ARCHAEODIET IN THE GREEK WORLD

Hesperia Supplements

The *Hesperia* Supplement series (ISSN 1064–1173) presents book-length studies in the fields of Greek archaeology, art, language, and history. Founded in 1937, the series was originally designed to accommodate extended essays too long for inclusion in the journal *Hesperia.* Since that date the Supplements have established a strong identity of their own, featuring single-author monographs, excavation reports, and edited collections on topics of interest to researchers in classics, archaeology, art history, and Hellenic studies.

Hesperia Supplements are electronically archived in JSTOR (www.jstor.org) where all but the most recent titles may be found. For order information and a complete list of titles, see the ASCSA website (www.ascsa.edu.gr).

Occasional Wiener Laboratory Series (OWLS)

OWLS is a subseries of *Hesperia* Supplements under the general editorship of Sherry C. Fox, former director of the Wiener Laboratory at the American School of Classical Studies at Athens. Although the topics covered are scientific in nature, reflecting the interests of the Wiener Laboratory, it is hoped that the books will appeal to a wide archaeological audience. Authors are therefore encouraged to avoid overly technical discussions, and clearly indicate why their work is important to researchers outside their immediate field.

1 L. A. Schepartz, S. C. Fox, and C. Bourbou, eds., *New Directions in the Skeletal Biology of Greece* (2009)

2 A. Papathanasiou, M. P. Richards, and S. C. Fox, eds., *Archaeodiet in the Greek World: Dietary Reconstruction from Stable Isotope Analysis* (2015)

Hesperia Supplement 49

ARCHAEODIET IN THE GREEK WORLD

Dietary Reconstruction from Stable Isotope Analysis

EDITED BY
ANASTASIA PAPATHANASIOU, MICHAEL P. RICHARDS, AND SHERRY C. FOX

The American School of Classical Studies at Athens
2015

Library of Congress Cataloging-in-Publication Data

Archaeodiet in the Greek world: dietary reconstruction from stable isotope analysis / edited by Anastasia Papathanasiou, Michael P. Richards, and Sherry C. Fox
p. cm. — (Hesperia supplement ; 49)
Includes bibliographical references and index.
ISBN 978-0-87661-549-2 (alk. paper)
1. Prehistoric peoples—Food—Greece. 2. Diet—Greece—History. 3. Food habits—Greece—History. 4. Human remains (Archaeology)—Greece. 5. Greece—Antiquities. 6. Stable isotopes. I. Papathanasiou, Anastasia. II. Richards, Michael P. (Michael Phillip). III. Fox, Sherry C.
GN799.F6A727 2015
394.1′20938—dc23 2014045487

CONTENTS

ILLUSTRATIONS

TABLES

CHAPTER I

Introduction

by Anastasia Papathanasiou and Sherry C. Fox

After water, food is the most critical element in human survival. Food procurement—encompassing the raw materials and the time invested in provisioning and preparation—continues to be of primary concern to people today as it was in the past, and is very closely linked with culture. As Samuel noted, "Nutritional, economic, belief systems and others, all can influence the preparation and consumption of a single type of food. The structures of diet and cuisine are shaped by such systems; they are a key part of cultural expression based on biological necessity."[1] The choices made concerning foodstuffs and their preparation appear to be a cultural universal among humans and unique when compared to non-human animals. In the archaeology of food, various lines of evidence have been used to discern choices involving foods and food preparation made by people in the past. Within this framework, this book aims to bridge biology and culture, biochemical analyses, and archaeological questions through a comprehensive presentation of palaeodietary reconstructions for a number of prehistoric and historic Greek sites, covering a broad spatial and cultural range. The data derive from individual evidence based on the biochemical signatures of the stable isotopes of two elements, carbon and nitrogen, that have survived in the skeletal material of people from the Greek past. This book provides the biochemical evidence for Greek archaeodiet in an attempt to help reconstruct the nutrition, subsistence, and economy of ancient Greece and the greater Greek world.

ARCHAEOLOGY AND DIET

The term *diet,* defined by R. W. Dennell as "what is eaten and drunk habitually,"[2] refers to both solids and liquids consumed over a period of time, including daily staples, drinks, medicines, and flavorings.[3] *Diet* is often used interchangeably and/or confused with the term *nutrition,* which refers not to the range of consumed foodstuffs, but to the utilization and value of the entire diet—the nourishment process—in terms of adequacy of food quantity, energy production, food quality, and balance of vitamins and minerals.[4] Additionally, the term *subsistence* is somewhat broader since it includes diet among other means exploited to meet vital needs such as food production and shelter.

1. Samuel 1996, p. 13.
2. Dennell 1979, p. 122.
3. Valamoti 2007, pp. 89, 91.
4. Sarpaki 2000, p. 116.

In every diet reconstruction there are two entities involved: the consumed items and the consumers. Consumables are the environmentally available foodstuffs, and their consumers make the cultural choice to incorporate certain foods from the available range in certain proportions into the diet. This choice is closely related to the entire socioeconomic structure and culture of human communities. Therefore, a dietary reconstruction deals not so much with the bioavailability of certain resources, but rather with what was actually consumed, to what extent, and why. Subsequently, what was actually consumed is vitally interwoven with specific health problems and diseases, lifestyles, food preparation techniques, the ecology of plants and animals, economic strategies, animal husbandry, fishing, farming, crop selection, trade, rituals, religion, and language, and it may be codified in a series of food consumption rules.

Issues that can be addressed by dietary reconstruction at the individual or the communal level include: what foodstuffs were consumed, what their relative proportions and importance were, what the nutritional value of the consumed foods was, how diet affected human health, and what dietary differences existed among the members of a community with respect to sex, age, and/or status. Additional questions could include which economic strategy was followed for the exploitation of the available resources, when and how new plant or animal species were introduced into the diet, what dietary or lifestyle changes were triggered by major cultural transitions, and how cultural transitions affected lifestyle and/or diet. Furthermore, we may ask to what extent religious practices created dietary taboos and affected the consumption of certain foodstuffs,[5] or even how close biochemical data are to textual, iconographic, or archaeological data and interpretations. Existing theories based on other data series can be challenged or confirmed by the biochemical evidence of past diet.

DIETARY RECONSTRUCTIONS

It is difficult to find evidence for these questions, particularly given the nature of the archaeological material. Greek archaeology, however, has a long and rich record of addressing them with the evidence at hand. In order to overcome the inherent connection between food consumption and the sociocultural framework, a variety of primary analytical approaches from a wide range of disciplines has been used for palaeodietary reconstruction, including archaeozoological and archaeobotanical remains, organic residues, adult stature, growth disturbances, pathological conditions on bones and teeth, dental microwear, ancient textual and artistic data, chemical analyses, DNA identification of residues, phytoliths, and coprolites. Secondary evidence for what people ate has come from the architectural study of hearths and olive and wine presses, as well as from household food preparation tools such as mortars and pestles, querns, and ceramic vessels for cooking. Much of this evidence has likely been lost over the years, but the study of the remaining vestiges, within the methodological limitations of each approach, can lead to an understanding of what people ate in the past.[6]

Prior to the introduction of isotopic studies in Greece, several other approaches to reconstruct diet along with its economic and social

5. Dennell 1979, p. 122; Sarpaki 2000, p. 116.

6. See also Chap. 3.

implications were used. Archaeological evidence for food in Greece, and how it was cultivated or raised, traded, and prepared, was reported as early as 1936 by Vickery,[7] who also included hunting and fishing in his monograph, *Food in Early Greece.* The Mediterranean diet, or as Renfrew[8] has coined it, the Mediterranean triad of grain, olive oil, and vine, has been a part of the ancient as well as modern Greek diet. Sarpaki[9] has aptly added legumes to the ancient Greek diet, which she describes as a tetrad. A number of books have appeared on the subject of diet in ancient Greece, covering a spectrum of issues that include subsistence strategies, culinary practices, agricultural production, consumption patterns, exploitation of domestic and wild resources, and ceramic assemblages.[10]

Within this framework, stable isotope analysis of osteological material is a biochemical approach, which can directly assess ancient diets. This method tracks the stable isotopes of carbon and nitrogen acquired through food, which provide the building material for bone. Bones and teeth are usually the only organic remains preserved in the archaeological record that may retain their original structure unaltered; they therefore constitute a record of individual and communal life history, diet, growth, development, and behavior.[11] As such, osteological remains are the focus of several interdisciplinary research fields, such as anthropology, archaeology, and chemistry. Palaeodietary reconstruction is one such application of isotopic composition of skeletal remains.[12]

Stable isotope analysis is based on the premise "you are what you eat."[13] Field and laboratory studies have shown that stable isotope ratios of carbon and nitrogen in tissues reflect similar ratios in the diet. As soon as foodstuffs are consumed, they are used as anatomical building blocks. Body tissues are composed of biomolecules, such as collagen, that are themselves made up of elements such as ingested carbon and nitrogen. Furthermore, carbon and nitrogen isotopic analysis of human tissue reliably reflects the isotopic composition of past diets, as collagen is resilient to contamination by external factors, and there are established controls for sample purity.

Katzenberg notes that "although stable isotope analysis may be viewed as a fairly technical research specialty, the results of such analyses make a significant contribution to the reconstruction of past human life,"[14] which is the ultimate goal of any archaeological investigation. Isotopic analysis has helped to clarify a number of important research questions, such as the date of introduction and the intensity of maize cultivation,[15] the relative importance of marine versus terrestrial resources in coastal populations, and the shift from hunting and gathering to food production.[16] The use of stable isotopes is now widespread in archaeology mainly because of their ability to provide insight into the social aspect of past cultures.

7. Vickery 1936.
8. Renfrew 1973.
9. Sarpaki 1992.
10. E.g., *Famine and Food Supply in the Graeco-Roman World: Responses to Risks and Crisis* (Garnsey 1988); *Eat Not This Flesh* (Simoons 1994); *Food in Antiquity* (Wilkins, Harvey, and Dobson 1995); "Loaves and Fishes" (Garvie-Lok 2001); *Food in the Ancient World from A to Z* (Dalby 2003); *Zooarchaeology in Greece* (Kotjabopoulou et al. 2003); *Food, Cuisine and Society in Prehistoric Greece* (Halstead and Barrett 2004); and *Cooking Up the Past: Food and Culinary Practices in the Neolithic and Bronze Age Aegean* (Mee and Renard 2007).
11. Larsen 1997, pp. 2–3; Ambrose and Krigbaum 2003.
12. See also Chap. 2.
13. DeNiro and Epstein 1981, p. 341; Ambrose and Norr 1993, p. 1.
14. Katzenberg 2000.
15. Buikstra and Milner 1991; Larsen et al. 1992; Ambrose and Norr 1993.
16. Tauber 1981; Norr 1991; Lubell et al. 1994; Papathanasiou 2003; Garcia-Guixé, Richards, and Subirà 2006.

HISTORY OF STABLE ISOTOPE ANALYSIS

The field began from "the domain of isotope geochemistry, particularly the radiocarbon dating of bone,"[17] and originated in the 1950s and 1960s when mass spectrometry became a popular technique. Significant stable isotopic variation was identified in plants and animals from different climates and environments (for reviews of the field, see Ambrose,[18] Ambrose and Krigbaum,[19] Katzenberg,[20] Koch et al.,[21] Pate,[22] Schoeninger and Moore[23]). Stable isotopes of carbon from bone collagen were the first to be used in archaeological research, followed by the stable isotopes of nitrogen.[24] Since then, the field has diversified considerably to include analyses of more elements and body tissues. Precision has also been refined, and several aspects of environmental, metabolic, or diagenetic effects that may alter the reconstruction of past diets have been clarified. Consequently, the chemical variation of the composition of hard tissues was applied to a vast array of archaeological materials from all cultural periods and geographical and environmental settings from all over the world, thus making possible the reconstruction of actual diets, subsistence strategies, and life histories.

Although in the Americas, Asia, and Africa, isotopic differences are large and values usually do not overlap, there is a different picture in Europe, where the general absence of tropical (C_4) foodstuffs and the paucity of marine values result in small variation, clustering, and overlapping of values within a relatively more confined range. Variation, however, does exist with respect to regional climatic factors and culturally induced food choices.[25] In the specific case of Greece, stable isotope analysis can define such dietary variation, which may relate to differences in sociocultural behaviors, access to animal protein resources, or climatic differentiation.

ISOTOPIC STUDIES IN GREECE

This volume represents both the direct assessment of past diet in Greece and the Greek world using chemical isotopic composition of osteological remains from archaeological contexts as well as the cultural and archaeological interpretations of those data. The papers that comprise this comprehensive volume were originally presented at a special session of the 16th European Meeting of the Paleopathology Association held on the Greek island of Santorini in 2006. The session, entitled "Stable Isotopic Studies in Greece," focused on palaeodietary research employing stable isotope analysis. Seven participants reviewed the present state and future perspectives of palaeodietary reconstruction in Greece through the lens of stable carbon and nitrogen isotope analysis. All of the presented papers except one are included in this book in expanded form. Two additional contributions have also been included. The aims of the 2006 session were to bring together researchers studying the archaeodiet of Greece through direct chemical evaluation, communicate and compare their results, discuss technical and interpretational problems, and design new research questions. It was also the goal of the session's organizers, as well as that of

17. Ambrose and Krigbaum 2003, p. 193.
18. Ambrose 1993.
19. Ambrose and Krigbaum 2003.
20. Katzenberg 2000.
21. Koch, Fogel, and Tuross 1994.
22. Pate 1994.
23. Schoeninger and Moore 1992.
24. DeNiro and Epstein 1981.
25. Van Klinken, Richards, and Hedges 2000.

the participants, to combine most of the stable isotope work conducted in Greece over the past 15 years into a single, comprehensive volume. In this way, a general perspective on the study of diet becomes apparent by comparing isotopic data from several sites of the same or different cultural and chronological periods, assessing Greek diets in conjunction with their environmental and socioeconomic implications, and addressing a number of long-standing archaeological questions in the field from prehistoric to medieval times, from a variety of geographic and social contexts within Greece and the Greek world.

The present volume may be considered a continuation of the monograph *Palaeodiet in the Aegean* (Wiener Laboratory Monograph 1), edited by Vaughan and Coulson in 2000.[26] In that volume, which resulted from a colloquium held in 1993, Vaughan stated on p. 5: "The reader will have noticed that, in contrast to research in other parts of the world, where stable isotope analysis of organic components of skeletal remains in palaeodietary work is an established approach, this technique is missing in most of the research in this volume, and indeed from the Aegean in general." The same observation had also been made by Jacobsen and Cullen in 1990[27] in Jane Buikstra's edited volume, *A Life in Science: Papers in Honor of J. Lawrence Angel.*[28]

This facet of Greek archaeology has changed, since not only are there now researchers engaged in stable isotope analysis, but the field of isotopic studies in general has made tremendous advances.[29] Indeed, Greece today is one of the most systematically examined and best-documented places in the world, offering:

1. dense coverage of the country's broad regional variation in climate and geomorphology;
2. a chronological and cultural depth that extends from the Mesolithic era to post-Byzantine times (Fig. 1.1, Table 1.1); and
3. application of stable isotopic techniques that extend beyond diet to include the investigation of cultural practices such as weaning, population movement, residence patterns, social differentiation, and economic strategies.

Isotopic research on Greek materials has been conducted in several laboratories both in Europe and in North America. Results complement each other regionally, temporally, and culturally, while the map is being filled with ever-increasing detail. Scholars reconstruct not only past human diet, but also geographical and ecosystem differences, economic models and strategies, social differentiation, variation through time for the same location, relation of burial practices with status, diet, and health, as well as dietary induced health conditions. This volume constitutes the first attempt to incorporate recent advances and the current status of palaeodietary reconstruction in Greece through isotopic studies. The contributions review the evidence with temporal breadth, geographical density, and social, cultural, and environmental variation. Most importantly, the volume maintains a bioarchaeological angle, which helps merge the biochemical data with major cultural issues.

26. Vaughan and Coulson 2000.
27. Jacobsen and Cullen 1990, p. 46.
28. Buikstra 1990.
29. Papathanasiou, Larsen, and Norr 1995; Garvie-Lok 2001; Triantaphyllou 2001; Papathanasiou 2003; Bourbou and Richards 2007; Richards and Hedges 2008; Richards and Vika 2008; Buikstra and Lagia 2009.

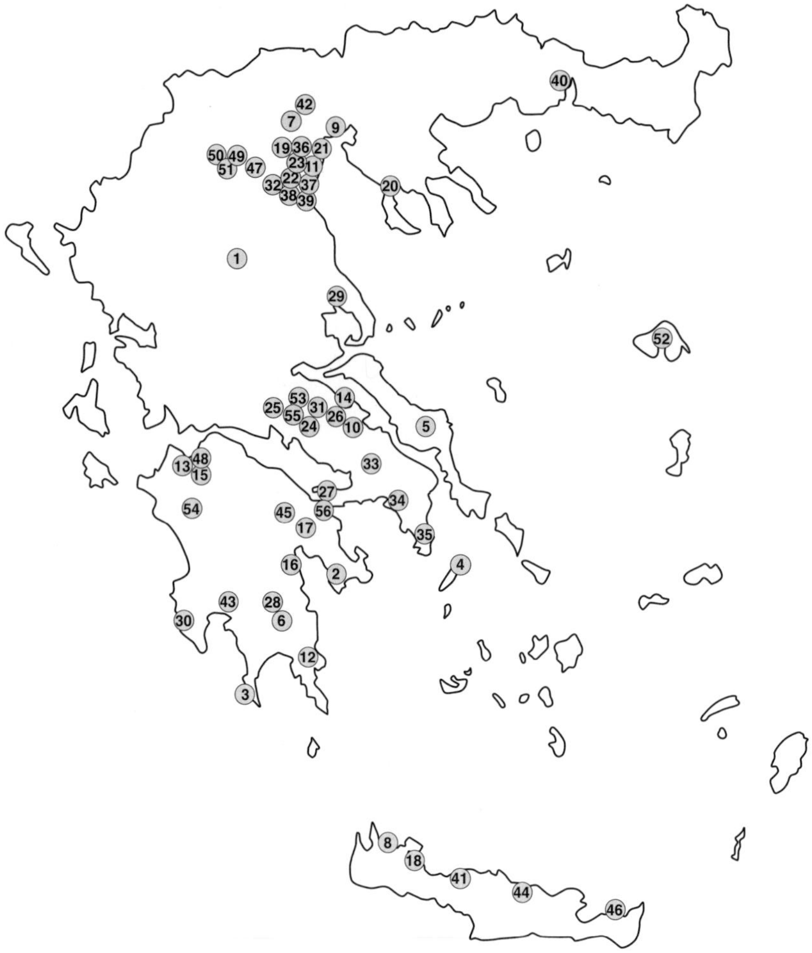

Abdera	40	Kalamaki	15	Kynos	14	Perahora	27	Sykia	12
Agia Triada	54	Kalapodi	55	Laurion	35	Petras	46	Tharrounia	5
Agios Mamas	20	Karitsa	38	Lerna	16	Pontokomi	50	Thebes	33
Alepotrypa Cave	3	Kastella	44	Makriyalos	11	Proskynas	10	Theopetra	1
Almyri	56	Kazanaki	29	Mavropigi	51	Pydna	36	Tragana	26
Armenoi	18	Kephala	4	Messene	43	Pylos	30	Treis Elies	23
Atalanti	31	Kladeri	39	Modi	24	Rymnio	32	Voudaini	48
Athens, Pl. Kotzia	34	Koilada	19	Mycenae	17	Servia	47	Xirolimni	49
Kerameikos		Kolaka	25	Mytilene	52	Sourtara	42	Zeli	53
Eleutherna	41	Korinos	21	Nea Nikomedeia	7	Spaliareika	13		
Franchthi Cave	2	Koufovouno	28	Nemea	45	Spathes	22		
Gerani	8	Kouveleiki	6	Olympus tumuli	37	Stavroupoli	9		

Figure 1.1. Map of Greece showing archaeological sites mentioned in this volume

The volume is organized in chronological sequence, starting with the Neolithic period and ending with the Byzantine era. A detailed methodology chapter precedes the presentation and analyses of the data for this volume. This contribution, Chapter 2 by Michael Richards, reviews the recent research advances in the field of stable isotope analysis of archaeological tissues, provides an in-depth presentation of the underlying biology

TABLE 1.1. CHRONOLOGY AND ABBREVIATIONS

Period	*Abbreviation*	*Chronology*
Mesolithic		8000–6800 B.C.
Early Neolithic	EN	6800–5800 B.C.
Middle Neolithic	MN	5800–5300 B.C.
Late Neolithic	LN	5300–4500 B.C.
Final Neolithic	FN	4500–3000 B.C.
Early Helladic I	EH I	3000–2500 B.C.
Early Helladic II	EH II	2500–2300 B.C.
Early Helladic III	EH III	2300–2100 B.C.
Middle Helladic	MH	2100–1550 B.C.
Late Helladic I	LH I	1550–1500 B.C.
Late Helladic II	LH II	1500–1400 B.C.
Late Helladic IIIA	LH IIIA	1400–1300 B.C.
Late Helladic IIIB	LH IIIB	1300–1190 B.C.
Late Helladic IIIC	LH IIIC	1190–1060 B.C.
Submycenaean		1060–1000 B.C.
Protogeometric		1000–900 B.C.
Geometric		900–750 B.C.
Late Geometric		750–700 B.C.
Archaic		700–500 B.C.
Classical		500–300 B.C.
Hellenistic		300 B.C.–A.D. 1
Early Roman		A.D. 1–250
Late Roman		A.D. 250–500
Byzantine		A.D. 500–1453
Alternate Bronze Age Terminology*		
Early Bronze Age	EBA	3000–2100 B.C.
Middle Bronze Age	MBA	2100–1800 B.C.
Late Bronze Age	LBA	1800–1100 B.C.
Early Iron Age	EIA	1100–900 B.C.
Early Minoan	EM	3000–2200 B.C.
Middle Minoan	MM	2200–1500 B.C.
Late Minoan	LM	1500–1100 B.C.

*Contemporary with Helladic depending on geographical location.

and chemistry on which this method is based, and discusses environmental influence on isotopic ratios. Richards describes in detail the methodology, laboratory procedures and protocols, and preparation processes used by most of the contributors of this volume. In the cases where different methodologies were applied, they are described in detail within each individual chapter. Richards also reviews the potentials and limitations of the interpretations of isotopic results, which is an essential step in any analysis for palaeodietary reconstruction.

The third chapter consists of two parts. The first part may serve as an introduction to all the papers referring to Neolithic and Bronze Age Greece. In this section, Anastasia Papathanasiou sets the background by describing the foodweb available for prehistoric Greek populations as it is understood from zooarchaeological and archaeobotanical studies. This section also stresses the importance of incorporating the stable isotope analysis of animal bone in conjunction with human bone in palaeodietary reconstructions. In the second part, Papathanasiou reviews the history and the development of stable isotope analysis for Neolithic and Bronze Age Greece over the last 15 years. She discusses the major data collected in her own research and incorporates published material in order to present the resulting current picture of prehistoric Greece as comprehensively as possible. She integrates faunal and botanical data and calls into question various archaeological hypotheses on the economies and subsistence of past societies, especially the ones that deal with the importance of fishing, meat consumption, and the very early indications of C_4 plants, specifically millet. To this end, information and inferences from other sources are confirmed through direct assessment of human osteological material, using animal osteological material as a reference. The chapter continues with a cross-laboratory comparison, proving the accuracy and repeatability of the discussed laboratory techniques. The chapter concludes with inter- and intra-site comparisons of regional, temporal, and social subgroups and their potential implications.

In the fourth chapter, Sevasti Triantaphyllou summarizes existing research from a large number of sites in the region of Macedonia in northern Greece. Triantaphyllou presents important new evidence from Neolithic, Bronze Age, and Early Iron Age (EIA) archaeological sites from Macedonia, expanding the existing human and faunal database. She incorporates dental data and pathological observations on the sampled human skeletons, comments on the regional variations of the analyzed sites, and uses the data to formulate interesting interpretations. Particularly, she underlines evidence for the presence of C_4 plants, as well as the importance of dairy products in conjunction with studies of ceramic typology and residues, and her results tie in nicely with Papathanasiou's in terms of low meat consumption and the negligible contribution of marine resources in Neolithic Greece.

In the fifth chapter, Efrossini Vika focuses on an Early and Middle Bronze Age assemblage from the site of Thebes in the central Greek mainland, delving into intra-site variation. She compares two groups of 12 individuals each, one from an Early Bronze Age (EBA) mass grave, and the other from a Middle Bronze Age (MBA) burial area. Her results indicate a C_3 plant diet with non-detectable fish protein that does not change through time. Vika reports several very low $\delta^{15}N$ values indicative of legume consumption or a possible vegan diet. A statistically significant difference appears, however, when comparing two urban groups with a third group, located 2 km away and closer to the sea. Her results indicate that significantly higher amounts of animal and marine protein seem to be incorporated into the diet of this third group. Finally, when comparing these results with those from her faunal samples, a C_4 signal was revealed, indicating the use of millet in Early Bronze Age Greece, perhaps used as animal fodder.

In the sixth chapter, Carina Iezzi focuses on a later time in the Bronze Age, in a previously unstudied region on the periphery of the Mycenaean world, east Lokris in central Greece. Iezzi introduces the parameters of ecology and geography and presents new isotopic evidence integrated meaningfully with archaeological, dental, and osteological indicators to show that independent lines of evidence correlate with regional differences in Mycenaean subsistence economies. These differences depend on the varied ecosystems and their environmental characteristics. She also points to the existence of sporadic C_4 evidence, probably millet, a negligible marine resource consumption in even coastal groups, and gendered dietary differences that reveal males were incorporating more animal dietary protein than females from the same context.

The following chapter by Eleni Panagiotopoulou and Anastasia Papathanasiou brings to attention the paucity of data concerning the Early Iron Age.[30] This lack of information is largely due to a long-standing academic neglect of the so-called Dark Age. The frequent practice of cremation in this period further hinders chemical analyses of osteological remains. The authors contribute new data for this little-studied time period by focusing on a cemetery of primary burials from central Greece. They document the diet of the Ayios Dimitrios population, which is characterized by non-significant contribution of C_4 and marine resources despite proximity to the sea, and they also document that there are no sex or age differences within the sample regarding access to animal dietary protein. Lastly, they explore weaning age and correlate the dietary reconstruction results with the health status of the group, highlighting the dietary and cultural differences compared with those from the earlier Mycenaean era.

The next chapter changes the focus to the Classical period. Anna Lagia reports new evidence from a time period that has been heavily scrutinized archaeologically but for which no isotopic analyses have thus far been undertaken. She focuses on the city of Athens during the Classical, Hellenistic, and Early Roman periods, and on two cemeteries, believed to represent differences in socioeconomic status, from the center of the *polis*. These two cemeteries are compared with each other and against a third cemetery in the periphery of Athens, which may have been used during the Classical period for the burial of enslaved mine workers. Using stable isotopic signatures as an avenue to economic history, Lagia compares the diet of the population across the three aforementioned chronological periods and across cemeteries in order to evaluate economic models that suggest the majority of the population lived at or near the subsistence level. The economic models also predict significant social differentiation in the population of the ancient city, while no significant changes in the economy occurred across time.

Lagia's results suggest a C_3 terrestrial diet with significant amounts of animal products and argue against an impoverished diet for the majority of the adult population. In terms of socioeconomic status, no major differences are observed in the diet, although individuals from the upper-class cemetery exhibit a significantly higher consumption of animal protein. Temporal comparison of dietary signatures supports the occurrence of significant changes in the diet through time. Two time periods are identified with major

30. Triantaphyllou 2001.

declines in protein consumption. The chapter closes with comparisons to other sites from the eastern Mediterranean and the Black Sea providing evidence for the existence of significant temporal and regional distinctions.

In Chapter 9, Cynthia Kwok and Anne Keenleyside focus on the Classical and Hellenistic eras, but on the border of the ancient Greek world, at the colony of Apollonia Pontica on the Black Sea. The chapter sheds light on a minimally studied research area, that of weaning and infant feeding strategies, and it contributes significant information to our understanding of breastfeeding and weaning processes in antiquity.[31] Kwok and Keenleyside study numerous individuals ranging from birth to 15 years of age and compare the results to archaeological data, palaeopathological evidence, and ancient literature. They conclude that weaning began between 6 months to 1 year of age and was completed by the age of 2 to 4 years, with some degree of variation. The weaning diet as well as the childhood diet consisted of C_3 plants and dairy products, and it differed from the adult diet, which included more meat and marine resources. This protein deficient diet increased children's susceptibility to developing nutritional deficiencies, such as iron, vitamin B_{12}, and folic acid, and this corresponds with the observed pathological lesions on the osteological material.

Chapter 10 by Chryssa Bourbou and Sandra Garvie-Lok focuses on the subsequent Byzantine period, from which very little is known about the diet. They review evidence from five sites covering the north, the south, and the islands of Greece, concentrating on both adult and weaning diet. They integrate isotopic evidence with a very detailed survey of both Byzantine literature and art, illustrating what is known historically. Bourbou and Garvie-Lok further investigate the relative importance of certain foods in the Byzantine diet, such as legumes, meat, and fish, and conclude that there was in fact a "Byzantine diet" based on the Mediterranean triad, although very small variation between different populations can be seen. The Byzantine diet consists primarily of C_3 grains with no substantial C_4 contribution, large amounts of domesticated animal protein mainly in the form of dairy products, very low amounts of marine resources, and an insignificant amount of legumes. Finally, they report evidence that weaning was practiced at a late age in contrast to contemporary Western Europe.

In the concluding chapter, Anastasia Papathanasiou and Michael Richards summarize the aforementioned studies and the approximately 1,000 samples presented in this volume from all time periods of prehistoric and historic Greece. They incorporate the results into broad and integrated conclusions addressing major archaeological questions, emphasizing the growing concern for uncovering the economic and social aspect of the past, and suggesting new research directions for stable isotopic studies of Greece. The authors address the key question of how these results are interpreted and how they are tied in with the economic–cultural context and current archaeological research objectives such as the transition to agriculture, the role of marine resources in prehistoric and historic diets, documentation of gender and status differences in access to resources, fluctuations of animal protein consumption through time, weaning age, and the role of C_4 plants in human and animal diet through time.

31. Garvie-Lok 2001; Bourbou and Richards 2007; Triantaphyllou et al. 2008.

QUESTIONS OF GREEK ARCHAEOLOGY ADDRESSED BY ISOTOPIC STUDIES

In summary, this volume, through stable isotope analysis, directly addresses the relative importance of particular food categories in the ancient Greek diet. These results complement what has been found from other lines of evidence from the archaeological record. To date, certain food remnants, such as animal bones, tend to preserve better than botanical remains and as such have led to an overestimation of their actual abundance and the importance of meat consumption in the reconstruction of past diets. By contrast, stable isotope studies, in conjunction with other lines of evidence, including faunal and botanical analyses, as well as historical literature and art, have led to more accurate interpretations of past diets and economies. Major questions in Greek archaeology have been assessed through this methodology, such as the intensity of domestication and the subsequent reduction in the variety of consumed species during the transition to agriculture; the testing of proposed economic models for Classical societies; the flourishing and decline in the standard of living along with the rise and fall of related cultures; the introduction of specific crops, such as millet; the estimation of the age of weaning during different periods in antiquity; the assessment of sex, age, and status differentiation in access to animal protein; and the relative accessibility and symbolism of particular food items and practices, such as fishing and hunting, which have been considered important throughout the prehistory and history of Greece and the Greek world. Additionally, stable isotopes studies used in correlation with skeletal evidence for stress and disease have helped specialists answer questions about past health. Finally, through surveying the approximately 1,000 samples presented in this volume, it becomes clear that the sum of individual diets reflects the diet of the population through time. This allows researchers to document long-term dietary changes and to quantify internal social variability.

It is also worth mentioning the great enthusiasm shared by the session's participants while exchanging ideas, comparing results, and receiving feedback. The contributors were enthralled by the convergence of this ongoing research and the general picture of archaeodiet in Greece that is beginning to emerge. The editors hope that this enthusiasm may be contagious for the readers of this volume. Indeed, as comprehensive as this volume might be during its preparation, by the time it appears in print it will surely have already been surpassed by current work since isotopic research in Greece is now growing fast, producing many intriguing results, and offering interesting new questions.

ACKNOWLEDGMENTS

Numerous people helped this volume reach its final form and the editors greatly appreciate their efforts. They include many former members of the Publications Office at the American School of Classical Studies at Athens, in particular Andrew Reinhard and Tracey Cullen. The comments of the external reviewers and those of the Publications Committee of the Managing Committee of the American School of Classical Studies at Athens were extremely helpful. We thank Kristina Killgrove for copyediting this volume as well as Tina Ross for drafting the excellent artwork and Kate Mertes for preparing the index. This volume would not have happened without the technical and moral support from George Stamoulis. Many more individuals could be acknowledged, but Malcolm H. Wiener deserves the greatest thanks for his steadfast support of the Wiener Laboratory at the American School of Classical Studies at Athens.

REFERENCES

Ambrose, S. H. 1993. "Isotopic Analysis of Paleodiets: Methodological and Interpretive Considerations," in *Investigations of Ancient Human Tissue: Chemical Analyses in Anthropology* (Food and Nutrition in History and Anthropology 10), ed. M. K. Sandford, Langhorne, Pa., pp. 59–130.

Ambrose, S. H., and J. Krigbaum. 2003. "Bone Chemistry and Bioarchaeology," *JAnthArch* 22, pp. 193–199.

Ambrose, S. H., and L. Norr. 1993. "Experimental Evidence for the Relationship of the Carbon Isotope Ratios of Whole Diet and Dietary Protein to Those of Bone Collagen and Carbonate," in *Prehistoric Human Bone: Archaeology at the Molecular Level,* ed. J. B. Lambert and G. Grupe, Berlin, pp. 1–37.

Bourbou, C., and M. P. Richards. 2007. "The Middle Byzantine Menu: Paleodietary Information from Isotopic Analysis of Humans and Fauna from Kastella, Crete," *International Journal of Osteoarchaeology* 17, pp. 63–72.

Buikstra, J. E., ed. 1990. *A Life in Science: Papers in Honor of J. Lawrence Angel* (Center for American Archeology Scientific Papers 6), Kampsville, Ill.

Buikstra, J. E., and A. Lagia. 2009. "Bioarchaeological Approaches to Aegean Archaeology," in *New Directions in the Skeletal Biology of Greece* (*Hesperia* Suppl. 43, OWLS 1), ed. L. A. Schepartz, S. C. Fox, and C. Bourbou, Princeton, pp. 7–29.

Buikstra, J. E., and G. R. Milner. 1991. "Isotopic and Archaeological Interpretations of Diet in the Central Mississippi Valley," *JAS* 18, pp. 319–329.

Dalby, A. 2003. *Food in the Ancient World from A to Z,* London.

DeNiro, M. J., and S. Epstein. 1981. "Influence of Diet on the Distribution of Nitrogen Isotopes in Animals," *Geochimica et cosmochimica acta* 45, pp. 341–351.

Dennell, R. W. 1979. "Prehistoric Diet and Nutrition: Some Food for Thought," *WorldArch* 11, pp. 121–135.

Garcia-Guixé, E., M. P. Richards, and M. E. Subirà. 2006. "Palaeodiets of Humans and Fauna at the Spanish Mesolithic Site of El Collado," *CurrAnthr* 47, pp. 549–556.

Garnsey, P. 1988. *Famine and Food Supply in the Graeco-Roman World: Responses to Risk and Crisis,* Cambridge.

Garvie-Lok, S. J. 2001. "Loaves and Fishes: A Stable Isotope Reconstruction of Diet in Medieval Greece" (diss. Univ. of Calgary).

Halsted, P., and J. C. Barrett, eds. 2004. *Food, Cuisine, and Society in Prehistoric Greece* (Sheffield Studies in Aegean Archaeology 5), Oxford.

Jacobsen, T. W., and T. Cullen. 1990. "The Work of J. L. Angel in the Eastern Mediterranean," in *A Life in Science: Papers in Honor of J. Lawrence Angel* (Center for American Archeology Scientific Papers 6), ed. J. E. Buikstra, Kampsville, Ill., pp. 38–51.

Katzenberg, M. A. 2000. "Stable Isotope Analysis: A Tool for Studying Past Diet, Demography, and Life History," in *Biological Anthropology of the Human Skeleton,* ed. M. A.

Katzenberg and S. R. Saunders, New York, pp. 305–328.

Koch, P. L., M. L. Fogel, and N. Tuross. 1994. "Tracing the Diets of Fossil Animals Using Stable Isotopes," in *Stable Isotopes in Ecology and Environmental Science* (Methods in Ecology), ed. K. Lajtha and R. H. Michener, Oxford, pp. 63–92.

Kotjabopoulou, E., Y. Hamilakis, P. Halstead, C. Gamble, and P. Elefanthi. 2003. *Zooarchaeology in Greece: Recent Advances* (*BSA* Studies 9), London.

Larsen, C. S. 1997. *Bioarchaeology: Interpreting Behavior from the Human Skeleton* (Cambridge Studies in Biological Anthropology 21), Cambridge.

Larsen, C. S., M. J. Schoeninger, N. J. van der Merwe, K. M. Moore, and J. A. Lee-Thorp. 1992. "Carbon and Nitrogen Stable Isotopic Signatures of Human Dietary Change in the Georgia Bight," *American Journal of Physical Anthropology* 89, pp. 197–214.

Lubell, D., M. Jackes, H. Schwarcz, M. Knyf, and C. Meiklejohn. 1994. "The Mesolithic–Neolithic Transition in Portugal: Isotopic and Dental Evidence of Diet," *JAS* 21, pp. 201–216.

Mee, C., and J. Renard, eds. 2007. *Cooking Up the Past: Food and Culinary Practices in the Neolithic and Bronze Age Aegean,* Oxford.

Norr, L. C. 1991. "Nutritional Consequences of Prehistoric Subsistence Strategies in Lower Central America" (diss. Univ. of Illinois, Urbana-Champaign).

Papathanasiou, A. 2003. "Stable Isotope Analysis in Neolithic Greece and Possible Implications on Human Health," *International Journal of Osteoarchaeology* 13, pp. 314–324.

Papathanasiou, A., C. S. Larsen, and L. Norr. 1995. "A Bioarchaeological Analysis of a Neolithic Ossuary from Alepotrypa Cave, Diros, Greece," *American Journal of Physical Anthropology* 38, Suppl. 20, p. 167 (abstract).

Pate, F. D. 1994. "Bone Chemistry and Paleodiet," *Journal of Archaeological Method and Theory* 1, pp. 161–209.

Renfrew, J. M. 1973. *Palaeoethnobotany: The Prehistoric Food Plants of the Near East and Europe,* London.

Richards, M. P., and R. E. M. Hedges. 2008. "Stable Isotope Evidence of Past Human Diet at the Sites of the Neolithic Cave of Gerani; the Late Minoan III Cemetery of Armenoi; Grave Circles A and B at the Palace Site of Mycenae; and Late Helladic Chamber Tombs," in *Archaeology Meets Science: Biomolecular Investigations in Bronze Age Greece. The Primary Scientific Evidence, 1997–2003,* ed. Y. Tzedakis, H. Martlew, and M. K. Jones, Oxford, pp. 220–230.

Richards, M. P., and E. Vika. 2008. "Stable Isotope Results from New Sites in the Peloponnese: Sykia, Kalamaki, and Spaliareka," in *Archaeology Meets Science: Biomolecular Investigations in Bronze Age Greece. The Primary Scientific Evidence, 1997–2003,* ed. Y. Tzedakis, H. Martlew, and M. K. Jones, Oxford, pp. 231–235.

Samuel, D. 1996. "Approaches to the Archaeology of Food," *Petits propos culinaires* 54, pp. 12–21.

Sarpaki, A. 1992. "The Palaeoethnobotanical Approach: The Mediterranean Triad or Is It a Quartet?" in *Agriculture in Ancient Greece. Proceedings of the Seventh International Symposium at the Swedish Institute at Athens, 16–17 May, 1990* (*ActaAth* 4°, 42), ed. B. Wells, Stockholm, pp. 61–76.

———. 2000. "The Study of Paleodiet in the Aegean: Food for Thought," in *Palaeodiet in the Aegean. Papers from a Colloquium Held at the 1993 Meeting of the Archaeological Institute of America in Washington D.C.* (Wiener Laboratory Monograph 1), ed. S. J. Vaughan and W. D. E. Coulson, Oxford, pp. 115–121.

Schoeninger, M. J., and K. Moore. 1992. "Bone Stable Isotope Studies in Archaeology," *Journal of World Prehistory* 6, pp. 247–296.

Simoons, F. J. 1994. *Eat Not This Flesh: Food Avoidances from Prehistory to the Present,* Madison.

Tauber, H. 1981. "^{13}C Evidence for Dietary Habits of Prehistoric Man in Denmark," *Nature* 292, pp. 332–333.

Triantaphyllou, S. 2001. *A Bioarchaeological Approach to Prehistoric Cemetery Populations from Central and Western Greek Macedonia* (*BAR-IS* 976), Oxford.

Triantaphyllou, S., M. P. Richards, C. Zerner, and S. Voutsaki. 2008. "Isotopic Dietary Reconstruction of Humans from Middle Bronze Age Lerna, Argolid, Greece," *JAS* 35, pp. 3028–3034.

Valamoti, S.-M. 2007. "Traditional Foods and Culinary Novelties in Neolithic and Bronze Age Northern Greece: An Overview of the Archaeobotanical Evidence," in *Cooking Up the Past: Food and Culinary Practices in the Neolithic and Bronze Age Aegean,* ed. C. Mee and J. Renard, Oxford, pp. 89–108.

van Klinken, G. J., M. P. Richards, and R. E. M. Hedges. 2000. "An Overview of Causes for Stable Isotopic Variations in Past European Human Populations: Environmental, Ecophysiological, and Cultural Effects," in *Biogeochemical Approaches to Paleodietary Analysis* (Advances in Archaeological and Museum Science 5), ed. S. H. Ambrose and M. A. Katzenberg, New York, pp. 39–63.

Vaughan, S. J., and W. D. E. Coulson, eds. 2000. *Palaeodiet in the Aegean. Papers from a Colloquium Held at the 1993 Meeting of the Archaeological Institute of America in Washington D.C.* (Wiener Laboratory Monograph 1), Oxford.

Vickery, K. F. 1936. *Food in Early Greece* (Illinois Studies in the Social Sciences 20.3), Urbana, Ill.

Wilkins, J., D. Harvey, and M. Dobson. 1995. *Food in Antiquity,* Exeter.

CHAPTER 2

Stable Isotope Analysis of Bone and Teeth as a Means for Reconstructing Past Human Diets in Greece

by Michael P. Richards

The use of the stable isotope ratios of carbon and nitrogen in bone collagen, and carbon in bioapatite in bone and tooth enamel, is a powerful method of reconstructing the past diets of human and animals. The method was first applied in archaeology in the late 1970s and early 1980s[1] and has been increasingly used over the past 20 years. It is now widely applied to many time periods and sites all over the world, and increasingly in Greece.[2] Here I give a short introduction to the method of isotopic analysis, especially as it relates to its use in reconstructing the diet of humans in prehistoric and historic Greece. This general presentation is intended to serve as an introduction to the methodology for the case studies that are presented in this volume.

BONE AS A SOURCE OF PALAEODIETARY INFORMATION

Our body tissues are composed of biomolecules that are themselves made up of elements such as carbon and nitrogen taken from food we have eaten over our lifetimes. In archaeological contexts, the most commonly preserved body tissues are bone and teeth, due to their composition, and therefore these are the two main tissues used in isotope palaeodietary research. Bone is composed mainly of mineral (calcium phosphate), and approximately 20% is protein (collagen). Currently, most isotope palaeodietary studies focus on bone collagen as the substrate for isotopic analysis of carbon and nitrogen, as collagen is often the best-preserved component of bone. Isotope analysis of bone mineral (bioapatite, where carbon and oxygen isotopes are measured) has also been undertaken, and some of the case studies presented here use this substrate. Below I explain the different aspects of diet that these two bone components reflect and discuss the methodological issues associated with the isotopic analysis of both.

1. E.g., Vogel and van der Merwe 1977; DeNiro and Epstein 1981; Tauber 1981.

2. E.g., Papathanasiou, Larsen, and Norr 2000; Papathanasiou 2001, 2003; Bourbou and Richards 2007; Richards and Hedges 2008; Richards and Vika 2008; Triantaphyllou et al. 2008.

CARBON AND NITROGEN ISOTOPE ANALYSIS OF BONE COLLAGEN

Collagen is a fairly robust biomolecule, and based largely on radiocarbon dating experiments, it has been shown to be the portion of bone that is least susceptible to contamination from carbon and nitrogen in the burial environment. Therefore, if we can extract intact collagen and measure its properties to ensure that it is well-preserved, we can be reasonably sure that it is largely the same molecule that was present when the human or animal was alive. After establishing that we have intact and well-preserved collagen, we measure the ratio of two different isotopes of carbon (^{13}C and ^{12}C, represented using delta notation, discussed below, as a $\delta^{13}C$ value) and nitrogen (^{15}N and ^{14}N, also represented using delta notation as a $\delta^{15}N$ value) in the collagen. We can use these values to determine the diets of the humans and animal. Below I describe the extraction and measurement procedures in more detail and explain how we can use these isotopic measurements to reconstruct past diets.

Collagen Extraction and Isotopic Measurement Procedures

The first step in isotope analysis for reconstructing diets is the extraction of collagen from bone samples. Usually between 200 and 1,000 mg of whole bone or bone powder are taken. The samples are then treated with hydrochloric acid, which removes the main mineral part of the bone. The mineral phase often contains carbon and nitrogen that has been incorporated into the bone after it was buried, and therefore this step removes the contaminants. This step is followed by most labs and generally employs the procedure first outlined by Longin.[3] After this initial step, the pretreatment procedures differ between labs. Some laboratories simply freeze-dry the remaining residue after this first step and then measure the carbon and nitrogen isotope values of that residue, while most other labs do further pre-treatment to better isolate the intact collagen and ensure that contaminants are removed. This further pre-treatment often involves the heating of the extract to over 57°C in a slightly acidic solution, which causes the triple helix of collagen to separate and the individual collagen strands to solubilize. The resulting solution is filtered. The most effective way of doing this is by using an ultrafilter, which isolates the >30 kDa fraction for further analysis,[4] although this is not a commonly used method. After extraction, the resulting solid, mostly collagen, is freeze-dried and measured in an isotope mass spectrometer. Most labs today use a continuous flow mass spectrometer where the sample is combusted in an elemental analyzer in a stream of helium, the CO_2 and N_2 gases are separated using gas chromatography, and the isotope ratios of these gases are measured in the mass spectrometer. At the same time, a measurement of the amount of carbon and nitrogen in the sample is made. These measurements are used to calculate the C:N ratio, which is an important indicator of whether the collagen was well preserved. Another indicator is the %C and %N in a sample. For well-preserved collagen samples, the C:N ratio should be between 2.9 and 3.6, and the %C and %N should be close to 45% and 15%, respectively.[5]

3. Longin 1971.
4. Brown et al. 1988.
5. DeNiro 1985; Ambrose 1990.

Bone Collagen as a Dietary Indicator

The isotope ratios of carbon and nitrogen that are produced by this method can be used to reconstruct the diet of the human or animal. From a number of important controlled animal feeding experiments in the early 1990s, it has been demonstrated that carbon isotope values mostly show the source of dietary protein and are not a measure of all of the carbon that a person has eaten.[6] Due to poorly understood processes in the body, as amino acids from food are digested and processed and as new amino acids are formed that are incorporated into bone collagen, there is a change, or fractionation, in the carbon isotope ratio between dietary protein and bone collagen of approximately 5 per mil (‰) with diets with sufficient amounts of protein.[7] Nitrogen is only present in dietary protein, so bone collagen nitrogen isotope values must reflect dietary protein isotope values. Similar to carbon, there is a poorly understood fractionation effect where bone collagen has a nitrogen isotope ratio of approximately 3‰ higher than dietary protein.[8] Collagen is constantly being formed and incorporated into bones as old collagen is removed, and this process of renewal, or turn-over, takes a considerable amount of time and differs between different bones, likely dependent on the amount of stress and use that the bones are subject to, as well as the age of the individual.[9] Therefore, it may take 20 to 30 years for the bone collagen in a femur to be completely replaced,[10] whereas a rib may take only five years.[11] The turn-over times for bone collagen are poorly known and likely differ between humans and among mammal species. Therefore, the best estimate of what the carbon and nitrogen isotope values are telling us is that they are a record of all of the dietary protein consumed over the past 10 to 20 years of life.

BONE BIOAPATITE AS A SOURCE OF ISOTOPE PALAEODIETARY INFORMATION

Bone mineral contains about 6% bioapatite, which is formed in equilibrium with blood bicarbonate.[12] Unlike collagen $\delta^{13}C$ values, which come primarily from dietary protein, bone (and tooth enamel) carbonate reflects the $\delta^{13}C$ value of the total diet from protein, lipids, and carbohydrates. Although there has been some observed variation in the fractionation between total diet $\delta^{13}C$ and bioapatite $\delta^{13}C$, generally there is approximately a 12‰ to 13‰ shift between diet and apatite $\delta^{13}C$ values.[13] Bone carbonate $\delta^{13}C$ values can be an important source of dietary information, especially as they can record signals from carbohydrates (e.g., sugar cane) and lipids that would generally be invisible to collagen isotope analysis, which mainly measures protein sources. Therefore, carbonate $\delta^{13}C$ has the potential to provide better information on the foods that are the main caloric contributors to diet (lipids and carbohydrates) and not just the source of dietary protein. Measurements of bone collagen and carbonate $\delta^{13}C$ values together can therefore provide a more complete picture of the total diet. Bone carbonate

6. Ambrose and Norr 1993.
7. Ambrose and Norr 1993; Kellner and Schoeninger 2007.
8. Schoeninger and DeNiro 1984; Ambrose 2000; Hedges and Reynard 2007.
9. Hedges et al. 2007.
10. Wild et al. 2000.
11. Stenhouse and Baxter 1979.
12. Lee-Thorp 2008.
13. Ambrose and Norr 1993; Lee-Thorp 2008.

is perhaps most useful in environments where C_4 plants (like maize and millet, as compared to C_3 plants like wheat and barley) are important and can be used to detect their dietary importance, especially if they are consumed only as carbohydrates or lipids and would therefore be almost invisible in bone collagen isotope analysis.

Preparation methods for bone carbonate isotope measurements are much simpler than for collagen. These require a relatively simple pre-treatment to first remove organics (NaOCl or H_2O_2), and then a relatively gentle acid treatment (e.g., acetic acid) to remove any recent carbonate from the burial environment.[14] Measurement of the $\delta^{13}C$ values is also fairly simple, and generally automated, with CO_2 being evolved from the pre-treated bone in a vacuum, after adding orthophosphoric acid.

There are major problems with bone carbonate measurements, as there is good evidence that these values can be significantly altered in the burial environment by the addition of new carbonate to the bone or the alteration of the existing bone mineral, which can lead to the incorporation of new carbonate into the bone structure.[15] There are a number of ways to judge the preservation quality of the bone mineral, including looking at its structure and at the size of the apatite crystals. These methods do not always detect contamination and alteration of the bone bioapatite $\delta^{13}C$ values, however, so most researchers use bone carbonate measurements with caution.

CARBON AND NITROGEN ISOTOPE VALUES IN THE BIOSPHERE

In order to interpret the carbon and nitrogen isotope values obtained from bone collagen and bioapatite, it is necessary to understand the range of values present in the biosphere and especially in different ecosystems such as terrestrial temperate and arid environments and aquatic freshwater and marine ecosystems. Below is a short introduction to the typical range of values observed in these environments.

Carbon Isotopes

There are two stable isotopes of carbon: the more plentiful ^{12}C (ca. 98.89% of all carbon) and the rarer ^{13}C (ca. 1.11%).[16] The $\delta^{13}C$ notation is used to express the ratio of ^{13}C to ^{12}C in a sample and is given in units of per mil (‰). The ratio of ^{13}C to ^{12}C in a sample is measured compared to the ratio in the Vienna Pee Dee Belemnite (VPDB) standard[17] according to the following equation:

$$\delta^{13}C = [[(^{13}C/^{12}C)\text{sample} / (^{13}C/^{12}C)\text{standard}] - 1] \times 1{,}000$$

Atmosphere and Plant $\delta^{13}C$ Values

Atmospheric CO_2 has an average global $\delta^{13}C$ value of -7 to -8‰,[18] although local variations in this value have been observed.[19] During photosynthesis, plants preferentially take up ^{12}C relative to ^{13}C from atmospheric CO_2, resulting in the plant tissues having a lower $\delta^{13}C$ value than atmospheric

14. Lee-Thorp 2008.
15. Schoeninger and DeNiro 1982.
16. Hoefs 1997.
17. Coplen 1994.
18. Hoefs 1997.
19. Van der Merwe 1989.

CO_2.[20] There are three photosynthetic pathways plants can follow; each varies in how they "fix" the carbon, and each can fractionate atmospheric CO_2 to different degrees.[21] C_3 (Calvin) photosynthetic pathway plants discriminate against ^{13}C the most and therefore have the most negative $\delta^{13}C$ values, ca. -27‰, while C_4 (Hatch-Slack) photosynthetic pathway plants discriminate less against ^{13}C and therefore have more positive $\delta^{13}C$ values averaging ca. -13‰.[22] Crassulean acid metabolism (CAM) pathway plants have $\delta^{13}C$ values intermediate between the two, depending on their growth conditions. Most plants in temperate environments are C_3 plants (e.g., trees, cereals), while many arid environment plants are C_4 (e.g., maize, millet, grasses). CAM plants are rare and include succulents like cacti.

Ocean Water and Marine Plant $\delta^{13}C$

Marine plants derive their carbon from the dissolved ocean bicarbonate, which is more enriched in ^{13}C than atmospheric carbon. Surface ocean dissolved bicarbonate has an average $\delta^{13}C$ value of about 2.2‰, but this gets closer to zero in deeper waters.[23] Marine plants use the same processes of fixing carbon as terrestrial plants (i.e., C_3 and C_4 pathways), but as ocean bicarbonate is more enriched in ^{13}C (approximately 10‰ higher in $\delta^{13}C$ than atmospheric CO_2), marine plants are about 10‰ higher in $\delta^{13}C$ than terrestrial plants.

Freshwater $\delta^{13}C$ Values

The dissolved carbonate in water in freshwater lakes and rivers can come from a number of sources. The carbonate could have been derived from atmospheric CO_2 and could also include carbonate from geological sources. The diverse nature of the origins of freshwater carbonate means that plants in those systems can have a wide range of $\delta^{13}C$ values. Dissolved atmospheric CO_2 usually predominates, however, so freshwater aquatic plants often have $\delta^{13}C$ values similar to terrestrial plants.[24]

Nitrogen Isotopes

There are two stable isotopes of nitrogen: ^{14}N (99.634%) and ^{15}N (0.366%).[25] The ratio of ^{15}N to ^{14}N in a sample is expressed in $\delta^{15}N$ notation and in per mil (‰). The ratio of ^{15}N to ^{14}N in a sample is compared to the ratio in the AIR (air) standard according to the following equation:

$$\delta^{15}N = [[(^{15}N/^{14}N)\text{sample} / (^{15}N/^{14}N)\text{standard}] - 1] \times 1{,}000$$

Atmosphere, Soil, and Plant $\delta^{15}N$ Values

Almost 80% of the atmosphere is composed of elemental nitrogen,[26] and atmospheric nitrogen has a similar $\delta^{15}N$ value all over the world.[27] The standard for $\delta^{15}N$, atmospheric nitrogen (AIR), is defined as having a value of 0‰, although the actual observed value of atmospheric nitrogen is often closer to 1.6 ± 1.4‰.[28]

Leguminous plants (e.g., lentils) derive their nitrogen partly from the atmosphere, where bacteria in the roots fix nitrogen by converting

20. Hoefs 1997.
21. O'Leary 1981.
22. Hoefs 1997.
23. Hoefs 1997.
24. Hoefs 1997.
25. Owens 1987.
26. Hoefs 1997.
27. DeNiro 1987; Owens 1987.
28. Owens 1987.

atmospheric inorganic nitrogen to organic nitrogen. They also derive nitrogen from the soil, where soil bacteria have already fixed atmospheric nitrogen. Non-leguminous plants obtain their nitrogen from the soil only. There is little fractionation of nitrogen when it is used by plants, so plant tissue $\delta^{15}N$ values are often similar to atmospheric $\delta^{15}N$ values (i.e., close to 0‰), but plant $\delta^{15}N$ values that differ greatly from atmospheric $\delta^{15}N$ values have been observed.[29] These differences may be related to differing soil conditions that result in unusual soil nitrogen $\delta^{15}N$ values, which are then reflected in plant $\delta^{15}N$ values. Unusual nitrogen values may be partly due to the large amounts of fractionation that occur with denitrification in soils.[30] Therefore, although the average values of most legumes and non-legumes are close to 1‰, researchers have observed values between -10 and +10‰.[31]

Marine $\delta^{15}N$ Values

Dissolved nitrogen in the ocean, due to complex processes of denitrification, has a higher $\delta^{15}N$ value than atmospheric $\delta^{15}N$, and dissolved deep ocean nitrate has $\delta^{15}N$ values of between 6‰–8‰.[32] Therefore, marine plants and organisms such as algae have higher $\delta^{15}N$ values than terrestrial plants and organisms.

INTERPRETING THE DIETARY SIGNAL FROM BONE CARBON AND NITROGEN ISOTOPE DATA IN GREEK ARCHAEOLOGY

Stable carbon and nitrogen isotope ratios of bone collagen and bone bioapatite carbon can tell us significant new information about the diets of humans and animals in prehistoric and historic Greece. Figure 2.1 is a theoretical plot of the bone collagen isotope values that might be observed in Greece.

The carbon isotope ratios are of particular interest for Greece, as they are well suited to distinguishing between marine and terrestrial proteins. This is of interest especially in earlier periods where marine foods may have played a significant role, such as the Mesolithic period. It is also intriguing to explore whether marine foods, in a country with a long coastline and numerous islands, were ever important in the diets of humans, and if this changed through time. Also of interest with carbon isotope measurements is the ability to distinguish C_4 plants, namely millet, in the diet from more common C_3 plants such as wheat and barley.

The nitrogen isotope value, when used in conjunction with the carbon value, can be used to distinguish between a $\delta^{13}C$ value that indicates either millet or marine food consumption. Marine food chains are much longer than terrestrial ones, so marine organisms, especially the ones consumed by humans, are higher up the food chain and consequently have higher $\delta^{15}N$ values, as there is an increase of approximately 3‰ in $\delta^{15}N$ value with each step of the food chain.[33] When C_4 or marine foods were not consumed, the $\delta^{15}N$ value can be used to distinguish between the consumption of animal and plant proteins, but this is really only possible where contemporary

29. Schwarcz and Schoeninger 1991.
30. Hoefs 1997.
31. Létolle 1980; Owens 1987.
32. Hoefs 1997, p. 123.
33. Schoeninger and DeNiro 1984; Ambrose 2000.

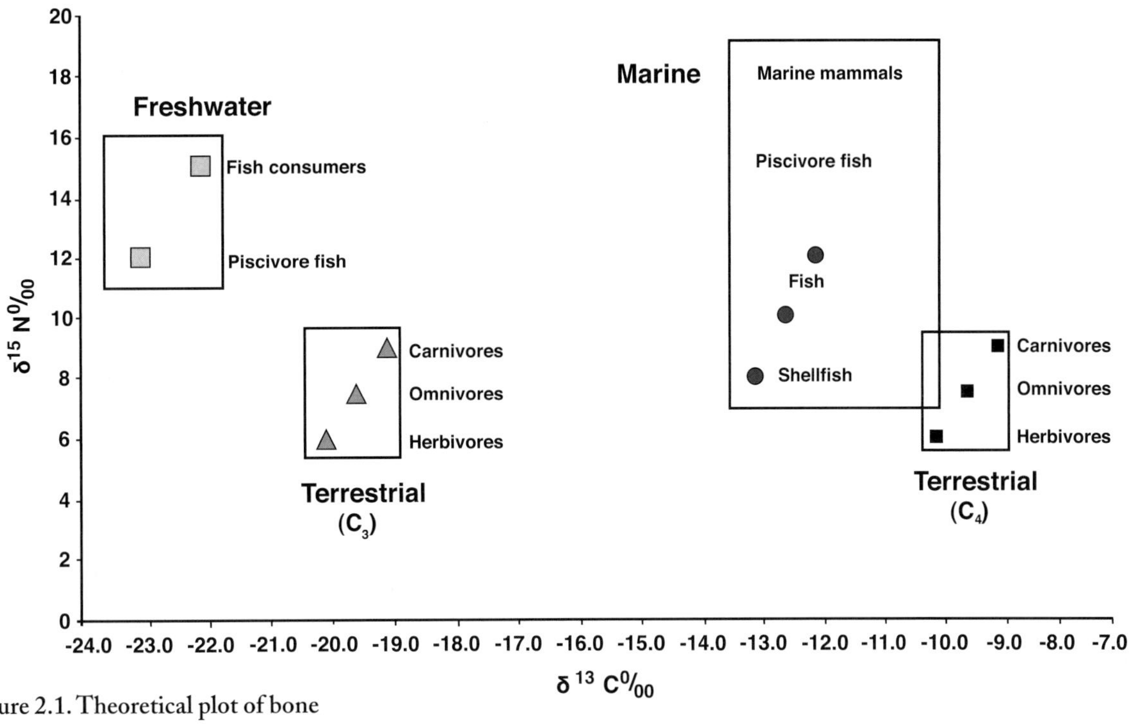

Figure 2.1. Theoretical plot of bone collagen isotope values in Greece

animals have been measured and the local food web established. Unfortunately, many of the isotopic studies undertaken in Greece do not include associated animal remains, mostly because the studies come from human burial sites such as cemeteries that do not contain animal remains. The number of background animal isotopic values is increasing for Greece, however, including a wealth of new data included in this volume. With these new data and ongoing isotope studies in Greece, we are now able to build a better picture of Greek diets at different time periods, look for status and sex differences, compare different regions, and begin to explore in much more depth how human diets have changed through time in the prehistoric and historic Aegean.

REFERENCES

Ambrose, S. H. 1990. "Preparation and Characterization of Bone and Tooth Collagen for Stable Carbon and Nitrogen Isotope Analysis," *JAS* 17, pp. 431–451.

———. 2000. "Controlled Diet and Climate Experiments on Nitrogen Isotope Ratios of Rats," in *Biogeochemical Approaches to Paleodietary Analysis* (Advances in Archaeological and Museum Science 5), ed. S. H. Ambrose and M. A. Katzenberg, New York, pp. 243–259.

Ambrose, S. H., and L. Norr. 1993. "Experimental Evidence for the Relationship of the Carbon Isotope Ratios of Whole Diet and Dietary Protein to Those of Bone Collagen and Carbonate," in *Prehistoric Human Bone: Archaeology at the Molecular Level,* ed. J. B. Lambert and G. Grupe, Berlin, pp. 1–37.

Bourbou, C., and M. P. Richards. 2007. "The Middle Byzantine Menu: Paleodietary Information from Isotopic Analysis of Humans and Fauna from Kastella, Crete," *International Journal of Osteoarchaeology* 17, pp. 63–72.

Brown, T. A., D. E. Nelson, J. S. Vogel, and J. R. Southon. 1988. "Improved Collagen Extraction by Modified Longin Method," *Radiocarbon* 30, pp. 171–177.

Coplen, T. B. 1994. "Reporting of Stable Hydrogen, Carbon, and Oxygen Isotopic Abundances," *Pure and Applied Chemistry* 66, pp. 273–276.

DeNiro, M. J. 1985. "Post-Mortem Preservation and Alteration of *in vivo* Bone Collagen Isotope Ratios in Relation to Paleodietary Reconstruction," *Nature* 317, pp. 806–809.

———. 1987. "Stable Isotopy and Archaeology," *American Scientist* 75, pp. 182–191.

DeNiro, M. J., and S. Epstein. 1981. "Influence of Diet on the Distribution of Nitrogen Isotopes in Animals," *Geochimica et Cosmochimica Acta* 45, pp. 341–351.

Hedges, R. E. M., J. G. Clement, C. D. L. Thomas, and T. C. O'Connell. 2007. "Collagen Turnover in the Adult Femoral Mid-Shaft: Modelled from Anthropogenic Radiocarbon Tracer Measurements," *American Journal of Physical Anthropology* 133, pp. 808–816.

Hedges, R. E. M., and L. M. Reynard. 2007. "Nitrogen Isotopes and the Trophic Level of Humans in Archaeology," *JAS* 34, pp. 1240–1251.

Hoefs, J. 1997. *Stable Isotope Geochemistry,* 4th ed., Berlin.

Kellner, C. M., and M. J. Schoeninger. 2007. "A Simple Carbon Isotope Model for Reconstructing Prehistoric Human Diet," *American Journal of Physical Anthropology* 133, pp. 1112–1127.

Lee-Thorp, J. A. 2008. "On Isotopes and Old Bones," *Archaeometry* 50, pp. 925–950.

Létolle, R. 1980. "Nitrogen-15 in the Natural Environment," in *Handbook of Environmental Isotope Geochemistry,* ed. P. Fritz and J. C. Fontes, Amsterdam, pp. 407–433.

Longin, R. 1971. "New Method of Collagen Extraction for Radiocarbon Dating," *Nature* 230, pp. 241–242.

O'Leary, M. H. 1981. "Carbon Isotopic Fractionation in Plants," *Phytochemistry* 20, pp. 553–567.

Owens, N. J. P. 1987. "Natural Variations in ^{15}N in the Marine Environment," *Advances in Marine Biology* 24, pp. 389–451.

Papathanasiou, A. 2001. *Bioarchaeological Analysis of Neolithic Alepotrypa Cave, Greece* (*BAR-IS* 961), Oxford.

———. 2003. "Stable Isotope Analysis in Neolithic Greece and Possible Implications on Human Health," *International Journal of Osteoarchaeology* 13, pp. 314–324.

Papathanasiou A., C. S. Larsen, and L. Norr. 2000. "Bioarchaeological Inferences from a Neolithic Ossuary from Alepotrypa Cave, Diros, Greece," *International Journal of Osteoarchaeology* 10, pp. 210–228.

Richards, M. P., and R. E. M. Hedges. 2008. "Stable Isotope Evidence of Past Human Diet at the Sites of the Neolithic Cave of Gerani; the Late Minoan III Cemetery of Armenoi;

Grave Circles A and B at the Palace Site of Mycenae; and Late Helladic Chamber Tombs," in *Archaeology Meets Science: Biomolecular Investigations in Bronze Age Greece. The Primary Scientific Evidence, 1997–2003,* ed. Y. Tzedakis, H. Martlew, and M. K. Jones, Oxford, pp. 220–230.

Richards, M. P., and E. Vika. 2008. "Stable Isotope Results from New Sites in the Peloponnese: Sykia, Kalamaki, and Spaliareka," in *Archaeology Meets Science: Biomolecular Investigations in Bronze Age Greece. The Primary Scientific Evidence, 1997–2003,* ed. Y. Tzedakis, H. Martlew, and M. K. Jones, Oxford, pp. 231–235.

Schoeninger, M. J., and M. J. DeNiro. 1982. "Carbon Isotope Ratios of Apatite from Fossil Bone Cannot Be Used to Reconstruct the Diets of Animals," *Nature* 297, pp. 577–578.

———. 1984. "Nitrogen and Carbon Isotopic Composition of Bone Collagen from Marine and Terrestrial Animals," *Geochimica et Cosmochimica Acta* 48, pp. 625–639.

Schwarcz, H., and M. Schoeninger. 1991. "Stable Isotope Analyses in Human Nutritional Ecology," *American Journal of Physical Anthropology* 34, pp. 283–321.

Stenhouse, M. J., and M. S. Baxter. 1979. "The Uptake of Bomb ^{14}C in Humans," in *Radiocarbon Dating. Proceedings of the Ninth International Conference, Los Angeles and La Jolla, 1976,* ed. R. Berger and H. E. Suess, Berkeley, pp. 324–341.

Tauber, H. 1981. "^{13}C Evidence for Dietary Habits of Prehistoric Man in Denmark," *Nature* 292, pp. 332–333.

Triantaphyllou, S., M. P. Richards, C. Zerner, and S. Voutsaki. 2008. "Isotopic Dietary Reconstruction of Humans from Middle Bronze Age Lerna, Argolid, Greece," *JAS* 35, pp. 3028–3034.

van der Merwe, N. J. 1989. "Natural Variation in ^{13}C Concentration and Its Effect on Environmental Reconstruction Using ^{13}C/^{12}C Ratios in Animal Bones," in *The Chemistry of Prehistoric Human Bone,* ed. T. D. Price, Cambridge, pp. 105–125.

Vogel, J. C., and N. J. van der Merwe. 1977. "Isotopic Evidence for Early Maize Cultivation in New York State," *AmerAnt* 42, pp. 238–242.

Wild, E. M., K. A. Arlamovsky, R. Golser, W. Kutschera, A. Priller, S. Puchegger, W. Rom, P. Steier, and W. Vycudilik. 2000. "^{14}C Dating with the Bomb Peak: An Application to Forensic Medicine," *Nuclear Instruments and Methods in Physics Research: Section B, Beam Interactions with Materials and Atoms* 172, pp. 944–950.

CHAPTER 3

Stable Isotope Analyses in Neolithic and Bronze Age Greece: An Overview

by Anastasia Papathanasiou

Stable isotope–based palaeodietary analyses of archaeological populations in the Americas, Asia, and Africa have been widely applied and have yielded results indicating isotopic differences that are large and values that usually do not overlap. In Europe, however, there is a different picture: the general absence of C_4 foodstuffs and the paucity of marine values result in small variation, clustering, and overlapping of the values within a relatively more confined area. Variation does exist, however, with respect to regional climatic factors and culturally influenced food choices.[1] In the specific case of prehistoric Greece, it is possible to define such dietary variation, which may relate to differences in sociocultural behaviors and access to animal protein resources or to climatic differentiation. Several broader archaeological questions can be answered through isotopic studies, as they are directly or indirectly related to dietary reconstruction, such as what formed the base of a prehistoric economy, the pace of the shift to agriculture, the use of secondary products, animal husbandry and pastoralism practices, the importance of fishing and hunting, the introduction of new plants, internal social differentiation, as well as regional, temporal, and cultural variations.

In the first part of this chapter, the food web of prehistoric Greece, as defined by archaeobotanical and zooarchaeological studies will be reviewed, and in the second part, the results and interpretations of stable isotope analyses will be presented, combining the archaeological evidence with chemical analyses.

THE FOOD WEB

Zooarchaeological and Archaeobotanical Studies of Prehistoric Greece

Palaeodietary analysis has never been an easy, straightforward process, mainly due to the biases and limitations inherent in the archaeological record. In prehistoric Greek archaeology, conventional methods such as archaeobotanical and zooarchaeological analyses have been employed by archaeologists to reconstruct past diets. Botanical and faunal specialists

1. Van Klinken, Richards, and Hedges 2000.

have often mentioned the problems and biases of such analyses, which are further elaborated below.

Most animal and plant remains in archaeological deposits are the result of complex human and nonhuman interactions with the environment. Bones and seeds retrieved from an archaeological site are the outcome of a series of predepositional, depositional, and postdepositional procedures that have contributed to the generation of the zooarchaeological and archaeobotanical records.[2] Thus, one of the primary roles of the specialist is to interpret these records in order to determine the animal and plant species consumed by the inhabitants of a settlement.

The relative quantities of different plant and animal species in the diet of a past population deviate from the relative quantities of the preserved material recovered archaeologically. Commonly, what is recovered is only a small proportion of the original resources, instead of bulk information about the site as a whole, with very little indication, if any, regarding how these resources were actually used.

In particular, quite a number of preservation and recovery biases arise in botanical analyses. These biases include the following:

1. Not all excavators systematically recover seeds, and the retrieval can be inadequate if the aperture of the sieve is not fine enough to allow for their recovery.
2. Not all plants are equally well preserved, so that relative quantities recovered from a site may not accurately reflect the importance (or by contrast, the insignificance) of the various plants once exploited at the site.[3]
3. It is difficult to determine whether recovered seeds represent a crop, a weed, a wild plant deliberately gathered for human consumption, or a plant consumed by animals; furthermore, seeds may represent: a foodstuff or fodder collected from the wild, a food resource comprised of deliberately sown plants, crop processing waste, or a combination of the above.[4]
4. Since charring and carbonization are the most common processes of preservation for botanical remains in prehistoric Greece, dense fruits or seeds are most likely to survive exposure to fire. Cereals, therefore, are more likely to be recovered, while legumes, fruits, nuts, and especially roots, leaves, and tubers, all are rarely preserved through carbonization and appear less frequently in the archaeological record.[5]

In faunal analyses, retrieval and interpretation problems include the following:

1. Species representation is a product of the type of recovery method employed (hand versus water flotation and/or sieving), with more meticulous methods yielding a broader spectrum of species, including small-sized and/or juvenile animals, and certain fragile body parts.
2. Large animal bones exhibit a different (i.e., greater) rate of survival than small bones and microfauna, such as birds, fish, reptiles, and mollusks.[6]

2. Dennell 1976; Lyman 1982; Jones 1983; Klein and Cruz-Uribe 1984; Reitz and Wing 1999, pp. 124–139.
3. Hansen 2000, p. 13.
4. Hansen 2000, p. 14.
5. Hansen 2000, p. 15.
6. Payne 1972.

3. Proportions of wild and domestic fauna are not directly comparable between sites due to differences in local habitats and in standards of bone retrieval during and after excavation.[7]
4. "Failure to recognize structured (intentionally arranged) deposits resulting from ceremonial consumption, or large concentrations of food refuse resulting from feasting," as opposed to common food remains characteristic of residential areas, leads to "misinterpretation of the scale, frequency, and biological and social implications of consumption of a certain food."[8]
5. Not all animal remains recovered at a site can be treated as food in a static manner without understanding the food production and consumption strategy of each society.
6. It is important to stress that "food and drink are socially defined as such, and their consumption is primarily a social activity."[9]
7. Human actions prior to or following consumption, such as extensive butchering or discard of food processing or cooking waste outside the confines of the settlement, may lead to underrepresentation of the zooarchaeological and archaeobotanical information.

In summary, although archaeobotany and zooarchaeology can provide positive identification of botanical and faunal species present at a site and occasionally reveal how these food resources were exploited, when these approaches are applied to dietary reconstruction, emphasis should be placed on understanding the relative proportions and dietary uses of the recovered species, determining the extent to which the recovered species were used as food resources, and quantifying discernable consumption patterns.

Archaeobotanical studies have generally demonstrated that from the Mesolithic period, people in Greece were exploiting wild cereals (oats and barley), legumes (lentils, peas, and bitter vetch), as well as almonds, pistachios, and pears (see Table 3.1 for the scientific names of the species mentioned).[10] In the Early Neolithic (EN), a new picture emerges as cereal resources change from oats to glume wheats, emmer, einkorn, and barley. The six-row variety of barley becomes dominant over wheat, especially in southern Greece, during the Late Neolithic (LN). Two-row barley occurs together with six-row barley in the Early Neolithic, but it is soon completely replaced by the more productive six-row variety.[11] Bread wheat also appears in the Neolithic, but it is rare, perhaps because it is a more labor-intensive crop.[12] Legumes include lentils as the most common species, followed by grass peas, peas, bitter vetch, chickling vetch, and chick peas. Horsebeans are introduced in the Late Neolithic when there is an overall increase in the use of legumes.[13] In addition, more fruits are added, such as blackberries, cornelian cherries, pears, elderberries, figs, almonds, and acorns.[14] As far as olives and grapes are concerned, wild varieties of both probably existed at Neolithic sites, but neither macrofossil nor palynological evidence supports their exploitation and domestication until the Bronze Age or perhaps slightly earlier during the Final Neolithic (FN).[15]

During the Bronze Age, emmer wheat and two-row barley are the two most common cereal species, followed by einkorn and six-row barley and bread wheat. Spelt wheat and possibly millet (a C_4 plant) are also

7. Halstead 1999.
8. Hamilakis 2000, p. 59.
9. Hamilakis 2000, p. 59.
10. Hansen 2000, pp. 15–17.
11. Hansen 1988, p. 44.
12. Renfrew 1973; Hansen 1988, p. 43.
13. Hansen 1988, p. 44.
14. Hansen 2000, p. 17; Valamoti 2004; 2007, p. 91.
15. Hansen 1988, pp. 44–45.

TABLE 3.1. PLANTS IDENTIFIED BY ARCHAEOBOTANICAL STUDIES FOR PREHISTORIC GREECE

Common Name	*Scientific Name*
lentils	*Lens culinaris*
peas	*Pisum sativum*
bitter vetch	*Vicia ervilia*
emmer	*Triticum dicoccum*
einkorn	*Triticum monococcum*
six-row barley	*Hordeum hexastichum*
two-row barley	*Hordeum distichum*
grass peas	*Lathyrus sativus*
chickling vetch	*Lathyrus cicera*
chick peas	*Cicer arietinum*
horsebeans	*Vicia faba*
blackberries	*Rubus fruticosus*
cornelian cherries	*Cornus mas*
pears	*Pyrus amygdaliformis*
elderberries	*Sambucus nigra* and *Sambucus ebulus*
figs	*Ficus carica*
almonds	*Prunus dulcis*
acorns	*Quercus* sp.
olives	*Olea europea* var. *oleaster*
wild grapes	*Vitis sylvestris*
domesticated grapes	*Vitis vinifera*

introduced. Among the legumes, lentils, peas, and bitter vetch are the most dominant species, followed by grass pea and fava beans. Grapes, olives, and figs similarly predominate, while other fruits and nuts are rarely identified,[16] and no aromatic or medicinal plants are known.[17]

Turning to zooarchaeological results, sheep and goats (ovicaprids; *Ovis/Capra*), especially sheep, seem to predominate over cattle *(Bos taurus)* and pig *(Sus scrofa domesticus)* in the Early Neolithic period. As the Neolithic progresses, the use of sheep and goats gradually gives way to a mixed model with cattle and, to a lesser extent, pigs, but ovicaprids continue to predominate throughout the entire Neolithic.[18] During the Early Neolithic, ovicaprids and cattle were exploited for their meat according to slaughter-age patterns, as the young slaughter ages indicate an unspecialized meat strategy.[19] From the Middle Neolithic (MN) onward, they were also exploited for their secondary products such as milk, wool, and (in the case of cattle) traction, while pigs were only raised for meat and were slaughtered at a young age.[20] The discernable ratio between wild and domestic animals indicates that the Neolithic economy of most settlements in Greece was based primarily on domestic animals, despite occasional increases in the proportion of game.[21] The predominant wild species include red deer

16. Hansen 2000, p. 18.
17. Valamoti 2004, pp. 132–133; 2007, p. 91.
18. Halstead 1981, 2006.
19. Halstead 2000; 2008, p. 240.
20. Curci and Tagliacozzo 2003, p. 130.
21. Halstead 1999.

(Cervus elaphus), fallow deer *(Dama dama),* aurochs *(Bos primigenius),* and wild boar *(Sus scrofa)* as well as a range of smaller species that appear in various frequencies reflective of the environment of a site.[22]

This trend continues in the Bronze Age when "wild animals are rarely represented and remain an insignificant component of subsistence (10% or less)."[23] Although hunting is frequently depicted in the iconography and epic traditions of the time, it seems that hunting played an institutional, social, and political role rather than having been simply a practical subsistence strategy.[24]

Inhabitants of Neolithic sites located close to the sea or in deltaic environments also collected marine resources from cockles (*Cerastoderma* spp.), limpets (*Patella* spp.), and topshells (Trochidae), along with a number of other shellfish such as oysters (Ostreidae), queen scallops (Pectinidae), rayed trough-shells (Mactridae), tellins (Tellinidae), and razor shells (Solenidae). Fish bones from coastal species also found in waters with low salinity, such as various sparids (Sparidae), mullets (*Mugil* spp.), eels *(Anguilla anguilla),* groupers *(Dicentrarchus labrax),* seabass (*Epinephelus* spp.), meagres *(Argyrosomus regius),* sharks and rays, which have been recovered in low quantities. Mesopelagic catches, such as tuna and mackerel (Scombridae), are occasional and sometimes found both at coastal and inland sites, recovered in the form of chunks, possibly indicating that they were salted. Seasonality studies indicate that the fishing activity would start in early spring and would continue through summer to autumn, with occasional preserved catches for winter consumption. Fishing and shellfish gathering would likely have been supplementary to pastoral and agricultural activities.[25]

Similarly, during the Bronze Age, fish, although frequently mentioned in the archaeological record,[26] are notably absent even at coastal Bronze Age sites.[27] The alimentary emphasis remains on domestic animals whose respective importance and prevalence varies over time. Cattle are more common in the Early Bronze Age, but as they declined thereafter, ovicaprids increased up to 60% of domestic animals in the archaeological assemblage. Pigs, exploited only for the primary product of their meat, remained at a low but constant frequency (17% to 24%). Over time, animals without significant secondary product potential become less important.[28]

Although a review of the zooarchaeological and archaeobotanical record indicates the relative representation of species as reflected in the archaeological record of Neolithic and Bronze Age Greece (both mainland and the islands), this representation may not correlate exactly with the foodstuffs consumed. It is important to understand to what extent the plant and animal species of the archaeological record actually constituted past human diet. To this end, biochemical analyses should be employed.[29] Before undertaking the task of reconstructing human diet, however, it is essential to identify the local ecology and the parameters and values of the food web (plants and animals) upon which humans relied.

22. Halstead 1999; Curci and Tagliacozzo 2003.
23. Cosmopoulos, Greenfield, and Ruscillo 2003, p. 151.
24. Hamilakis 2003.
25. Theodoropoulou 2007a, 2007b.
26. Doumas 1992; Powell 1996.
27. Cosmopoulos, Greenfield, and Ruscillo 2003, p. 151.
28. Cosmopoulos, Greenfield, and Ruscillo 2003, p. 151.
29. See Chap. 2.

Stable Isotope Analysis of Animal Bone Samples

Human bone composition reflects the particular plants and animals consumed in a specific environment, as the isotopic composition of human body tissues is indicative of the plant and animal foods consumed by individuals during tissue formation. In order to comprehend human diet, therefore, a reference baseline needs to be established consisting of the isotopic values of the consumed foodstuffs (see also Chap. 2). Although animal bones in direct association with human bones are rarely available for analysis in isotopic studies,[30] they nevertheless are essential for the interpretation of the human isotopic values from a site. Two further points should be made. First, herbivore values represent the isotopic values of the animal protein consumed by humans, as well as the value of the plants upon which they fed (by subtracting one trophic level for their own ^{15}N enrichment). Second, carnivore values can be compared to human values, because both animals and humans consumed the same animal protein source.[31] Additional conclusions can be drawn from animal values with regard to the use of the animals themselves and their feeding strategies.

For animal isotopes to prove useful in dietary reconstructions, there is one prerequisite already alluded to above: the animal isotopes must originate ideally from the same context or, at least, the same site as the human population under study, so as to be representative of the same geographic and environmental context.

STABLE ISOTOPE ANALYSIS: RESULTS AND INTERPRETATIONS

Background Studies

Biochemical analysis of bone for understanding prehistoric diet in Greece was first conducted by Angel and Bisel,[32] who utilized trace element (calcium, strontium, zinc, manganese, phosphorus) analysis on human and animal bone samples while controlling for possible contamination. In the same study, the authors attempted a dietary reconstruction for the inhabitants of prehistoric Franchthi Cave. Their results, in which they identify fish, meat, and bread wheat, were heavily criticized by Cook,[33] since these findings do not fit with the archaeological evidence from the site, and problems of contamination were obvious due to diagenesis, the trace element exchange between the analyzed bone material and its environment (soil and grave goods). Although trace element analysis is still occasionally used today, most bioarchaeologists employ stable isotope analysis for palaeodietary reconstruction, since stable isotopes provide a direct measure of past human diet and can be controlled for contamination.

Neolithic Period: Human Bone Samples

In the early 1990s, a number of researchers began using stable isotopes to assess ancient Greek diet, food-producing strategies, economy, and health. Papathanasiou, during an analysis of human osteological material

30. Hedges and Reynard 2007.
31. Müldner and Richards 2005, pp. 44–45; Hedges and Reynard 2007, pp. 1242–1246.
32. Angel and Bisel 1986.
33. Cook 2000, pp. 102–103.

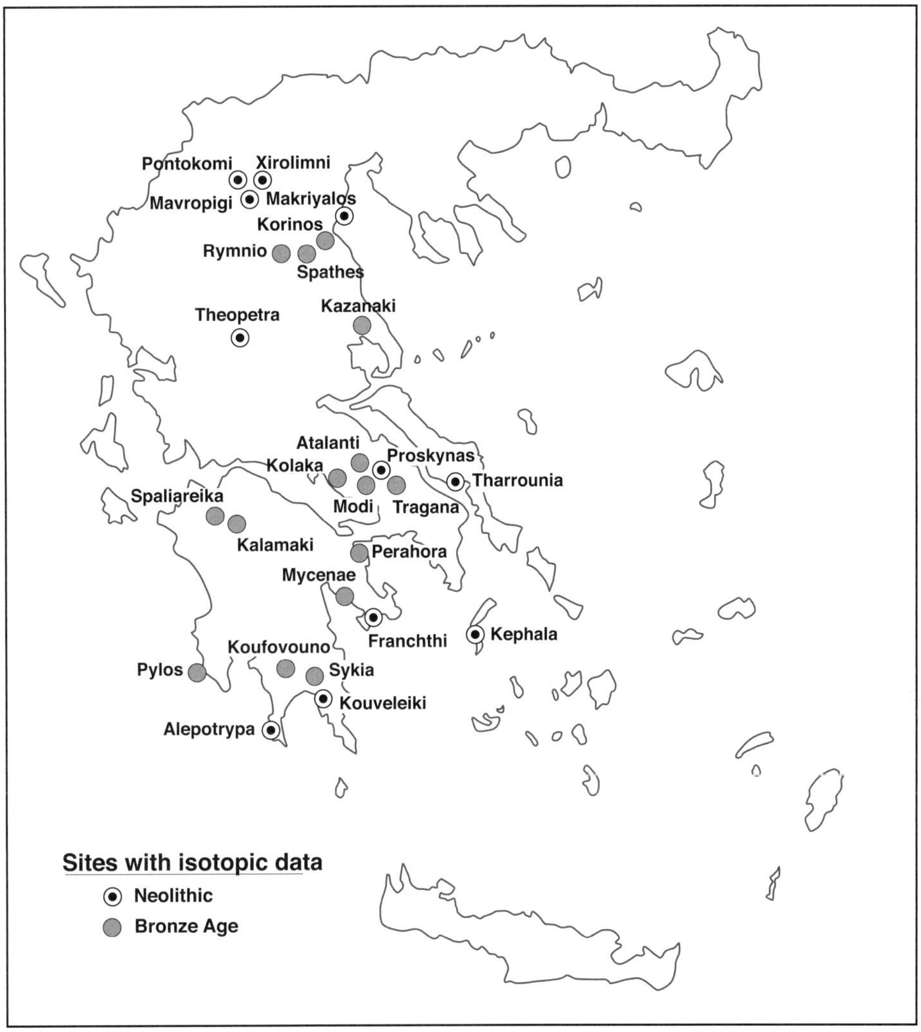

Figure 3.1. Neolithic and Bronze Age sites with isotopic data discussed in the text

from Neolithic Alepotrypa Cave in southern Greece, observed a great prevalence of anemic conditions. In an effort to investigate the etiology of anemia, and more specifically to determine if it was of genetic (as had been assumed up to that point) or of nutritional origin, isotope analysis was employed beginning with a pilot study of 11 human and 10 animal samples in 1993 at the University of Illinois at Urbana-Champaign.[34] The results pointed to a rather poor diet based mainly on C_3 terrestrial resources with insignificant consumption of marine resources despite accessibility to the latter.[35] This initial study was then broadened to compare coastal and inland Neolithic sites from all over Greece in order to verify or refute the initial results. These analyses were conducted at the University of Florida.

The study's first results, published in 1999, and later in 2001 and 2003,[36] referred to three coastal Neolithic sites (Alepotrypa, Franchthi, and Kephala) and three inland sites (Theopetra, Tharrounia, and Kouveleiki). All sites included in the study are shown in Figure 3.1. A total of 101 samples were examined, most of which were very well preserved and gave valid values (see Table 3.2 and Fig. 3.2). In general, all the $\delta^{13}C$ values of both coastal and inland populations cluster together. The differences between these values are statistically insignificant, but the overall pattern

34. Papathanasiou 1994, pp. 54–57; Papathanasiou, Larsen, and Norr 1995; 2000, pp. 221–223.
35. Papathanasiou, Larsen, and Norr 2000, pp. 221–223.
36. Papathanasiou 1999a, 2001, 2003.

TABLE 3.2. MEAN STABLE ISOTOPE VALUES FOR HUMANS AT PREHISTORIC SITES

Site	*$\delta^{13}C \pm 1\sigma$ (‰ VPDB)*	*$\delta^{15}N \pm 1\sigma$ (‰ AIR)*	*N*
NEOLITHIC SITES			
Alepotrypa	-19.89 ± 0.32	7.28 ± 0.86	25
Franchthi	-18.83 ± 0.57	8.74 ± 0.90	10
Kephala	-18.58 ± 0.54	9.22 ± 1.13	4
Theopetra	-20.05 ± 0.37	7.65 ± 0.55	12
Tharrounia	-19.99 ± 0.22	8.04 ± 0.67	20
Kouveleiki	-19.84 ± 0.04	8.09 ± 0.33	2
Proskynas	-19.72 ± 0.70	7.50 ± 1.57	10
Makriyalos	-19.86 ± 0.33	6.78 ± 0.87	22
Xirolimni	-20.22 ± 0.01	8.30 ± 0.10	2
Mavropigi	-20.40 ± 0.20	7.45 ± 0.63	10
Pontokomi	-19.90	10.31	1
BRONZE AGE SITES			
Mycenae chamber	-19.20 ± 0.14	7.76 ± 1.18	10
Mycenae A	-18.66 ± 0.53	10.43 ± 1.07	9
Mycenae B	-19.23 ± 0.56	10.07 ± 0.97	6
Pylos tholos	-19.75 ± 0.38	9.56 ± 0.55	4
Pylos chamber	-19.82 ± 0.31	8.80 ± 0.58	26
Sykia	-19.38 ± 0.26	8.17 ± 0.65	6
Spaliareika	-20.07 ± 0.27	8.39 ± 0.94	7
Kalamaki	-19.97 ± 0.57	7.82 ± 0.85	15
Spathes	-18.54 ± 0.44	7.64 ± 0.72	5
Rymnio	-17.64 ± 0.78	8.61 ± 1.17	4
Korinos	-19.20 ± 0.56	8.74 ± 1.40	4
Makriyalos	-19.61 ± 0.38	7.35 ± 0.71	6
Perahora	-19.26 ± 0.34	9.05 ± 0.57	19
Koufovouno	-19.81 ± 0.26	7.88 ± 0.38	5
Tragana	-19.33 ± 1.28	9.30 ± 0.95	4
Atalanti	-20.45 ± 0.88	7.73 ± 0.83	4
Kolaka	-18.30 ± 1.64	7.83 ± 1.66	4
Modi	-19.25 ± 0.31	7.30 ± 1.33	4
Kazanaki	-19.51 ± 0.11	11.40 ± 2.07	2

that emerges represents primarily a C_3 terrestrial diet. The $\delta^{15}N$ values are generally low and indicate that very little of the dietary protein had a marine origin. Although there is significant clustering, the variation that exists will be discussed later in this chapter.

Besides collagen, additional isotopic evidence comes from the carbon isotope ratios in the bone apatite carbonate (as outlined in Chap. 2) for

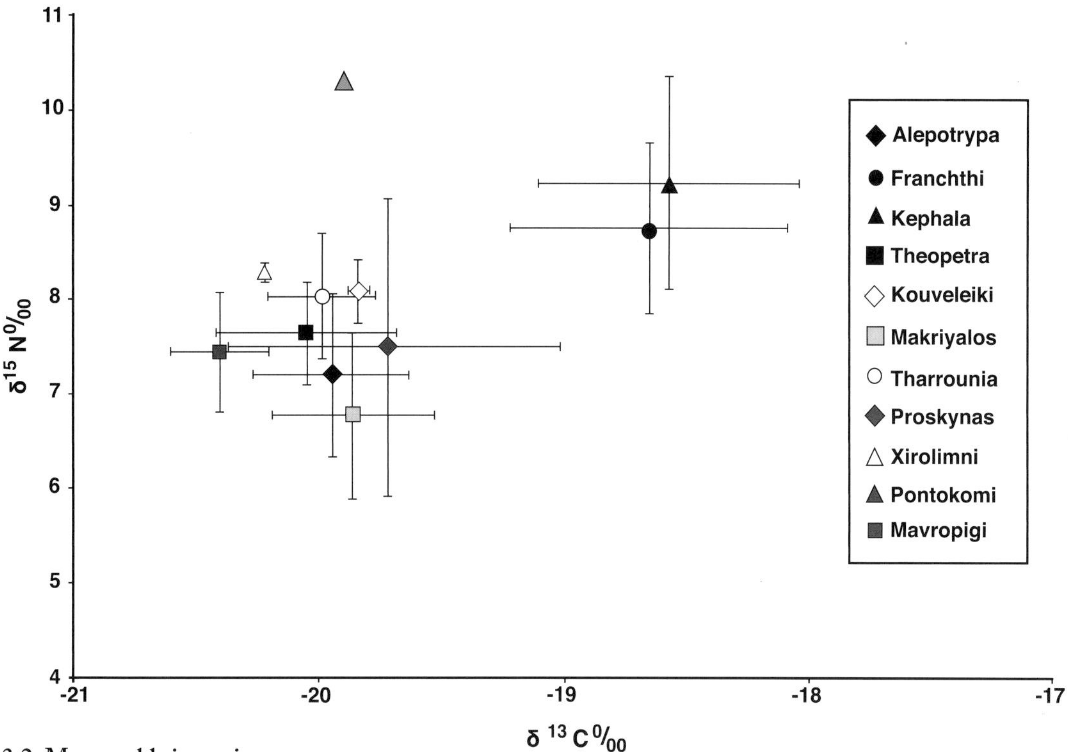

Figure 3.2. Mean stable isotopic values for humans from Neolithic Greek sites

a group of 11 samples referred to above as the "pilot study." The mean difference between human bone carbonate and collagen values falls into the intermediate range, which suggests that protein and energy were derived from foodstuffs having similar isotopic compositions. Both carbonate and collagen analyses point consistently toward C_3 resources. Furthermore, the very small range of values (4‰ in δ^{15}N and 2‰ in δ^{13}C) implies a high degree of homogeneity in the diet for the populations under study.

The above results reaffirmed the hypothesis that Neolithic Greece, regardless of geographical location, was occupied by agricultural groups with a land-based economy and a diet consisting of only occasional or periodic exploitation of near-shore marine protein resources. Greek Neolithic populations focused their diets on C_3 plants (wheat, barley, and legumes), dairy products, and possibly meat. They lacked significant amounts of marine protein, however, since their nitrogen values are not as positive as would be expected in a diet incorporating marine foods.

Subsequently, another study was conducted on Early Neolithic peoples of northern Greece for the sites of Xirolimni, Pontokomi, and Mavropigi to compare Early versus Late Neolithic populations and clarify the Early Neolithic shift to agriculture in Greece. Twenty-five samples from 25 different individuals were analyzed at the Max Planck Institute for Evolutionary Anthropology in Leipzig. The mean value for δ^{13}C at Xirolimni is -19.80‰ ± 0.27‰ (n = 14), at Mavropigi -20.40‰ ± 0.45‰ (n = 10), and at Pontokomi -19.90‰ (n = 1), while the mean value for δ^{15}N at Xirolimni is 8.69‰ ± 0.39‰, at Mavropigi 7.45‰ ± 0.80‰, and at Pontokomi 10.31‰.[37] Both the δ^{13}C values and δ^{15}N values cluster in the

37. Papathanasiou and Richards 2011.

range of values that are indicative of a primarily C_3 terrestrial diet with no marine input, as would be anticipated given the large distance of the sites from the sea. These Early Neolithic populations, as did the Later Neolithic populations, focused their diet on domesticated C_3 plants, indicating a very rapid and complete transition to an agricultural economy and diet.[38] When the above results are compared against the means of other Greek Neolithic sites from later periods, there is a difference in the $\delta^{15}N$ values, especially at Xirolimni, which suggests that this population incorporated more terrestrial animal protein in their diet, either from domesticated animal or wild game, than populations in later periods. The difference is small but statistically significant using Student's *t*-test ($t = 2{,}220$, $p<0.00003$).

A degree of variation is observed, however, as statistically significant differences exist between the mean $\delta^{15}N$ values for Alepotrypa Cave (7.3‰) on the one hand, and the coastal sites of Franchthi Cave (8.7‰) and Kephala (9.2‰) on the other, at levels of $p<0.0005$ ($t = 1{,}408$) and $p<0.0353$ ($t = 18$), respectively, based on Student's *t*-test. Also, a statistically significant difference was found when comparing the mean values of Franchthi and Kephala against all of the other inland sites. Franchthi and Kephala depart from a mainly C_3 terrestrial pattern by exhibiting a more varied diet. Although these results were based on C_3 foodstuffs, they show more significant amounts of meat and dairy products, and possibly marine foods. No other discernible differences in isotopic values—in terms of gender and social status, as expressed in comparisons between males and females, adults and subadults, or among different burial types—were detected for the Neolithic period.

The foregoing results are supported by another study, conducted by Triantaphyllou, on the Neolithic site of Makriyalos in northern Greece.[39] Triantaphyllou tested 22 samples, which generally exhibit the same terrestrial C_3 pattern. This pattern has also been observed in the dietary analysis of 14 samples from Proskynas in central Greece, a Neolithic site extending into the Middle Bronze Age, where mean values of $\delta^{13}C$ at -19.7‰ and $\delta^{15}N$ at 7.5‰ indicate a terrestrial diet focused on C_3 plants.[40]

Neolithic Period: Animal Bone Samples

To further refine the picture of human diet in Neolithic Greece, a number of faunal samples of Neolithic animal bones were selected from the same sites of Alepotrypa, Franchthi, Theopetra, Xirolimni, Mavropigi, and Kephala, from which human bone had already been tested. These samples represented as many species as possible, including wild and domestic animals. Bones from herbivores, carnivores, omnivores, and fish were sampled. The species identification, number, and provenance of the samples are shown in Tables 3.3 and 3.4, and in Figure 3.3.

The results for most of the aforementioned animals fall within the expected range. With respect to $\delta^{13}C$ values, nearly all of them, except for a few samples discussed below, fall within the C_3 range (from -22.9‰ to -18.8‰) and therefore show a high degree of homogeneity. Fish values are very positive in both nitrogen and carbon.[41]

38. Papathanasiou 2011, pp. 88–95.
39. Triantaphyllou 2001; see also Chap. 4.
40. Papathanasiou, Zachou, and Richards 2009, pp. 229–230.
41. Vika and Theodoropoulou 2012.

TABLE 3.3. MEAN STABLE ISOTOPE VALUES FOR ANIMALS AT NEOLITHIC SITES

	Bovid	*Sheep/Goat*	*Deer*	*Pig*	*Wild Pig*	*Dog*	*Fox*	*Fish*	*Cat*	*Badger*
Alepotrypa										
$\delta^{15}N \pm 1\sigma$ (‰ AIR)	4.41 ± 0.54	5.91 ± 0.53	4.32	4.61 ± 0.07	–	–	6.95	8.97 ± 3.39	–	–
$\delta^{13}C \pm 1\sigma$ (‰ VPDB)	-18.97 ± 6.24 (n = 7)	-21.72 ± 0.59 (n = 4)	-21.45 (n = 1)	-20.82 ± 0.03 (n = 3)	–	–	-19.41 (n = 1)	-11.78 ± 2.93 (n = 3)	–	–
Kephala										
$\delta^{15}N \pm 1\sigma$ (‰ AIR)	6.47 ± 0.02	7.02 ± 2.25	5.47	–	–	8.49	–	–	–	–
$\delta^{13}C \pm 1\sigma$ (‰ VPDB)	-20.70 ± 0.11 (n = 2)	-20.18 ± 0.21 (n = 2)	-19.99 (n = 1)	–	–	-18.89 (n = 1)	–	–	–	–
Makriyalos										
$\delta^{15}N \pm 1\sigma$ (‰ AIR)	–	–	3.24 ± 1.25	5.25 ± 0.76	4.20 ± 0.84	–	–	–	–	–
$\delta^{13}C \pm 1\sigma$ (‰ VPDB)	–	–	-21.67 ± 0.39 (n = 5)	-20.49 ± 0.27 (n = 5)	-20.38 ± 0.44 (n = 5)	–	–	–	–	–
Franchthi										
$\delta^{15}N \pm 1\sigma$ (‰ AIR)	–	5.53 ± 1.08	5.28 ± 1.22	6.52 ± 0.98	–	7.68 ± 0.03	–	8.27 ± 0.37	7.42	–
$\delta^{13}C \pm 1\sigma$ (‰ VPDB)	–	-19.58 ± 1.23 (n = 4)	-21.20 ± 0.39 (n = 4)	-19.95 ± 0.04 (n = 2)	–	19.07 ± 0.02 (n = 3)	–	-14.75 ± 0.73 (n = 2)	-19.47 (n = 1)	–
Theopetra										
$\delta^{15}N \pm 1\sigma$ (‰ AIR)	4.94 ± 0.01	5.49 ± 0.25	4.91	5.79 ± 3.84	3.89	7.48	–	–	–	4.61 ± 10.67
$\delta^{13}C \pm 1\sigma$ (‰ VPDB)	-20.18 ± 0.63 (n = 2)	-19.75 ± 0.35 (n = 4)	-20.80 (n = 1)	-20.35 ± 1.02 (n = 2)	-20.00 (n = 1)	-18.81 (n = 1)	–	–	–	-19.26 ± 0.45 (n = 2)
Mavropigi										
$\delta^{15}N \pm 1\sigma$ (‰ AIR)	6.06 ± 0.01	4.67 ± 0.21	4.60 ± 0.57	6.29 ± 0.39	–	–	6.70	–	–	–
$\delta^{13}C \pm 1\sigma$ (‰ VPDB)	-20.06 ± 0.79 (n = 2)	-20.32 ± 0.05 (n = 3)	-20.84 ± 0.55 (n = 3)	-20.73 ± 0.58 (n = 3)	–	–	-19.50 (n = 1)	–	–	–
Xirolimni										
$\delta^{15}N \pm 1\sigma$ (‰ AIR)	–	5.15 ± 0.60	–	4.73	–	–	–	–	–	–
$\delta^{13}C \pm 1\sigma$ (‰ VPDB)	–	-19.54 ± 0.22 (n = 5)	–	-19.54 (n = 1)	–	–	–	–	–	–

TABLE 3.4. STABLE ISOTOPE VALUES FOR ANIMALS AND MODERN FOODSTUFFS FROM NEOLITHIC SITES

Sample	*Site*	*Material*	*% Coll.*	*% N*	*% C*	*C:N*	*$\delta^{13}C$ (‰ VPDB)*	*$\delta^{15}N$ (‰ AIR)*
MAX PLANCK INSTITUTE								
APT-1	Alepotrypa	*Bos*	0.80	13.51	41.17	3.55	-17.26	5.13
APT-2	Alepotrypa	Red deer	2.82	15.81	44.09	3.25	-21.45	4.32
APT-3	Alepotrypa	Goat	2.71	15.50	43.88	3.30	-21.32	5.21
APT-6	Alepotrypa	Tuna	0.47	9.42	27.30	3.38	-13.66	7.38
APT-7	Alepotrypa	Pig	2.40	14.99	43.88	3.41	-20.81	4.74
APT-8	Alepotrypa	Fox	3.43	15.44	44.50	3.36	-19.41	6.95
APT-11	Alepotrypa	*Bos*	2.69	15.21	43.71	3.35	-21.27	3.82
APT-12	Alepotrypa	Bass	1.38	15.36	43.30	3.29	-10.32	8.55
FRA-1	Franchthi	Tuna	–	–	–	–	–	–
FRA-2	Franchthi	Tuna	–	–	–	–	–	–
FRA-3	Franchthi	Tuna	0.20	–	–	–	–	–
FRA-4	Franchthi	Bass	–	–	–	–	–	–
FRA-5	Franchthi	Fallow deer	–	–	–	–	–	–
FRA-6	Franchthi	Dog	0.71	15.03	42.69	3.31	-18.97	7.64
FRA-7	Franchthi	Feline	0.31	14.52	42.93	3.45	-19.47	7.42
FRA-8	Franchthi	Cervid	–	–	–	–	–	–
FRA-9	Franchthi	*Equus*	0.09	–	–	–	–	–
FRA-10	Franchthi	Cervid	0.16	–	–	–	–	–
FRA-11	Franchthi	Cervid	0.19	10.03	29.46	3.42	-21.44	5.39
FRA-12	Franchthi	Cervid	0.37	3.87	11.41	3.44	-20.40	3.83
FRA-13	Franchthi	*Equus*	–	–	–	–	–	–
FRA-14	Franchthi	Dog	0.03	–	–	–	–	–
FRA-15	Franchthi	Cervid	0.16	–	–	–	–	–
FRA-16	Franchthi	Cervid	0.31	–	–	–	–	–
FRA-17	Franchthi	Cervid	–	–	–	–	–	–
FRA-18	Franchthi	Deer	0.38	4.72	14.92	3.68	-20.56	6.41
FRA-19	Franchthi	Cervid	0.30	–	–	–	–	–
FRA-20	Franchthi	Cervid	–	–	–	–	–	–
FRA-21	Franchthi	Dog	0.77	14.61	43.31	3.46	-19.03	7.87
FRA-22	Franchthi	Pig	0.15	14.10	42.12	3.49	-20.08	5.82
FRA-23	Franchthi	Red deer	0.45	4.42	13.30	3.51	-21.87	5.37
FRA-24	Franchthi	Cervid	–	–	–	–	–	–
FRA-25	Franchthi	Cervid	0.31	6.20	18.11	3.41	-21.07	6.52
FRA-26	Franchthi	Pig	0.19	–	–	–	–	–
FRA-27	Franchthi	Tuna	–	13.49	41.00	3.55	-14.14	8.70
FRA-28	Franchthi	Sheep	0.38	14.44	42.01	3.40	-19.81	5.89
FRA-29	Franchthi	Cattle	–	–	–	–	–	–
FRA-30	Franchthi	Red deer	0.23	4.30	13.55	3.67	-21.39	4.19
FRA-31	Franchthi	Canid	2.23	14.34	41.01	3.34	-19.21	7.54
FRA-32	Franchthi	Sheep/goat	0.42	6.21	18.65	3.50	-20.70	4.77
FRA-33	Franchthi	Pig	1.15	11.72	33.52	3.33	-19.81	7.22

TABLE 3.4—*Continued*

Sample	*Site*	*Material*	*% Coll.*	*% N*	*% C*	*C:N*	*δ¹³C (‰ VPDB)*	*δ¹⁵N (‰ AIR)*
FRA-34	Franchthi	Sheep/goat	2.14	15.24	42.81	3.28	-19.75	4.62
FRA-35	Franchthi	Sheep/goat	2.46	15.27	43.26	3.31	-18.05	6.83
FRA-36	Franchthi	Tuna	0.28	5.27	17.22	3.81	-15.35	7.84
Xan1	Xirolimni	Pig	1.44	14.95	41.54	3.24	-19.54	4.73
Xan2	Xirolimni	Goat	0.83	13.71	39.35	3.35	-19.46	4.47
Xan3	Xirolimni	Sheep/goat	2.09	14.90	41.48	3.25	-19.77	6.04
Xan4	Xirolimni	Sheep/goat	3.76	16.92	46.30	3.19	-18.78	5.64
Xan5	Xirolimni	Sheep/goat	0.98	14.96	40.89	3.19	-20.04	5.37
Xan6	Xirolimni	Sheep/goat	2.59	15.38	41.76	3.17	-19.63	4.23
M1	Mavropigi	Pig	1.22	14.54	39.92	3.20	-19.86	6.82
M2	Mavropigi	Pig	0.45	14.11	39.17	3.24	-21.05	6.45
M3	Mavropigi	Pig	1.35	14.84	40.65	3.20	-21.29	5.60
M4	Mavropigi	*Capreolus*	1.29	14.67	40.86	3.25	-21.61	5.45
M5	Mavropigi	*Cervus elaphus*	0.78	14.82	40.93	3.22	-20.14	4.00
M6	Mavropigi	*Cervus elaphus*	1.00	12.98	35.88	3.23	-20.77	4.35
M7	Mavropigi	*Ovis/capra*	1.35	14.54	40.73	3.27	-20.45	4.82
M8	Mavropigi	*Ovis/capra*	0.69	14.63	40.61	3.24	-20.08	5.02
M9	Mavropigi	*Ovis/capra*	0.83	13.98	38.88	3.25	-20.45	4.15
M10	Mavropigi	*Ursus*	0.34	–	–	–	–	–
M11	Mavropigi	*Bos*	0.40	14.75	43.00	3.40	-20.69	6.06
M12	Mavropigi	*Bos*	2.08	14.30	39.40	3.22	-19.43	6.07
M13	Mavropigi	*Bos*	0.35	9.78	30.41	3.63	-19.06	6.34
M14	Mavropigi	*Vulpes*	1.99	15.26	42.07	3.22	-19.45	6.68
T 212	Theopetra	Sheep	4.26	15.38	42.36	3.21	-20.45	5.68
T 94B	Theopetra	Sheep	1.16	15.20	42.15	3.24	-19.97	4.74
T 252	Theopetra	Goat	1.19	15.31	43.02	3.28	-19.52	5.70
T 54	Theopetra	Goat	0.28	14.68	41.44	3.29	-19.07	5.82
T 70	Theopetra	Cattle	1.58	15.37	42.05	3.19	-20.74	4.86
T 237	Theopetra	Cattle	0.58	14.70	42.15	3.35	-19.62	5.01
T 275	Theopetra	Pig	0.77	4.37	12.16	3.25	-21.06	4.40
T 206	Theopetra	Pig	0.77	15.17	41.99	3.23	-19.63	7.17
T 94A	Theopetra	Boar	1.25	2.32	6.23	3.13	-20.00	3.89
T 98	Theopetra	Red deer	0.00	–	–	–	–	–
T 248	Theopetra	Red deer	0.76	14.67	42.89	3.41	-20.80	4.91
T 79	Theopetra	Badger	0.82	4.18	11.12	3.10	-18.78	6.92
T 507	Theopetra	Badger	1.09	14.40	40.41	3.27	-19.73	2.30
T 102	Theopetra	Dog	2.62	15.50	42.95	3.23	-18.81	7.48
UNIVERSITY OF ILLINOIS								
DA12	Alepotrypa	Bovid mandible	4.61	10.45	28.30	3.16	-20.13	5.46
DA13	Alepotrypa	Bovid rib	13.33	14.50	39.21	3.15	-20.52	3.36
DA14	Alepotrypa	Pig maxilla	8.43	13.24	35.94	3.17	-21.00	4.79
DA15	Alepotrypa	Pig maxilla	4.66	11.34	30.94	3.18	-20.66	4.30

TABLE 3.4—*Continued*

Sample	*Site*	*Material*	*% Coll.*	*% N*	*% C*	*C:N*	*$\delta^{13}C$ (‰ VPDB)*	*$\delta^{15}N$ (‰ AIR)*
DA16	Alepotrypa	Sheep	2.21	5.18	15.25	3.44	–	–
DA17	Alepotrypa	Sheep	8.17	12.30	33.67	3.19	-21.18	5.59
DA18	Alepotrypa	Fish	2.12	6.50	17.43	3.13	-11.36	10.99
DA19	Alepotrypa	Fish	7.68	11.20	29.50	3.07	-13.30	–
DA20	Alepotrypa	Fish	4.61	3.22	9.13	3.31	–	–
UNIVERSITY OF FLORIDA								
AP116	Alepotrypa	Sheep-goat	6.28	9.21	28.44	3.60	-21.53	6.91
AP117	Alepotrypa	Cow	14.20	14.61	43.76	3.50	-19.03	4.38
AP118	Alepotrypa	Cow	3.07	9.80	29.77	3.55	-20.47	4.10
AP119	Alepotrypa	Cow	3.11	9.88	29.22	3.45	-14.15	4.60
AP120	Alepotrypa	Goat	4.28	11.12	33.05	3.47	-22.85	5.95
AP157	Kephala	Sheep-goat	8.81	13.88	40.10	3.37	-19.85	5.96
AP158	Kephala	Sheep-goat	1.31	2.14	6.21	3.39	-20.50	8.08
AP159	Kephala	Cattle	1.10	1.74	5.65	3.79	-20.93	6.38
AP160	Kephala	Cattle	12.39	14.05	40.55	3.37	-20.47	6.56
AP161	Kephala	pig	1.18	0.65	1.84	3.30	-29.25	–
AP162	Kephala	Pig	1.07	1.71	5.79	3.95	–	–
AP163	Kephala	Dog	17.90	15.19	44.07	3.39	-18.89	8.49
AP164	Kephala	Deer	2.26	8.15	24.09	3.45	-19.99	5.47
AP204	modern	Almond	–	–	–	–	-26.24	-0.79
AP205	modern	Walnut	–	–	–	–	-23.92	0.67
AP206	modern	Chestnut	–	–	–	–	-28.12	-0.77
AP207	modern	Wild chestnut	–	–	–	–	-28.09	3.02
AP208	modern	Olive	–	–	–	–	-27.41	-1.65
AP209	modern	Acorn	–	–	–	–	-24.17	-2.58
AP210	modern	Wild acorn	–	–	–	–	-23.51	-0.97
AP211A	modern	Berries (A), skin only	–	–	–	–	-25.79	-2.12
AP211B	modern	Berries (A), skin + seeds	–	–	–	–	-26.71	-2.23
AP212	modern	Berries (B)	–	–	–	–	-26.42	0.52
AP213	modern	Fig	–	–	–	–	-26.06	1.72
AP214	modern	Black fig	–	–	–	–	-24.41	0.89
AP215	modern	Wild pear	–	–	–	–	-23.54	–
AP216	modern	Apple	–	–	–	–	-25.15	–
AP217	modern	Wild apple	–	–	–	–	-25.00	–
AP218	modern	Quince	–	–	–	–	-25.26	–
AP219	modern	Pomegranate	–	–	–	–	-23.62	-0.22
AP220	modern	Rabbit	–	–	–	–	-24.72	2.91
AP221	modern	Chicken	–	–	–	–	-21.97	6.65

*Wherever no value is reported in the table, either the quality of the collagen was not good and the sample was not run at all, or the analysis did not produce a value.

**Modern foodstuffs were analyzed according to the University of Florida protocol for such items.

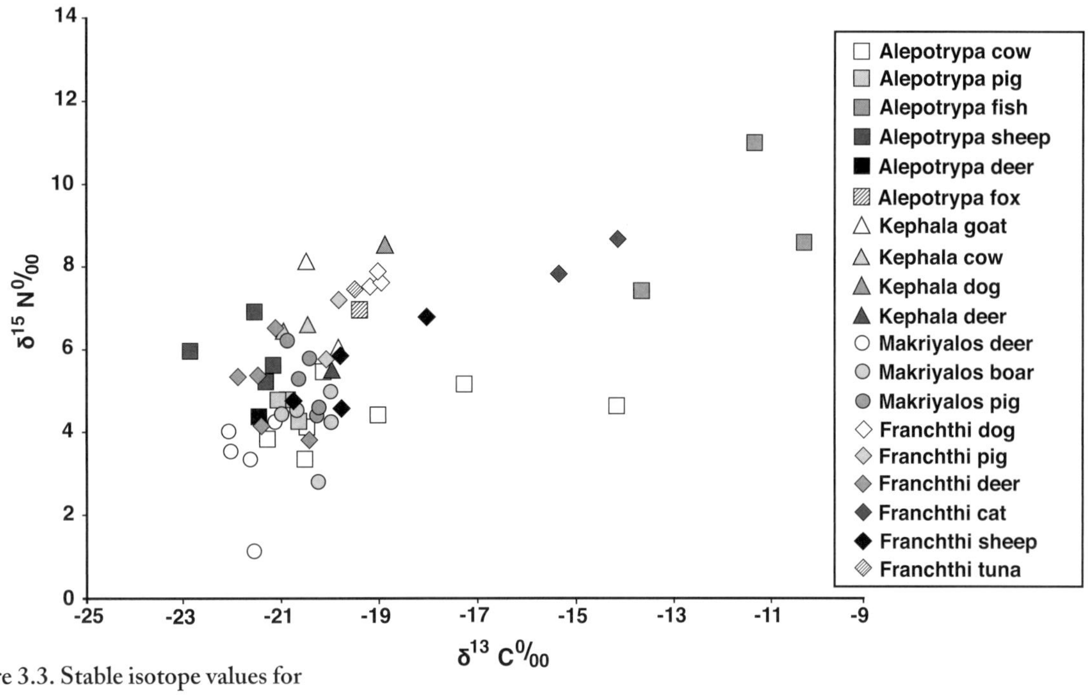

Figure 3.3. Stable isotope values for Neolithic Greek animal bones

With respect to δ^{15}N values, carnivore (dog, fox) values are higher than most human ones, while herbivores are lower in the food chain. Intermediate values are observed for pigs. If we subtract a 3‰ trophic level enrichment from the δ^{15}N values of the herbivores, we reach the 0‰–2‰ expected range for the plants that comprise the herbivores' diet. Human values are more positive than pigs and very close to those of carnivores, thereby indicating significant amounts of animal protein in the human diet. It is impossible, however, to isotopically distinguish different protein sources such as milk and meat. Furthermore, these high values should be interpreted as probably indicating large amounts of dairy foods instead of meat in the diet, as isotopic results must be consistent with archaeological data.[42] Additionally, the logistics of pastoralism and the use of grazing land versus arable land for growing crops support the aforementioned conclusion.

Domestic pigs show two trends, although the sample supporting this pattern is small. Values for pigs at Alepotrypa are very close to herbivore values, which suggests that they fed within areas such as gardens and cultivated areas close to habitation sites.[43] Most domestic pigs at the other sites show intermediate values between wild boars and humans, which suggests they acquired a large proportion of their dietary protein from animal products, most probably through feeding on waste or human refuse, implying that they were kept in confined areas close to the houses.[44]

A noteworthy point is that deer, cattle, and sheep/goat at Kephala exhibit higher δ^{15}N values than the equivalent species from other sites, and humans at Kephala express higher nitrogen values than at all other sites. Comparing the δ^{15}N values of all herbivores (deer, cattle, sheep/goat) from all sites, the values range from 1.1‰ to 6.8‰, whereas Kephala herbivores range between 5.5‰ and 8.1‰ (see Table 3.5). The average values from the sites range between 3.2‰ to 5.5‰, while the average value at Kephala is 6.5‰. The differences between Kephala and the other sites, except Franchthi, are statistically significant based on Student's *t*-test (p values range between 0.002 [$t = 318$] and 0.045 [$t = 13.4$]).

42. Hedges and Reynard 2007, p. 1242.
43. Halstead 2006.
44. Müldner and Richards 2005.

TABLE 3.5. NITROGEN ISOTOPE VALUES FOR HERBIVORES FROM NEOLITHIC SITES

Site	*Range (‰ AIR)*	*Mean*	*SD*	*N*
Kephala	5.47–8.08	6.49	0.97	5
Franchthi	3.83–6.83	5.40	1.00	8
Alepotrypa	3.36–5.46	4.40	0.46	8
Xirolimni	4.23–6.04	5.15	0.60	5
Theopetra	4.74–5.82	5.30	0.23	6
Makriyalos	1.10–4.24	3.24	1.57	5
Mavropigi	4.00–5.45	4.63	0.31	6

One possible explanation for this finding might be that the ecosystem of Kea Island, where the site of Kephala is located, is characterized by significant nitrogen enrichment. This may be why the Kephala human samples show significantly higher animal protein consumption. In other words, a shifted dietary picture may reflect the values of the specific ecosystem as well as regional diachronic fluctuations in the isotopic baseline.[45] The same could be true for Franchthi Cave, where the $\delta^{15}N$ values are also higher but the difference is not statistically significant (*t*-test, $t = 1$, $p>0.48$) and not as pronounced as with Kephala. However, further analysis is required to validate this interpretation and to identify the underlying reasons for this pattern, which may be due to different causes. Manuring of fields is a common explanation for the aforementioned phenomenon,[46] a practice that would elevate $\delta^{15}N$ values for plants and, consequently, for the animals feeding on them. Yet, there is no evidence for this practice having occurred in some Neolithic sites and not in others. Moreover, manuring may not explain the elevated deer nitrogen values at Kephala. Furthermore, there is evidence that the effect of manuring in palaeosoils is not significant.[47]

A second explanation points to regional differences in aridity, since it has been shown that in hot, arid environments, animals suffering from water stress may have significantly higher $\delta^{15}N$ values, especially water-conserving species such as goat and sheep.[48] This latter explanation may fit our data better, since it correlates with variation in rainfall in the aforementioned areas.

Another noteworthy result from this analysis is the very positive carbon signal for two cows from Alepotrypa Cave (samples APT-1 and AP119), indicating a C_4 diet (such as millet) for these cows (-17.3‰ and -14.2‰) if we use a cutoff point of -19‰ $\delta^{13}C$ to indicate the consumption of C_4 plants.[49] These two samples certainly come from different cows and were from different contexts. The first sample, when analyzed, was considered to be an outlier or intrusion from a much later period, since it was a single, unique occurrence. After the second sample was identified, it seemed probable that these animals were grazing consistently on C_4 plants, and the possibility arises that a C_4 plant was used as fodder as early as the Final Neolithic period. In addition, the nitrogen values for sheep and goat (samples AP158 at Kephala; AP116 at Alepotrypa) seem to be close to the boundary between a C_3 and C_4 diet, which, if they are not misidentified, could indicate herding of these animals over a large area—a practice that would have allowed them to exploit broader, isotopically varied plant biomasses.[50]

45. Hedges and Reynard 2007, p. 1242.
46. Bogaard et al. 2007.
47. Bogaard et al. 2007.
48. Ambrose 1986, 1991; van Klinken, Richards, and Hedges 2000; Hedges and Reynard 2007.
49. Pearson et al. 2007.
50. Richards et al. 2003, p. 75; Pearson et al. 2007.

The foregoing discussion emphasizes the importance of complementing human dietary analysis with associated animal samples from the same site in order to achieve the baseline needed for more precise dietary reconstructions. More human and animal bones from these and other Neolithic sites throughout Greece are currently being analyzed to corroborate the current findings, to verify trends observed that may lack an adequate sample size, and to continue adding pieces to the puzzle for all of Neolithic Greece.

Bronze Age

The first dietary studies for Bronze Age Greece were undertaken by Michael Richards and published in *Minoans and Mycenaeans: Flavours of Their Time.*[51] Bronze Age sites nicely cluster, exhibiting the same C_3 terrestrial diet with no significant marine protein consumption. Increased variability during the Bronze Age can be detected, however. As shown in Figure 3.4, values of individuals buried in chamber tomb cemeteries all group together, yet at Mycenae, individuals from Grave Circles A and B demonstrate significantly higher levels of animal and fish protein in their diets.[52] It is important to note that the Grave Circle burials considerably predate the chamber tombs and may reflect earlier dietary diversity. This difference is statistically significant (*t*-test, $t = 179{,}124$, $p<0.00003$), indicating socially variable consumption practices, and it suggests that status differences existed and were reflected in the diet. In another Mycenaean context from Achaia (Kalamaki), Vika observed a correlation between individuals buried in more elaborate tombs and elevated $\delta^{15}N$ values, but the sample size was extremely small.[53] Additionally, at Middle Helladic (MH) III and Late Helladic (LH) I Lerna, Triantaphyllou observes elevated $\delta^{15}N$ values in selected individuals.[54] Another parallel comes from the periphery of the Mycenaean world, from coastal Thessaly (Kazanaki, Volos), where a small, very rich tholos tomb came to light. In this case, despite the burial practices followed—including secondary burning of the grave contents including human bone, which can produce very poor collagen preservation—the two valid samples from this tholos exhibit the highest nitrogen values ($\delta^{15}N$ of 12.9‰ and 9.9‰) in prehistoric Greece.[55]

With this observed correlation in mind, Schepartz and Papathanasiou undertook a study at Pylos, the second most important Mycenaean palace.[56] The analysis of human skeletal remains from Pylos by Schepartz confirmed that variability exists in the health status of different subgroups. Specifically, individuals from tholos tombs—contexts that are among the richest in artifacts, the most complex architecturally and ritually, and generally considered to belong to members of elite families—exhibit better skeletal and dental health (at a statistically significant level, χ^2 test, $\chi^2 = 3.84$, $p<0.05$) than individuals buried in the more numerous but less elaborate chamber tombs.[57] In addition, across tomb types, females in chamber tombs exhibit poorer dental and skeletal health when compared to their male counterparts. The results indicate that there are significant differences in dental health of males and females at Pylos, and that these differences are also correlated with tomb type. Similar sex differences in dental health were observed at the Mycenaean Athenian Agora[58] and at Middle to Late Helladic Lerna.[59]

51. Richards and Hedges 1999.
52. Richards and Hedges 1999.
53. Vika 2002.
54. Triantaphyllou et al. 2008, pp. 3031–3033.
55. Adrymi-Sismani and Alexandrou 2009; Papathanasiou 2009.
56. Papathanasiou et al. 2012.
57. Schepartz, Miller-Antonio, and Murphy 2009.
58. Schepartz, Miller-Antonio, and Murphy 2009.
59. Triantaphyllou et al. 2008.

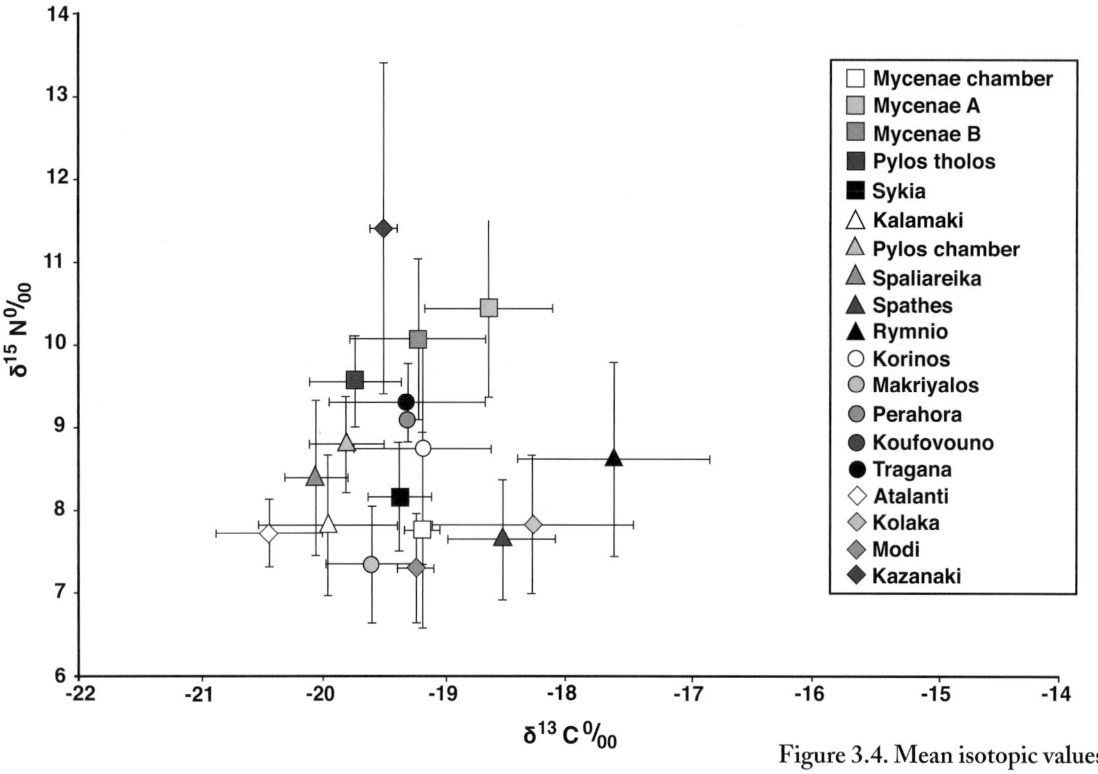

Figure 3.4. Mean isotopic values for humans from Bronze Age Greek sites

To test whether dietary factors may have been responsible for these differences, a sample of 95 individuals from Pylos was used, representing different status, age, and gender subgroups. Bone preservation was very poor, with only 63 of the 95 samples yielding valid results.[60] Once again, all samples cluster in the same C_3 terrestrial diet, but when different subgroups are compared increased variability is revealed. The individuals from tholos tomb contexts, supposedly associated with higher status, exhibit greater consumption of animal resources—at a statistically significant level (*t*-test, $t = 2143$, $p<0.003$), despite the small sample size—and exhibit some of the highest values of the entire sample. The difference becomes more marked when only adult individuals from tholoi and chamber tombs are compared. In addition, when comparing males and females from the chamber tombs, a statistically significant difference (*t*-test, $t = 132$, $p<0.005$) in the amount of animal protein consumed is observed, with females showing less animal protein in their diet. These isotopic data clearly correlate with the results of the osteological analysis.

In general, although Pylos exhibits less variability and much smaller differences than Mycenae, nevertheless the variability is present. Furthermore, this variability is statistically significant, and the dietary data clearly verify the hypotheses and results of the osteological analysis of the sample. The immediate conclusions are that the life experiences were different for Mycenaean males and females, and that individuals buried in tholos tombs seem to be distinct from people buried in chamber tombs, since the former demonstrated better nutrition and a better health status. Another distant parallel comes from Armenoi (Late Minoan Crete), where, again, males incorporated significantly more animal protein in their diets compared to females.[61]

60. Papathanasiou et al. 2012, pp. 147–149.

61. Richards and Hedges 1999, p. 214.

Other dietary studies of Bronze Age Greece include those published by Lagia and colleagues[62] for MBA (Middle Helladic) Koufovouno in the southern Peloponnese, where a typical C_3 resource diet was detected, indicating little animal derived protein; by Triantaphyllou and colleagues for 39 samples from Middle Helladic Lerna,[63] where a similar C_3 terrestrial diet was observed but with the inclusion of more animal protein; by Ingvarsson-Sundström and colleagues[64] for 19 samples from Middle Helladic Asine, reaching the same conclusions; and by Petroutsa and colleagues[65] for EBA (Early Helladic [EH]) Perahora near Corinth, where 19 samples with slightly elevated $\delta^{15}N$ values were interpreted as showing a mix of terrestrial and marine resources. A small sample of eight individuals from the Middle Helladic site of Proskynas also points in the same direction,[66] as does a sample from Makriyalos of six individuals.[67]

For the Late Bronze Age (LBA) Mycenaean period, several isotopic analyses have been conducted including those at Spaliareika, Voudaini, and Kalamaki in Achaia;[68] Sykia in Laconia;[69] East Lokris in central Greece;[70] and Korinos, Spathes, and Rymnio in northern Greece.[71] Results from all sites cluster closely together in what amounts to a homogenous C_3 terrestrial diet. Similar dietary signals have also been detected for Classical[72] and Byzantine populations in Greece.[73]

Statistical Inferences from Temporal and Regional Comparisons

Since the temporal and regional contexts of the Neolithic and Bronze Age are very broad, it would be a reasonable step to compare between the subphases of the aforementioned time periods as well as between geographical areas. The resulting regional and temporal subdivisions do not represent artificial partitions but exhibit distinct cultural characteristics. To mention only a few of the major differentiations, in the Late and Final Neolithic more regionalization and exploitation of marginal environments is observed, and it is only during the LBA and only in southern Greece that prominent palatial centers emerge with all the socioeconomic consequences that this emergence entails. In order to test for differences between these temporal and regional subdivisions, a number of statistical tests were conducted.

First, the Early and Middle Neolithic periods were compared to the Late and Final Neolithic for differences in the mean $\delta^{13}C$ and $\delta^{15}N$ values. Regarding the $\delta^{15}N$ values, it must be noted that all the samples may include neonates, as the ages of the individuals from which the samples were taken are not specified. The bias that may be introduced in the nitrogen values by the presence of nursing infants, whose nitrogen signatures are higher than that of their mothers because they are consuming their mothers' tissues, however, pertains to all sites and time periods. Sites included in the Early and Middle Neolithic sample are Franchthi Cave ($n = 7$), Xirolimni ($n = 14$), Mavropigi ($n = 10$), and Pontokomi ($n = 1$), while the Late and Final Neolithic sample consisted of Alepotrypa Cave ($n = 25$), Kephala ($n = 4$), Theopetra ($n = 12$), Tharrounia ($n = 20$), Kouveleiki ($n = 2$), Proskynas ($n = 7$), Makriyalos ($n = 22$), and Franchthi Cave ($n = 4$). A Student's t-test comparison between the two samples indicated that, while there is no statistically significant difference in the $\delta^{13}C$ values, there is a statistically

62. Lagia, Petroutsa, and Manolis 2007, pp. 321–323.
63. Triantaphyllou et al. 2008, pp. 3031–3033.
64. Ingvarsson-Sundström, Richards, and Voutsaki 2009.
65. Petroutsa, Richards, and Manolis 2007, pp. 292–294.
66. Papathanasiou, Zachou, and Richards 2009, pp. 229–230.
67. Triantaphyllou 2001.
68. Vika 2002.
69. Papathanasiou 1999b, 2002.
70. Iezzi 2005; Petroutsa 2007.
71. Triantaphyllou 2001.
72. Vika 2007; see also Chap. 5.
73. Garvie-Lok 2001; Bourbou and Richards 2007.

significant difference (t = 2,220, p<0.0003) in the mean value of $\delta^{15}N$ (7.5‰ vs. 8.6‰), with Early and Middle Neolithic populations incorporating more animal protein in their diets. As our samples are not site-specific but adequately representative of the two subperiods, it can be inferred that a significant decrease in animal protein consumption occurred in the Late and Final Neolithic, which can be correlated with the cultural characteristics of this period and the colonization of marginal environments.

Two separate samples were formed for the EBA/MBA and the LBA. The EBA/MBA sample was formed with data from Proskynas (n = 8), Perahora (n = 19), Makriyalos (n = 6), Koufovouno (n = 15), Lerna (n = 39), and Asine (n = 19), while the LBA sample included Pylos (n = 39), Mycenae (n = 25), Tragana (n = 4), Atalanti (n = 4), Kolaka (n = 4), Modi (n = 4), Korinos (n = 4), Rymnio (n = 4), Spathes (n = 5), Kalamaki (n = 15), Spaliareika (n = 7), Sykia (n = 6), Kazanaki (n = 2), Kalapodi (n = 14), and Voudaini (n = 24). Statistical analysis does not discern any significant differences (t-test, t = 2, p>0.20) between the EBA/MBA and the LBA populations, although the cultural differences are profound. Additionally, although significant intra-population variations exist based on gender and status, they do not appear when all the samples are lumped together, because there is not much variability in the general Bronze Age diet. There is, however, a statistically significant difference in the $\delta^{15}N$ values (8.5‰ vs. 7.5‰) (t-test, t = 955, p<0.001) between the Late/Final Neolithic and the EBA/MBA samples, suggesting that animal protein consumption increased again. In the LBA, the mean value of $\delta^{15}N$ further increases to 8.6‰ but, as mentioned above, the difference is not statistically significant. However, if the LBA sample is restricted to only the palatial centers that include tholos burials (Pylos, Mycenae, and Kazanaki), thus excluding the non-elite tombs, the mean $\delta^{15}N$ value becomes 9.3‰, which is significantly higher than the EBA and MBA values. This result implies increased access to animal protein for those groups associated with the palatial center, a result that correlates well with the cultural characteristics of a stratified society that arose during this time period.

In a regional perspective, when comparing the nitrogen values for northern and southern EBA and MBA Greece, it seems that there is no statistically significant difference although significant cultural differences exist. Again, the effect of the palatial centers is hidden by the rural sites, which constitute the majority of the samples and exhibit remarkable homogeneity. Conversely, except for nitrogen variation, when comparing the carbon values for northern and southern Greece, there is a statistically significant difference (p<0.0003), which may imply more widespread C_4 consumption in northern Greece.

Cross-Checking

Cross-checking of the technique of isotopic analysis in a comparative study involving multiple laboratories was conducted. In order to achieve this, the same samples were reprocessed independently in two different laboratories for the purpose of verifying the replicability of the process and the results. Six samples from Alepotrypa Cave, analyzed in 1993 by Papathanasiou

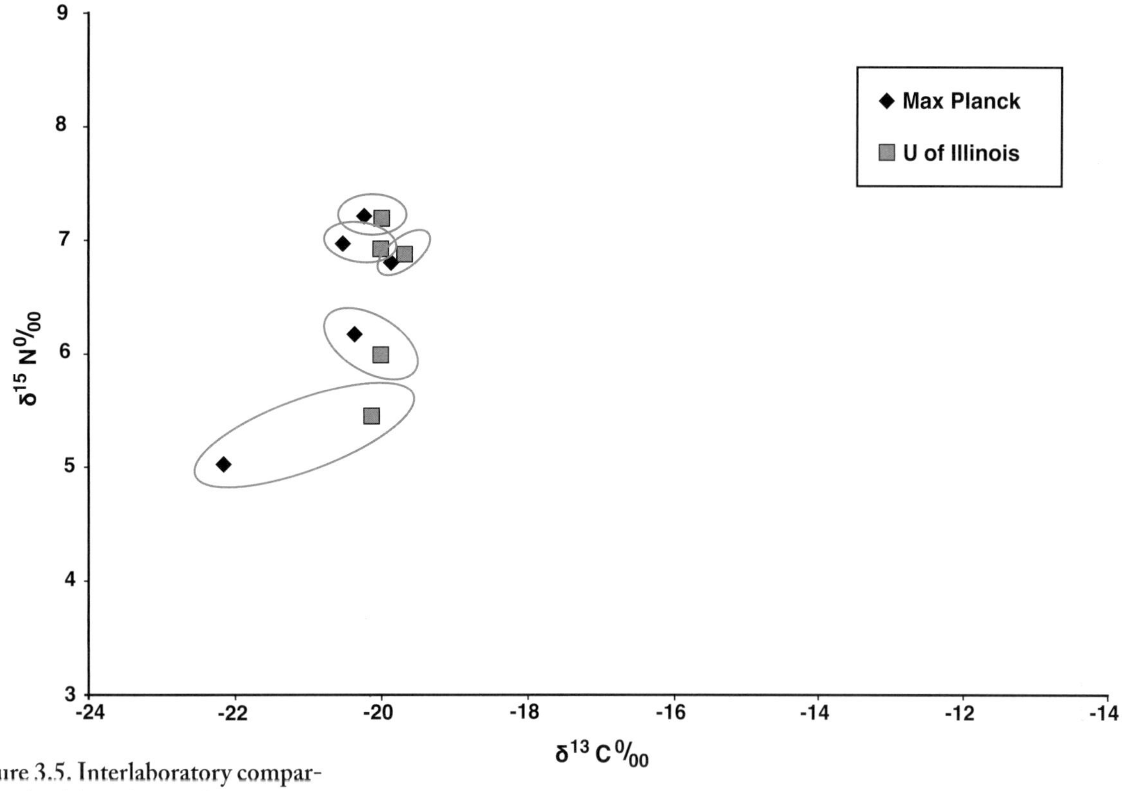

Figure 3.5. Interlaboratory comparison of stable carbon and nitrogen isotopes of human collagen from Alepotrypa Cave

and Norr at the University of Illinois at Urbana-Champaign, yielded a surplus of pure collagen that had subsequently been preserved in the original vials. These vials were taken to Richards' laboratory at the Max Planck Institute in Leipzig, where they were reanalyzed in 2007 according to the technique used there. The results are presented in Figure 3.5 and clearly replicate each other. After analyzing the data with Student's *t*-test, no statistically significant difference exists between the two sets. Although the two analyses were performed in different laboratories using slightly different techniques, all the results appear identical and within acceptable limits; therefore, both the techniques themselves and their respective interpretations are validated.

Judging from the results of this inter-laboratory comparison, archaeologists can safely place more confidence in stable isotope analysis knowing—even if its outcome requires them to reconsider their existing interpretations—that isotope analysis offers a positive, direct assessment of past diet that is free of diagenetic effects and postmortem contamination. Indeed, isotope analysis is a tight, standardized, systematically controlled technique that provides safe, positive evidence. There is no need to double-check the results of stable isotope analysis as a standard procedure. If further doubts persist, especially when traditional beliefs are challenged, results can be directly compared in different laboratories. By integrating isotope analysis into the existing spectrum of archaeological investigative techniques and interpreting chemical results and archaeological data together, it is possible to gain a greater understanding of past dietary patterns after considering all related parameters including geological, ecological, chemical, representational, and retrieval factors that pertain to any archaeological material.

Discussion

For the Neolithic period in Greece, a primarily C_3 terrestrial diet has been isotopically documented. The Neolithic diet, according to archaeological, archaeobotanical, and zooarchaeological evidence, consisted of plants and grains that represented everyday foodstuffs, while meat and, later, dairy foods were also included. The relative importance of each food category, as well as intra-population variation, is discussed below.

Neolithic Age

Domestic Animals

Animal protein also featured in the Neolithic Greek diet, but stable isotope analysis cannot determine these proportions exactly. Since meat and dairy products have the same isotopic signature, the $\delta^{15}N$ values cannot differentiate the exact source of the animal protein.[74] The use of secondary products, including dairy foods, is positively documented for the Bronze Age, but the extent that dairy products were used in the Neolithic is still an open research question. Recent analysis of pottery residues from southeastern Europe has shown that dairy products were consumed at least at some sites as early as the Middle Neolithic.[75] The zooarchaeological evidence from Greece indicates an emphasis on meat consumption, at least for the Early and Middle Neolithic, though in some cases the use of dairy products has been suggested.[76] Analysis of pottery residues has shown that milk may also have been consumed in the Middle Neolithic,[77] but, again, the extent of its consumption remains unknown. It would make more sense to assume that the animal protein was derived from both milk and meat. According to the logistics of pastoralism, large herds are costly to keep and could not have been supported by the available grazing land in Neolithic times, which would be labor-intensive and at the expense of farming land.[78]

Meat was coming mainly from domestic animals, in particular ovicaprids, which dominate the faunal record throughout all time periods, and, to a lesser degree, from cattle and pigs. However, meat from domestic animals, except for being a more expensive resource since livestock would have been costly to raise, is also more difficult to preserve over time. Thus, the question about the pattern and frequency of meat consumption arises.[79] It has been suggested, based on archaeological evidence from various Neolithic sites,[80] that meat-sharing of fresh carcasses, depending on the size of the animal, was taking place between several households or extended social groups—usually in the form of feasting. Occasionally, a portion of fresh meat from livestock may have been preserved for piecemeal consumption over an extended period of time.[81] Although there are isolated episodes of excessive carnivory, like the feasting case at Makriyalos, such examples remain unique in the archaeological record of Neolithic Greece.

Wild Animals: Fish

Wild animals, on the other hand, form only a minor part of Neolithic faunal assemblages and, therefore, of the Neolithic diet, since hunting was no longer a primary subsistence strategy. Instead, it fulfilled non-dietary, religious, political, symbolic, and ideological needs in the same way observed elsewhere throughout Europe after the adoption of agriculture.[82]

74. DeNiro and Epstein 1981; Hedges and Reynard 2007.
75. Evershed et al. 2008.
76. Halstead 1999.
77. Evershed et al. 2008.
78. Halstead 1981, 1994, 2000, 2008.
79. Halstead 2007.
80. Pappa et al. 2004; Cavanagh 2007; Halstead 2007; Urem-Kotsou and Kotsakis 2007.
81. Cavanagh 2007, pp. 116–117; Halstead 2007.
82. Curci and Tagliacozzo 2003; Hamilakis 2003; Boyle 2006.

Among wild resources, the idea of a Mediterranean diet based on seafood has been extensively addressed. Even if more systematic fishing was taking place in the northwestern European Mesolithic period, similarly to the Atlantic and Baltic Mesolithic sites (a possibility that remains highly questionable and has yet to be verified through isotope analyses), there is a consensus that during the Neolithic period, fishing was declining in importance even at coastal sites, becoming subsidiary to animal husbandry and agriculture.[83] This apparent trend continues into the Bronze Age.[84] Powell reports a smaller amount of fish bones from Greek Neolithic sites, and that the fish are of small to moderate size with only a few large individuals.[85] High trophic level fish and marine mammals, which would provide a very positive nitrogen signal, were not part of the Mediterranean diet, while mollusks and small fish of low trophic levels, such as sardines and anchovies, were more common. The latter, however, exhibit $\delta^{13}C$ and $\delta^{15}N$ values very close to those of terrestrial resources, which means that it is difficult to detect them isotopically.[86] Even so, if prehistoric people had consumed significant quantities of the aforementioned small fish, it would be possible to identify it through slightly elevated $\delta^{13}C$ and $\delta^{15}N$ levels. Low trophic level fish, however, have also low nutritional yields, resulting in a small fraction by weight in the total diet. Thus, a combination of factors may mask prehistoric fish consumption. Prehistoric diets may have included some fish, but it is difficult to detect as it would comprise less than 10% of the total diet by weight, and the isotopic values would be closer to a C_3 terrestrial diet.[87]

Furthermore, Stratouli has argued that neither the boats nor the nets available during the Neolithic period would have allowed deep-water fishing.[88] Garcia-Guixé and colleagues, as well as Cavanagh,[89] argue that Mediterranean and Aegean fish could never have been a staple since these seas are much less productive compared to the Atlantic. The reconstruction of minimum numbers of fish, shellfish, and meat yields of Greek Neolithic sites[90] suggests that the contribution of marine resources was not insignificant but in no way comparable in magnitude to coastal Mesolithic western and northern Europe. A seasonality study[91] suggests that fishing would likely have taken place from spring to autumn, coinciding with agricultural activities, and would have provided a seasonal input of marine protein, unless some quantities were preserved for later consumption. In any case, fish need to comprise a considerable proportion of the diet (at least 10%) relative to other foodstuffs before their effect appears in the isotopic analysis. Apparently, the populations under study did not cross this threshold. However, the role of fishing as well as hunting still remains to be assessed in terms of the symbolic importance related to gender, power, and harvesting of the wild within a proto-sedentary way of life.[92]

Cereals

Cereals such as glume wheats and barley, as well as legumes such as lentils, peas, grass peas, and bitter vetch,[93] supplied the bulk of carbohydrates and plant protein to the Neolithic diet, while domesticated animals and their secondary products supplied the majority of animal protein. Cereals could be eaten uncooked; pounded into the form of bulgur[94] to make porridges, stews, and soups; or ground into flour to be either baked in various breads

83. Powell 2003, pp. 79–82; Cavanagh 2007.
84. Cosmopoulos, Greenfield, and Ruscillo 2003.
85. Powell 2003.
86. Garvie-Lok 2001; Lagia, Petroutsa, and Manolis 2007.
87. See also Chap. 11.
88. Stratouli 1996.
89. Garcia-Guixé, Richards, and Subirà 2006, pp. 553–554; Cavanagh 2007, p. 111; Gallant 1985, p. 31.
90. Theodoropoulou 2007a, 2007b.
91. Theodoropoulou 2007a, 2007b.
92. Hamilakis 2003.
93. Valamoti 2004, 2007.
94. Hansen 2000; Valamoti 2004, 2007.

or made into dough for pies that may have included other ingredients such as vegetables, legumes, oils, dairy products, or even meat.[95] Indeed, cereals are often found archaeologically to have been stored in dense concentrations as surplus, thereby indicating they were important staples that merited special care. Legumes appear in various contexts within individual houses, supplementing the main staple of cereals[96] and playing a significant role in the overall diet.[97]

The pattern observed in Neolithic Greece follows the general pattern observed with the transition to the Neolithic period. A number of isotopic studies of Mesolithic Europe have shown a great reliance on marine foods by coastal Mesolithic populations in Denmark,[98] the United Kingdom,[99] Portugal,[100] and the area of the Baltic Sea.[101] With the advent of agriculture in the Neolithic period, however, a shift in diet has been documented by isotopic studies, even at a number of the aforementioned sites, and a pattern emerges in which the terrestrial diet predominates.[102] This same pattern has been observed for Mesoamerican populations.[103] In the Mesolithic Mediterranean, on the other hand, it seems, although based on a very small number of samples, that this shift was not so sharp, since Mediterranean Mesolithic populations never relied so heavily (up to 100% elsewhere) on marine resources.[104] With the adoption of agriculture, Neolithic farming populations in areas that are unrelated geographically, chronologically, and culturally exhibit an over-reliance on domesticated species. Dietary variation and quality decrease, substituted by a greater investment in energy-rich plant foods and domesticated animal products.

Cereals became a staple as they could yearly yield a predictable amount of food of high calorific value. Furthermore, a surplus could be created that could be stored and preserved for later consumption or exchange. These attributes turned cereals into a dependable resource for further population increase, a major issue for Neolithic societies. Agricultural populations insist on this strategy even when it has adverse effects on their health.[105] The complex network of social processes that led to sedentary ways of life and the intensive exploitation of wild cereals and legumes resulted in what is conventionally described as the Neolithic Revolution.

Bronze Age

The sample size for Bronze Age Greece, although not extensive, covers all subperiods, the two most important palatial centers of the Mycenaean world, and a great number of smaller suburban sites from a broad geographical area. The same trend of predominantly terrestrial subsistence strategies, as indicated by chemical, archaeological, faunal, and botanical evidence, continues through the Bronze Age accompanied by an increasing importance in secondary products and a declining importance in hunting and fishing. Moreover, isotopic studies indicate that status and gender differences do arise during the later phase of this period and are reflected in the dental health and diet—especially at palatial centers (Mycenae and Pylos), with high-status males from tholos tombs exhibiting the highest levels of animal protein in their diet as well as better dental health. The sex and status based hierarchy appears to be representative of a more

95. Hansen 2000.
96. Valamoti 2004, 2007.
97. See also Chap. 11.
98. Tauber 1981, 1983, 1986.
99. Schulting and Richards 2001, 2002a, 2002b.
100. Lubell et al. 1994.
101. Liden 1995.
102. Tauber 1981; Lubell et al. 1994; Liden 1995.
103. Norr 1995.
104. Richards et al. 2001; Garcia-Guixé, Richards, and Subirà 2006.
105. Larsen 1997, p. 282.

widespread Mycenaean system as demonstrated by the isotope data, comparative archaeological studies, and archaeological and textual evidence.[106]

Potential explanations for these variations in dental health and diet may include either physiological or behavioral factors. Earlier tooth eruption in females and pregnancy have been suggested to correlate with poorer dental health by dental health studies, but they have not been linked with positive evidence.[107] Behavioral factors include: (a) differential access to meat and/or carbohydrates, with carbohydrates resulting in more dental caries and tooth loss; (b) meal frequency, with individuals who have fewer and larger meals exhibiting better dental health; and (c) oral hygiene.[108] Studies of prehistoric Native Americans from southern California (Santa Barbara Channel Islands for 4000 to 400 B.P.) and pygmy groups of central Africa (Aka, Efe, Mbuti vs. Bantu in the mid-1980s) strongly correlate frequent snacking and consumption of carbohydrates, as opposed to meat, with higher dental caries and tooth loss rates in females. Similarly, for Bronze Age Greece, different dietary habits for males and females may account for the observed differences in dental health and diet, if we assume that females spent more time preparing meals and probably had more snacks based on carbohydrates. Such a pattern of frequent female snacking could be linked to higher rates of caries in ancient Greece as in other cultures.[109] It cannot be positively affirmed, however, that elite women in Bronze Age Greece were intimately involved in food preparation and, therefore, the poor dental health of females from tholos tombs might still require further explanation.

Another factor to be considered is meat consumption, and possible differential participation in feasting, a main venue for meat consumption during the Bronze Age, which occurred in a range of forms, from very formalized ritual to casual feasting.[110] In their most formal form, Mycenaean feasts promoted the prestige of the palace and social solidarity and secured the loyalty of followers.[111] These feasts were performed on a grand scale that, as archaeological evidence suggests, would have required approximately two tons of meat, while at the same time they were characterized by cultural elaboration in a highly nonegalitarian social context.[112] Feasts and sacrifices were frequent occurrences in Mycenaean palaces, participation in which was one of the main duties of the ruling individuals[113]—a fact that may explain their elevated $\delta^{15}N$ values indicative of greater animal protein and perhaps marine food consumption. In addition, more frequent participation in feasting may have been "a predominantly male activity in a warrior society,"[114] as written (Homeric epics) and archaeological evidence suggests men were more active in military and political affairs. Such gender-biased participation could account for the dietary differences between Mycenaean males and females that appear in isotopic analyses.

106. Schepartz, Miller-Antonio, and Murphy 2009.

107. See discussion and related bibliography in Larsen 1997, pp. 75–76.

108. Walker and Erlandson 1986; Walker and Hewlett 1990.

109. Walker and Erlandson 1986; Walker and Hewlett 1990; Schepartz, Miller-Antonio, and Murphy 2009.

110. Sherratt 2004, pp. 301–302; Wright 2004b, p. 160.

111. For related bibliography see the Hesperia Special Issue on the Mycenaean Feast, Wright 2004a, 2004b.

112. Halsted and Isaakidou 2004; Stocker and Davis 2004.

113. Palaima 2004; Sherratt 2004; Wright 2004a.

114. Sherratt 2004, p. 315; Wright 2004a, p. 127, Wright 2004b p. 158.

C_4 Plants

Worth noting with respect to the C_4 signals in the Neolithic isotopic record is the possibility of the presence of millet. While millet does appear at EBA Kastanas and probably in the Middle Bronze Age Ayios Mamas,[115] there is no evidence known to date for millet grown locally in Greece during the Neolithic period. The very first occurrence of millet in any form comes from Olynthos in northern Greece, where a miniature pot filled with millet seeds was found in destruction debris dated by pottery finds to the Final Neolithic.[116] Valamoti suggests this single find may represent seeds brought as treasured goods by "a bride from a distant village in the north" or a traveler.[117] In stable isotope studies, the C_4 signal has also been detected at Bronze Age Rymnio by Triantaphyllou and was attributed to millet.[118] However, there is no isotopic evidence for its systematic cultivation and consumption, and it only appears sporadically in most prehistoric sites in either human or faunal samples. The two samples described herein from Alepotrypa Cave are the earliest chemically recognized C_4 signals in Greece (7000 to 5000 B.P.). Millet in the Neolithic period has been detected outside Greece at Çatal Höyük where it dates as early as 8000 to 7000 B.P.[119] Moreover, detection of C_4 in cattle from the Neolithic archaeological record may only imply that cattle were free-ranging over large distances, which gave them access to multiple environmental zones and allowed them to systematically include C_4 plants, including seaweed, in their diet. The possibility also exists that they were fed on a different fodder than sheep and goats.[120] The sample, however, is extremely small and cannot support further discussion.[121]

CONCLUSION

The foregoing discussion is an overview of all stable isotope results for Greek Neolithic and Bronze Age sites published to date, as well as a presentation of new data. These new data—which complement previous analyses with respect to sample size, temporal and spatial distribution, and variation in gender, status, and ecosystem—provide an integrated picture with finer distinctions and greater variability. Isotopic studies were initially undertaken to directly assess human diet in Neolithic Greece for two main reasons: (1) to explain the high percentage of observed anemic conditions of non-genetic origin; and (2) to supplement and test findings from faunal analyses and archaeological research that indicate significant access to, and consumption of, animal and fish protein.

In contrast to conventional methods of dietary reconstruction, stable isotope analyses indicate a diet for both Neolithic and Bronze Age Greece based primarily on C_3 plant resources, with smaller amounts of animal protein (meat, dairy products), and a statistically insignificant consumption of marine protein. The 17 prehistoric sites reviewed here all fall within the same range, exhibiting a generally similar C_3 diet with minimal variation. This variation is indicative of climatic differentiation, as well as socioculturally governed dietary patterns.

115. Kroll 1983, pp. 121–124; 2003.
116. Mylonas 1929 as referenced in Valamoti 2007, p. 99; Valamoti 2007, p. 99.
117. Valamoti 2004, 2007.
118. Triantaphyllou 2001.
119. Richards et al. 2003; Pearson et al. 2007.
120. Pearson et al. 2007.
121. See also Chap. 11.

Future research should initially continue to investigate the relative position of the values of different animal species, especially those of herbivores, in order to establish the reference point for a more accurate assessment of human values. Secondly, new research should identify certain variations within the ecosystem of the sites discussed above, determine whether certain sites are actually enriched in nitrogen, and, if so, find the cause for this phenomenon. Third, it would be worthwhile to examine the Mesolithic and very early Neolithic diet in Greece to determine whether a shift in diet occurs following this cultural transition, or whether the Neolithic diet resembles that of the Mesolithic (it must be noted, however, that collagen preservation is not very promising for this time period). Finally, exploration of the possible use of millet and evaluation of indicators for its introduction in Greece as early as the Neolithic would be useful.

Stable isotopic analysis has much to offer both now and in the future. Animal isotopic data can potentially contribute information to explicate aspects of animal husbandry and animal feeding strategies in the past. More extensive human sampling will enable us to detect inter- and intra-site variations related to age, sex, and status, while incorporation of juvenile individuals will help us estimate weaning patterns in prehistory. Dietary reconstruction is a well-founded, complex area of research that greatly benefits from scholars with a variety of backgrounds and approaches working together. Isotopic studies produce important, reliable results that, when taken in conjunction with biological, archaeological, textual, archaeobotanical, and zooarchaeological data, can lead to useful, viable interpretations and greater understanding of dietary patterns in ancient populations.

ACKNOWLEDGMENTS

I would like to thank all of the archaeologists of the corresponding sites for providing access to the skeletal materials utilized in this study. I would also like to thank Anna Lagia and George Stamoulis for their comments, suggestions, and technical advice, as well as Tatiana Theodoropoulou and Soultana-Maria Valamoti for the constructive discussions. Finally, I would like to thank the three anonymous reviewers for their constructive comments and their valuable remarks. Funding for this project was provided by two grants from the Wiener Laboratory of the American School of Classical Studies at Athens.

REFERENCES

Adrymi-Sismani, V., and M. Alexandrou. 2009. "The Mycenaean Tholos Tomb at Kazanaki–Funerary Rituals," in *Proceedings of the 2nd Archaeological Meeting of Thessaly and Central Greece,* Volos, pp. 133–149.

Ambrose, S. H. 1986. "Stable Carbon and Nitrogen Isotope Analysis of Human and Animal Diet in Africa," *Journal of Human Evolution* 15, pp. 707–731.

———. 1991. "Effects of Diet, Climate, and Physiology on Nitrogen Isotope Abundances in Terrestrial Foodwebs," *JAS* 18, pp. 293–317.

Angel, J. L., and S. C. Bisel. 1986. "The Human Skeletal Material from Franchthi Cave" (manuscript on file, Program in Classical Archaeology, Indiana University, Bloomington).

Bogaard, A., T. H. E. Heaton, P. Poulton, and I. Merbach. 2007. "The Impact of Manuring on Nitrogen Isotope Ratios in Cereals: Archaeological Implications for Reconstruction of Diet and Crop Management Practices," *JAS* 34, pp. 335–343.

Bourbou, C., and M. P. Richards. 2007. "The Middle Byzantine Menu: Paleodietary Information from Isotopic Analysis of Humans and Fauna from Kastella, Crete," *International Journal of Osteoarchaeology* 17, pp. 63–72.

Boyle, K. V. 2006. "Neolithic Wild Game Animals in Western Europe: The Question of Hunting," in *Animals in the Neolithic of Britain and Europe* (Neolithic Studies Group Seminar Papers 7), ed. D. Serjeantson and D. Field, Oxford, pp. 10–23.

Cavanagh, W. 2007. "Food Preservation in Greece during the Late and Final Neolithic Periods," in Mee and Renard 2007, pp. 109–122.

Cook, D. C. 2000. "Skeletal Evidence for Nutrition in Mesolithic and Neolithic Greece: A View from Franchthi Cave," in Vaughan and Coulson 2000, pp. 99–104.

Cosmopoulos, M. B., H. J. Greenfield, and D. Ruscillo. 2003. "Animal and Marine Remains from the New Excavations at Eleusis: An Interim Report," in Kotjabopoulou et al. 2003, pp. 145–152.

Curci, A., and A. Tagliacozzo. 2003. "Economic and Ecological Evidence from the Vetrebrate Remains of the Neolithic Site of Makri (Thrace–Greece)," in Kotjabopoulou et al. 2003, pp. 121–131.

DeNiro, M. J., and S. Epstein. 1981. "Influence of Diet on the Distribution of Nitrogen Isotopes in Animals," *Geochimica et cosmochimica acta* 45, pp. 341–351.

Dennell, R. W. 1976. "The Economic Importance of Plant Resources Represented on Archaeological Sites," *JAS* 3, pp. 229–247.

Doumas, C. 1992. *Οι τοιχογραφίες της Θήρας,* Athens.

Evershed, R. P., S. Payne, A. G. Sherratt, M. S. Copley, J. Coolidge, D. Urem-Kotsou, K. Kotsakis, M. Özdoğan, A. E. Özdoğan, O. Nieuwenhuyse, P. M. M. G. Akkermans, D. Bailey, R.-R. Andeescu, S. Campbell, S. Farid, I. Hodder, N. Yalman, M. Özbaraşan, E. Biçakci, Y. Garfinkel, T. Levy, and M. M. Burton. 2008. "Earliest Date for Milk Use in the Near East and Southeastern Europe Linked to Cattle Herding," *Nature* 455, pp. 528–531.

Gallant, T. W. 1985. *A Fisherman's Tale: An Analysis of the Potential Productivity of Fishing in the Ancient World* (Miscellanea Graeca 7), Gent.

Garcia-Guixé, E., M. P. Richards, and M. E. Subirà. 2006. "Palaeodiets of Humans and Fauna at the Spanish Mesolithic Site of El Collado," *CurrAnthr* 47, pp. 549–556.

Garvie-Lok, S. J. 2001. "Loaves and Fishes: A Stable Isotope Reconstruction of Diet in Medieval Greece" (diss. Univ. of Calgary).

Halstead, P. 1981. "Counting Sheep in Neolithic and Bronze Age Greece," in *Pattern of the Past: Studies in Honour of David Clarke,* ed. I. Hodder, G. Isaac, and N. Hammond, Cambridge, pp. 307–339.

———. 1994. "The North–South Divide: Regional Paths to Complexity in Prehistoric Greece," in *Development and Decline in the*

Mediterranean Bronze Age (Sheffield Archaeological Monographs 8), ed. C. Mathers and S. Stoddart, Sheffield, pp. 195–219.

———. 1999. "Neighbours from Hell? The Household in Neolithic Greece," in *Neolithic Society in Greece* (Sheffield Studies in Aegean Archaeology 2), ed. P. Halstead, Sheffield, pp. 77–95.

———. 2000. "Land Use in Postglacial Greece: Cultural Causes and Environmental Effects," in *Landscape and Land Use in Postglacial Greece* (Sheffield Studies in Aegean Archaeology 3), ed. P. Halstead and C. Frederick, Sheffield, pp. 110–128.

———. 2006. "Sheep in the Garden: The Integration of Crop and Livestock Husbandry in Early Farming Regimes of Greece and Southern Europe," in *Animals in the Neolithic of Britain and Europe* (Neolithic Studies Group Seminar Papers 7), ed. D. Serjeantson and D. Field, Oxford, pp. 42–55.

———. 2007. "Carcasses and Commensality: Investigating the Social Context of Meat Consumption in Neolithic and Early Bronze Age Greece," in Mee and Renard 2007, pp. 25–48.

———. 2008. "Between a Rock and a Hard Place: Coping with Marginal Colonisation in the Later Neolithic and Early Bronze Age of Crete and the Aegean," in *Escaping the Labyrinth: The Cretan Neolithic in Context* (Sheffield Studies in Aegean Archaeology 8), ed. V. Isaakidou and P. D. Tomkins, Oxford, pp. 229–257.

Halstead, P., and V. Isaakidou. 2004. "Faunal Evidence for Feasting: Burnt Offerings from the Palace of Nestor at Pylos," in *Food, Cuisine, and Society in Prehistoric Greece* (Sheffield Studies in Aegean Archaeology 5), ed. P. Halstead and J. C. Barrett, Oxford, pp. 136–154.

Hamilakis, Y. 2000. "The Anthropology of Food and Drink Consumption and the Aegean Archaeology," in Vaughan and Coulson 2000, pp. 55–63.

———. 2003. "The Sacred Geography of Farming: Wild Animals, Social Power, and Gender in Early Farming Societies," in Kotjabopoulou et al. 2003, pp. 239–247.

Hansen, J. M. 1988. "Agriculture in the Prehistoric Aegean: Data Versus Speculation," *AJA* 92, pp. 39–52.

———. 2000. "Palaeoethnobotany and Palaeodiet in the Aegean Region: Notes on Legume Toxicity and Related Pathologies," in Vaughan and Coulson 2000, pp. 13–27.

Hedges, R. E. M., and L. M. Reynard. 2007. "Nitrogen Isotopes and the Trophic Level of Humans in Archaeology," *JAS* 34, pp. 1240–1251.

Iezzi, C. A. 2005. "Regional Differences in the Health Status of Late Bronze Age Mycenaean Populations from East Lokris, Greece" (diss. State Univ. of New York at Buffalo).

Ingvarsson-Sundström, A., M. P. Richards, and S. Voutsaki. 2009. "Stable Isotope Analysis of the Middle Helladic Population from Two Cemeteries at Asine: Barbouna and the East Cemetery," *Mediterranean Archaeology and Archaeometry* 9, pp. 1–14.

Jones, G. 1983. "The Ethnoarchaeology of Crop Processing: Seeds of a Middle-Range Methodology," *Archaeological Review from Cambridge* 2, pp. 17–26.

Klein, R. G., and K. Cruz-Uribe. 1984. *The Analysis of Animal Bones from Archeological Sites* (Prehistoric Archeology and Ecology), Chicago.

Kotjabopoulou, E., Y. Hamilakis, P. Halstead, C. Gamble, and P. Elefanthi, eds. 2003. *Zooarchaeology in Greece: Recent Advances* (*BSA* Studies 9), London.

Kroll, H. J. 1983. *Kastanas: Ausgrabungen in einem Siedlungshügel der Bronze- und Eisenzeit Makedoniens, 1975–1979* 1: *Die Pflanzenfunde* (Prähistorische Archäologie in Südosteuropa 2), Berlin.

———. 2003. "Rural Plenty: The Result of Hardwork—Rich Middle Bronze Age Plant Remains from Agios Mamas, Chalkidiki," in *Troia and the Troad: Scientific Approaches* (Natural Science in Archaeology), ed. G. A. Wagner, E. Pernicka, and H.-P. Uerpmann, Berlin, pp. 293–301.

Lagia, A., E. I. Petroutsa, and S. K. Manolis. 2007. "Health and Diet during the Middle Bronze Age in the Peloponnese: The Site of Koufovouno," in Mee and Renard 2007, pp. 313–328.

Larsen, C. S. 1997. *Bioarchaeology: Interpreting Behavior from the Human Skeleton* (Cambridge Studies in Biological Anthropology 21), Cambridge.

Liden, K. 1995. "Megaliths, Agriculture, and Social Complexity: A Diet Study of Two Swedish Megalith Populations," *JAnthArch* 14, pp. 404–417.

Lubell, D., M. Jackes, H. Schwarcz, M. Knyf, and C. Meiklejohn. 1994. "The Mesolithic–Neolithic Transition in Portugal: Isotopic and Dental Evidence of Diet," *JAS* 21, pp. 201–216.

Lyman, R. L. 1982. "Archaeofaunas and Subsistence Studies," *Advances in Archaeological Method and Theory* 5, pp. 331–393.

Mee, C., and J. Renard. 2007. *Cooking Up the Past: Food and Culinary Practices in the Neolithic and Bronze Age Aegean*, Oxford.

Müldner, G., and M. P. Richards. 2005. "Fast or Feast: Reconstructing Diet in Later Medieval England by Stable Isotope Analysis," *JAS* 32, pp. 39–48.

Mylonas, G. E. 1929. *Excavations at Olynthus* I: *The Neolithic Settlement*, Baltimore.

Norr, L. 1995. "Interpreting Dietary Maize from Bone Stable Isotopes in the American Tropics: The State of the Art," in *Archaeology in the Lowland American Tropics: Current Analytical Methods and Recent Applications*, ed. P. W. Stahl, Cambridge, pp. 198–223.

Palaima, T. G. 2004. "Sacrificial Feasting in the Linear B Documents," *Hesperia* 73, pp. 217–246.

Papathanasiou, A. 1994. "Bioarchaeological Inferences from a Neolithic Ossuary at Alepotrypa Cave, Diros, Greece" (M.S. thesis Purdue University).

———. 1999a. "A Bioarchaeological Analysis of Health, Subsistence, and Funerary Behavior in the Eastern Mediterranean Basin: A Case Study from Alepotrypa Cave, Diros, Greece" (diss. Univ. of Iowa).

———. 1999b. "Συκιά Μολάων Λακωνίας," *ArchDelt* 54, B΄2, pp. 1009–1011.

———. 2001. *Bioarchaeological Analysis of Neolithic Alepotrypa Cave, Greece* (*BAR-IS* 961), Oxford.

———. 2002. "A Study of the Human Skeletal Remains of the Mycenaean Tombs from Bronze Age Sykia, Laconia," *American Journal of Physical Anthropology* 117 (Suppl. 34), pp. 122–123 (abstract).

———. 2003. "Stable Isotope Analysis in Neolithic Greece and Possible Implications on Human Health," *International Journal of Osteoarchaeology* 13, pp. 314–324.

———. 2009. "The Human Osteological Material from the Mycenaean Tholos at Kazanaki, Volos," in *Proceedings of the 2nd Archaeological Meeting of Thessaly and Central Greece*, Volos, pp. 151–161.

———. 2011. "Health, Diet, and Social Implications in Neolithic Greece from the Study of Human Osteological Material," in *Human Bioarchaeology of the Transition to Agriculture,* ed. R. Pinhasi and J. T. Stock, Chichester, pp. 87–106.

Papathanasiou, A., C. S. Larsen, and L. Norr. 1995. "A Bioarchaeological Analysis of a Neolithic Ossuary from Alepotrypa Cave, Diros, Greece," *American Journal of Physical Anthropology* 38, Suppl. 20, p. 167 (abstract).

———. 2000. "Bioarchaeological Inferences from a Neolithic Ossuary from Alepotrypa Cave, Diros, Greece," *International Journal of Osteoarchaeology* 10, pp. 210–228.

Papathanasiou, A., and M. P. Richards. 2011. "Ανθρωπολογικά κατάλοιπα από τις πρώιμες θέσεις Μαυροπηγής, Ξηρολίμνης και Ποντοκώμης της Αρχαιότερης Νεολιθικής στη Δυτική Μακεδονία," *Το Αρχαιολογικό Έργο στην Άνω Μακεδονία,* Aiani, pp. 257–274.

Papathanasiou, A., L. A. Schepartz, M. P. Richards, and E. Malapani. 2012. "Bioarchaeological Evidence for Social Differentiation in the Health and Diet of Mycenaean Pylos," in *Proceedings of the 2nd ARCH-RNT Symposium, University of the Peloponnese, Kalamata 2011,* pp. 143–151.

Papathanasiou, A., E. Zachou, and M. P. Richards. 2009. "Bioarchaeological Analysis of the Human Osteological Material from Proskynas, Lokris," in *New Directions in the Skeletal Biology of Greece* (*Hesperia* Suppl. 43, OWLS 1), ed. L. A. Schepartz, S. C. Fox, and C. Bourbou, Princeton, pp. 223–235.

Pappa, M., P. Halstead, K. Kotsakis, and D. Urem-Kotsou. 2004. "Evidence for Large-Scale Feasting at Late Neolithic Makriyalos, N. Greece," in *Food, Cuisine, and Society in Prehistoric Greece,* ed. P. Halstead and J. C. Barrett, Oxford, pp. 16–44.

Payne, S. 1972. "Partial Recovery and Sample Bias: The Results of Sieving Experiments," in *Papers in Economic Prehistory: Studies,* ed. E. S. Higgs, Cambridge, pp. 49–64.

Pearson, J. A., H. Buitenhuis, R. E. M. Hedges, L. Martin, N. Russell, and K. C. Twiss. 2007. "New Light on Early Caprine Herding Strategies from Isotope Analysis: A Case Study from Neolithic Anatolia," *JAS* 34, pp. 2170–2179.

Petroutsa, E. I. 2007. "Μελέτη της διατροφής σε πληθυσμούς της Ελλάδας της Εποχής του Χαλκού" (diss. Univ. of Athens).

Petroutsa, E. I., M. P. Richards, and S. K. Manolis. 2007. "Stable Isotope Analysis of Human Remains from the Early Helladic Site of Perachora, Korinth, Greece," in Mee and Renard 2007, pp. 290–296.

Powell, J. 1996. *Fishing in the Prehistoric Aegean* (*SIMA-PB* 137), Jonsered.

———. 2003. "Fishing in the Mesolithic and Neolithic–The Cave of Cyclops, Youra," in Kotjabopoulou et al. 2003, pp. 75–84.

Reitz, E. J., and E. S. Wing. 1999. *Zooarchaeology,* Cambridge.

Renfrew, J. 1973. "Agriculture," in *Neolithic Greece,* ed. D. R. Theocharis, Athens, pp. 147–164.

Richards, M. P., and R. E. M. Hedges. 1999. "How Chemical Analysis of Human Bones Can Tell Us the Diets of People Who Lived in the Past (and Other Contributions)," in *Minoans and Mycenaeans: Flavours of Their Time. National Archaeological Museum, 12 July–27 November 1999,* ed. Y. Tzedakis and H. Martlew, Athens, p. 214.

Richards, M. P., R. E. M. Hedges, I. Walton, S. Stoddart, and C. Malone. 2001. "Neolithic Diet at the Brochtorff Circle, Malta," *EJA* 4, pp. 253–262.
Richards, M. P., J. A. Pearson, T. I. Molleson, N. Russell, and L. Martin. 2003. "Stable Isotope Evidence of Diet at Neolithic Çatalhöyük, Turkey," *JAS* 30, pp. 67–76.
Schepartz, L. A., S. Miller-Antonio, and J. M. A. Murphy. 2009. "Differential Health Among the Mycenaeans of Messenia: Status, Sex, and Dental Health at Pylos," in *New Directions in the Skeletal Biology of Greece* (*Hesperia* Suppl. 43, OWLS 1), ed. L. A. Schepartz, S. C. Fox, and C. Bourbou, Princeton, pp. 155–174.
Schulting, R. J., and M. P. Richards. 2001. "Dating Women and Becoming Farmers: New Palaeodietary and AMS Dating Evidence from the Breton Mesolithic Cemeteries of Téviec and Hoëdic," *JAnthArch* 20, pp. 314–344.
———. 2002a. "Finding the Coastal Mesolithic in Southwest Britain: AMS Dates and Stable Isotope Results on Human Remains from Caldey Island, South Wales," *Antiquity* 76, pp. 1011–1025.
———. 2002b. "The Wet, Wild, and the Domesticated: The Mesolithic–Neolithic Transition on the West Coast of Scotland," *EJA* 5, pp. 147–189.
Sherrat, S. 2004. "Feasting in Homeric Epic," *Hesperia* 73, pp. 301–337.
Stocker, S. R., and J. L. Davis. 2004. "Animal Sacrifice, Archives, and Feasting at the Palace of Nestor," *Hesperia* 73, pp. 179–198.
Stratouli, G. 1996. "Die Fischerei in der Ägäis während des Neolithikums. Zur Technik und zum potentiellen Ertrag," *PZ* 71, pp. 1–27.
Tauber, H. 1981. "^{13}C Evidence for Dietary Habits of Prehistoric Man in Denmark," *Nature* 292, pp. 332–333.
———. 1983. "^{13}C Dating of Human Beings in Relation to Dietary Habits," *Journal of the European Study Group on Physical, Chemical and Mathematical Techniques applied to Archaeology* 8, pp. 365–375.
———. 1986. "Analysis of Stable Isotopes in Prehistoric Populations," in *Innovative Trends in Prehistoric Anthropology,* ed. B. Herrmann, Berlin pp. 31–38.
Theodoropoulou, T. 2007a. "L'exploitation des ressources aquatiques en Egée septentrionale aux périodes pré- et protohistoriques" (diss. Université de Paris 1, Panthéon-Sorbonne).
———. 2007b. "*La mer dans l'assiette:* L'exploitation alimentaire des faunes aquatiques en Egée pré- et protohistorique," in Mee and Renard 2007, pp. 72–88.
Triantaphyllou, S. 2001. *A Bioarchaeological Approach to Prehistoric Cemetery Populations from Central and Western Greek Macedonia* (*BAR-IS* 976), Oxford.
Triantaphyllou, S., M. P. Richards, C. Zerner, and S. Voutsaki. 2008. "Isotopic Dietary Reconstruction of Humans from Middle Bronze Age Lerna, Argolid, Greece," *JAS* 35, pp. 3028–3034.
Urem-Kotsou, D., and K. Kotsakis. 2007. "Pottery, Cuisine, and Community in the Neolithic of North Greece," in Mee and Renard 2007, pp. 225–246.
Valamoti, S. M. 2004. *Plants and People in Late Neolithic and Early Bronze Age Northern Greece: An Archaeobotanical Investigation* (*BAR-IS* 1258), Oxford.
———. 2007. "Traditional Foods and Culinary Novelties in Neolithic and Bronze Age Northern Greece: An Overview of the Archaeobotanical Evidence," in Mee and Renard 2007, pp. 89–108.
van Klinken, G. J., M. P. Richards, and R. E. M. Hedges. 2000. "An Overview of Causes for Stable Isotopic Variations in Past European Human Populations: Environmental, Ecophysiological, and Cultural Effects," in *Biogeochemical Approaches to Paleodietary Analysis* (Advances in Archaeological and Museum Science 5), ed. S. H. Ambrose and M. A. Katzenberg, New York, pp. 39–63.
Vaughan, S. J., and W. D. E. Coulson, eds. 2000. *Palaeodiet in the Aegean. Papers from a Colloquium Held at the 1993 Meeting of the Archaeological Institute of America in Washington D.C.* (Wiener Laboratory Monograph 1), Oxford.
Vika, E. 2002. "From Diet to Society: Stable Isotope Analysis and Its Cultural Context in Bronze Age Peloponnese, Greece" (M.S. thesis, Univ. of Bradford).
———. 2007. "A Diachronic Bioarchaeological Approach to the Society of Thebes, Greece. Paleopathological Investigations and Paleodietary Reconstruction Using $\delta^{13}C$, $\delta^{15}N$ and $\delta^{32}S$ Isotope Analysis" (diss. Univ. of Bradford).
Vika, E., and T. Theodoropoulou. 2012. "Re-Investigating Fish Consumption in Greek Antiquity: Results from $\delta^{13}C$ and $\delta^{15}N$ Analysis from Fish Bone Collagen," *JAS* 39, pp. 1618–1627.
Walker, P. L., and J. M. Erlandson. 1986. "Dental Evidence for Prehistoric Dietary Change on the Northern Channel Islands, California," *AmerAnt* 51, pp. 375–383.
Walker, P. L., and B. S. Hewlett. 1990. "Dental Health Diet and Social Status Among Central African Foragers and Farmers," *American Anthropologist* n.s. 92, pp. 383–398.
Wright, J. C. 2004a. "The Mycenaean Feast: An Introduction," *Hesperia* 73, pp. 121–132.
———. 2004b. "A Survey of Evidence for Feasting in Mycenaean Society," *Hesperia* 73, pp. 133–178.

CHAPTER 4

Stable Isotope Analysis of Skeletal Assemblages from Prehistoric Northern Greece

by Sevasti Triantaphyllou

The reconstruction of the diets of past peoples has been a focus of human bone studies over the course of the last few decades.[1] In addition to the macroscopic recording of dental lesions, more elaborate analytical methods such as stable isotope analysis have now been widely adopted for the identification of dietary patterns.[2] Until recently, however, dietary inferences in prehistoric Greece were based solely on the evidence of archaeobotanical and zooarchaeological analyses.[3]

A relatively high number of prehistoric settlements in northern Greece have employed systematic flotation and sieving during intensive excavation activities over the last 15 years. The rescue excavation of Neolithic Makriyalos, for example, has shed new light on several aspects of the Neolithic period in northern Greece such as settlement patterns, subsistence management, and consumption practices.[4] It is important to note that in this area of the Aegean, the discussion on palaeodiet has moved forward very early from the mere scientific identification of species consumed to the exploration of the social practices involved in subsistence strategies and the act of food procurement, preparation and consumption.[5]

The archaeobotanical data recovered from Neolithic and Bronze Age sites in northern Greece has been limited to Early Neolithic Nea Nikomedeia,[6] Middle Neolithic Servia,[7] Late Neolithic Sitagroi[8] and Olynthos,[9] and Bronze Age Kastanas,[10] until a few years ago. The above dataset has been enormously enriched in the last 15 years by the work of Valamoti on Neolithic sites such as, for example, Arkadikos, Apsalos, Mandalo, Makriyalos, Makri, and Dikili Tash, but also Early Bronze Age Archondiko.[11] Also, Margaritis[12] and Kotjamani[13] have added new information on archaeobotanical evidence from the area with their studies on LN Stavroupoli and EN Revenia, respectively. Assemblages are dominated by a combination of domestic cereal and pulse crops, which make up a carbohydrate and protein pair, that included einkorn, emmer, naked barley, lentils, grass pea, and bitter vetch during the Neolithic period, and einkorn, emmer, hulled barley, flax, and acorns during the Bronze Age.[14] Wild crops such as nuts and fruits appear to have been increasingly introduced to the diet of prehistoric populations in northern Greece from the LN period onward.[15] These latter plant categories may have been consumed seasonally after special preparation and preservation recipes were developed in order

1. Larsen 1997, pp. 270–301.
2. Ambrose 1993; Richards et al. 1998; Richards et al. 2003.
3. Jones 1987; Halstead 1987, 1992, 1994; Vaughan and Coulson 2000; Kotjabopoulou et al. 2003.
4. Pappa et al. 2013.
5. E.g., Valamoti 2003, 2004, 2007b; Urem-Kotsou 2007; Tsoraki 2007, 2008; Veropoulidou 2011.
6. Van Zeist and Bottema 1971.
7. Hubbard 1979.
8. Renfrew 1971.
9. Mylonas 1929.
10. Kroll 1983.
11. For a review on plant remains from prehistoric northern Greece, see Valamoti 2004, 2007b.
12. Margaritis 2002, 2004.
13. Kotjamani 2009.
14. Valamoti 2004, p. 13; 2007b, pp. 95–100.
15. Valamoti 2004, p. 13; 2007b, p. 97.

to make them available. Fruits, in particular, may have been used for the production of beverages as has been recently recognized by the evidence of grape pips and wine pressings at FN Dikili Tash.[16]

The domesticated animals identified include sheep, goat, cow, pig, and dog.[17] Evidence from Kastanas suggests a highly diversified subsistence system during the EBA that was succeeded in the LBA by a more specialized and extensive system of agriculture and stock-raising. The end of the Bronze Age was marked by the adoption of a more balanced regime, namely, intensive, small-scale cultivation of a wide variety of crops and new interest in gathering and especially hunting.[18] High juvenile mortality curves from all periods among sheep, goat, and cattle would suggest a strategy of animal husbandry targeted primarily for meat consumption. During the Bronze Age, however, an increase in the survivorship of male sheep and cattle at Sitagroi in eastern Macedonia may reflect a gradual increase in the use of the animals for their wool and also for traction.[19] More recently,[20] new evidence based on chemical analyses on the organic residues from pottery would imply a limited use of secondary products, and, in particular, milk, as early as the LN, associated perhaps only with certain households as suggested by Kotsakis and colleagues.[21]

Fallow, red, and roe deer are well represented in Neolithic and Bronze Age Macedonia, and their more frequent occurrence in Bronze Age assemblages has been attributed variously to progressive woodland clearance for the introduction of the horse, making the capture and transport of game easier, and thus leading to the growth of hunting. Hunting may be associated perhaps with special groups of people who could afford being involved in activities often considered as an elite type of sport.[22]

Furthermore, the contribution of seafood has been recently suggested in northern Greece by Veropoulidou and Theodoropoulou for the consumption of mollusks and fish, respectively.[23] It is noteworthy that mollusks appear to have been consumed in high quantities but in a rather low variety with regard to the species collected during the Neolithic, while the reverse pattern occurs in the Bronze Age, when mollusks were possibly comprising a smaller component to the diet. Overall, there is sufficient evidence in northern Greece to support a more regular consumption of seafood, particularly of mollusks such as *Cerastoderma glaucum,* which is the most abundant species identified and which particularly favors estuarine environments such as the river shorelines of the Thermaic Gulf in central Macedonia.[24]

Having thus considered very briefly the dietary regime of northern Greece provided mainly by the archaeobotanical and the zooarchaeological evidence, this paper aims to synthesize information derived from different

16. Valamoti et al. 2007.
17. Halstead 1994, p. 201.
18. Becker 1986, p. 291.
19. Bökönyi 1986; Halstead 1994, p. 201.
20. Evershed et al. 2008.
21. Kotsakis et al. 2008.
22. Halstead 1994, p. 200; and more recently, see Nikolaidou 2010, with evidence of lion bones from LBA Toumba Thessalonikis.
23. Theodoropoulou 2007; Veropoulidou 2011.
24. Veropoulidou 2011, p. 450.

levels of the study of the human remains as follows: (1) to present the results of further carbon and nitrogen stable isotope analyses conducted recently on human bone samples from central Macedonia; (2) to discuss the overall results of stable isotopic analysis conducted to date in this region of the prehistoric Aegean; and (3) to integrate them into the picture already provided by previous dietary reconstructions of past populations based on both archaeobotanical and zooarchaeological evidence along with organic residues from pottery. In order to achieve the above targets, it is important first to briefly review the history of human bone studies in northern Greece.

PREVIOUS HUMAN BONE STUDIES

Until 15 years ago, studies of prehistoric skeletal remains in Greece were limited to assemblages from the southern mainland, Crete, and the Cyclades. Recent work in northern Greece, however, has begun to fill in the gap and offers new insight into aspects of the reconstruction of the life history and the manipulation of the deceased from the Neolithic period to the EIA.[25] In comparing different skeletal populations from northern Greece, no constant pattern emerges with respect to demographic profiles and health status that appears consistent with any archaeological period or regional setting.[26] Instead, local conditions defined by environmental and social constraints seem to have affected past quality of life in Greek Macedonia, as reconstructed in this case study of prehistoric human skeletal assemblages. Overall, the present study shows a tendency toward declining levels of oral and physical health among LBA and EIA groups in Greek Macedonia.

A study of stable isotopes of carbon and nitrogen was conducted in central and western Greek Macedonia in the late 1990s by Triantaphyllou and van Klinken at the Oxford Research Laboratory for Archaeology and History of Art.[27] The aims of this initial work were: (1) to distinguish between human skeletal assemblages of the same region by selecting groups belonging for the most part to the same chronological period—the LBA; and (2) to investigate possible variation through time between human skeletal series settled at the same site. Thirteen permanent posterior teeth were analyzed from LBA Korinos, Spathes, and Rymnio[28] with regard to the first aim. Another 52 samples from human femoral midshafts and 15 samples of animal bones were provided from the site of Makriyalos dating from the Neolithic through Classical periods for testing with regard to the second aim.[29]

Another large-scale sampling for stable isotopes of carbon and nitrogen took place in 2003 at the EBA/MBA cemetery of Koilada located in the upland basin of Kitrini Limni in western Macedonia.[30] Fifty-four human and 12 faunal bone samples from the nearby EBA/MBA site of Polemistra, western Macedonia, were analyzed in order to explore: (1) general dietary patterns; (2) differences in isotopic signals between population subgroups defined by age or sex; and (3) the effect of breastfeeding and weaning on subadults.[31]

25. Triantaphyllou 1998a, 1998b, 2001, 2002, 2003, 2004, 2008, 2010; Antikas and Antikas 2002; Grammenos and Triantaphyllou 2004; Ziota and Triantaphyllou 2004; Karliabas, Bessios, and Triantaphyllou 2005.

26. Triantaphyllou 2001, p. 142; 2003, p. 219.

27. Triantaphyllou 2001, pp. 134–139.

28. Triantaphyllou 2001, p. 134, table 7.9.

29. Triantaphyllou 2001, p. 134, tables 7.10, 7.11.

30. Ziota 1995, 1996, 1998a, 1998b, 2007; Ziota and Triantaphyllou 2004.

31. Triantaphyllou 2010; Triantaphyllou, Richards, and Ziota in preparation.

TABLE 4.1. CHRONOLOGY FOR PREHISTORIC MACEDONIA*

Archaeological Phase	*Years* B.C.
Late Neolithic	5400/5300–4700/4500
Early Bronze Age	3300/3100–2300/2200
Middle Bronze Age	2300/2200–1700/1500
Late Bronze Age	1700/1500–1100
Early Iron Age	1100–900

*After Andreou, Fotiadis, and Kotsakis 1996

MATERIALS AND METHODS

The prehistoric skeletal assemblages that provided samples for the present stable isotope analysis cover a broad chronological range, from the Late Neolithic period to the Early Iron Age (see Table 4.1). The five sites that produced the human assemblages used in this study lie in various environmental zones, ranging from the coasts of the Thermaic Gulf in central Macedonia to the lowland and upland areas of Mount Olympus (Fig. 4.1).

Late Neolithic Stavroupoli, located on the coast, has both tell and flat-extended site components and provided: five primary inhumations containing two subadults (3–4 years and 13–15 years), two females (30–40 and 40–50 years), and one male individual (19–20 years); one cremation belonging to an adult individual of indeterminate sex; and 117 scattered bones.[32] Spathes, an LBA cemetery of the 13th–12th centuries B.C., is located on a steep slope of Mount Olympus at 1,000–1,100 m AMSL (above mean sea level) and consists of 34 graves arranged in rows. This cemetery is mainly characterized by its "Mycenaeanizing"-type shaft graves, burial practices, and related artifacts.[33] Treis Elies, located a short distance from Spathes, dates to the transitional phase between the end of the LBA and the beginning of the EIA.[34] The EIA Kladeri and Karitsa tumuli,[35] located in the lowlands of southern Pieria, are particularly distinctive, as group burials seem to indicate that family links may have defined local social order.[36]

The main aim of the present stable isotope analysis is to add more information on prehistoric diet derived from skeletal assemblages in northern Greece by expanding the existing sample to include newly excavated skeletal assemblages from LN Stavroupoli and the EIA Kladeri and Karitsa tumuli ("Olympus mountain tumuli"). In addition to the five human bone samples measured in the study by Triantaphyllou and van Klinken mentioned above,[37] LBA Spathes provides six new human bone samples.

Due to poor documentation, serious problems of retrieval, and post-excavation factors, however, the present sampling strategy could not take advantage of archaeological information derived from these sites, including chronological sub-phases, grave types, burial practices, and associated artifacts. Altogether, 21 individuals from five different skeletal assemblages, mainly adults of both sexes, provided the following rib fragments: Stavroupoli

32. Triantaphyllou 2002, pp. 829–846; 2004, pp. 613–624.

33. Poulaki-Pantermali 1986a, 1986b, 1987a, 1987b, 1990, 1991.

34. Poulaki-Pantermali 1988, 1989, 1992.

35. Poulaki-Pantermali 1985, 1986b, 1990.

36. Triantaphyllou 2001, p. 57.

37. Triantaphyllou 2001, table 7.9.

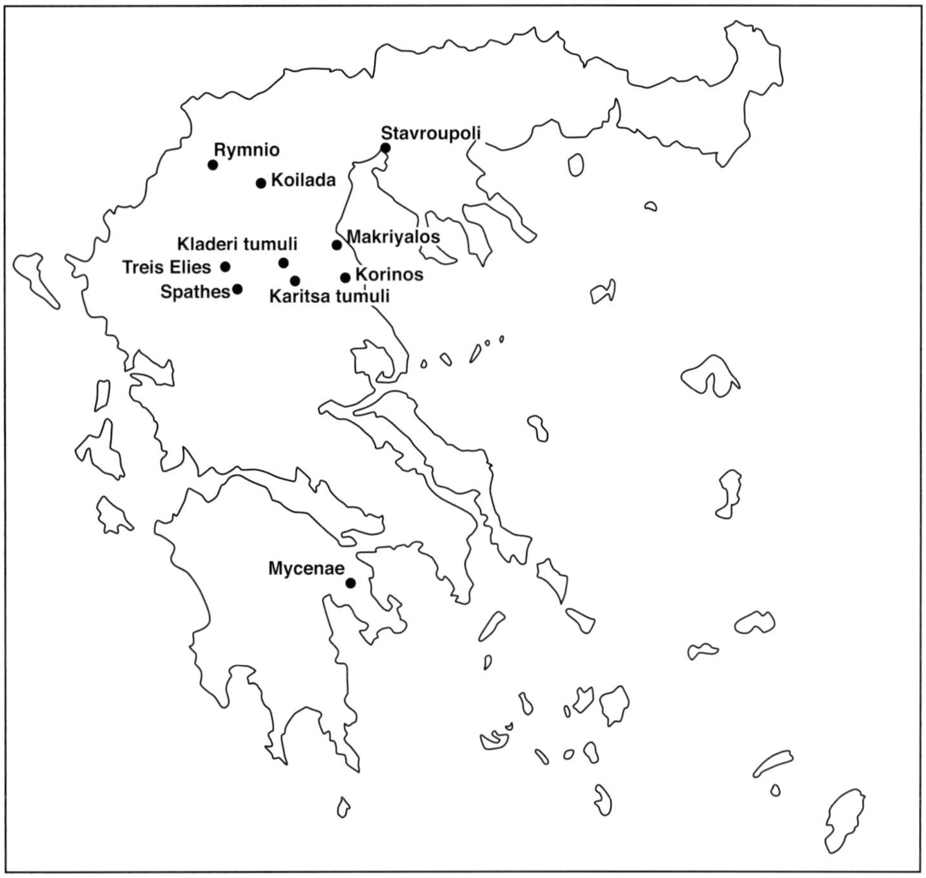

Figure 4.1. Map of Greece showing sites mentioned in the text

(n = 2), Spathes (n = 6), Treis Elies (n = 6), Karitsa tumuli (n = 2), and Kladeri tumuli (n = 5).

Table 4.2 lists all human skeletons sampled and gives their age and sex as assessed by macroscopic examination.[38] Only adults (individuals over 18 years of age) were sexed, with identification made on the basis of pelvic[39] and cranial morphologies,[40] as well as long bone measurements.[41] Age at death for subadults was estimated by examining tooth formation and eruption, epiphyseal closure, and measurement of long bones.[42] Age estimates for adults were based on pubic symphyseal morphology, changes in the auricular surface of the os coxae, and ectocranial suture closure.[43] Three age groups were defined for adults based on the above criteria: young adult, 18 to 30 years; middle adult, 30 to 40 years; and older adult, 40 to 50 years.

In addition to assessment of sex and age, information derived from the macroscopic analysis of the dentition, including evidence of dental disease, was examined in order to provide a background against which the results of stable isotope analyses could be compared. The relationship between diet and the oral health status of past populations has been previously debated, since other factors such as genetic variability could also affect dental conditions.[44] Nevertheless, it is broadly accepted that both environmental conditions and social factors (which include collection, production, and consumption of particular foodstuffs) play a major role in determining oral health status.

38. Triantaphyllou 2001.
39. Phenice 1969; Kelley 1978; Krogman and İşcan 1986; MacLaughlin and Bruce 1986; Lovell 1989; Sutherland and Suchey 1991.
40. Ascádi and Nemeskéri 1970.
41. Berrizbeitia 1989; Holland 1991; Holman and Bennett 1991.
42. Scheuer and Black 2000.
43. Buikstra and Ubelaker 1994.
44. Wood et al. 1992.

TABLE 4.2. STABLE ISOTOPE VALUES FOR HUMANS FROM PREHISTORIC NORTHERN GREECE

S-EVA	*Site*[1]	*Skeleton No.*	*Sex*	*Age (years)*[2]	*Date*[3]	*Sample Weight (g)*	*% Coll.*	*$\delta^{13}C$ (‰ VPDB)*	*$\delta^{15}N$ (‰ AIR)*	*% C*	*% N*	*C:N*
1604	STAV	1	?	Y child (3–4)	LN	4.5	1.50	-19.2	9.3	35.32	12.80	3.21
1606	STAV	3	F	MA (30–40)	LN	5.8	1.64	-19.9	9.3	33.60	12.07	3.24
1612	SPA	11 (2)	F	YA (20–25)	LBA	3.5	1.16	-19.3	8.1	40.89	14.72	3.24
1613	SPA	12 (1)	F	MA (25–35)	LBA	31	7.58	-19.4	7.7	44.14	16.22	3.17
1614	SPA	12 (2)	M	MA (30–40)	LBA	14.1	5.68	-19.2	8.5	43.91	15.97	3.20
1615	SPA	13 (1)	F	YA (25–30)	LBA	16.2	7.09	-19.3	8.4	43.95	16.05	3.19
1616	SPA	14 (2)	F	MA (30–40)	LBA	13.1	4.81	-19.6	7.8	42.84	15.55	3.21
1617	SPA	29 (East)	F	YA (18–25)	LBA	17.2	5.22	-19.2	7.3	45.03	16.57	3.17
1618	TREL	8	M	YA (24–30)	LBA/EIA	8.5	2.05	-16.3	7.5	44.21	15.70	3.28
1619	TREL	11	M	YA (18–20)	LBA/EIA	3.5	1.30	-18.2	8.0	42.18	14.32	3.43
1620	TREL	13	M	YA (18)	LBA/EIA	3	1.13	-15.1	9.5	37.17	12.71	3.41
1621	TREL	31	M	YA (25–30)	LBA/EIA	12.4	3.89	-17.9	8.1	42.98	15.67	3.19
1622	TREL	44 (2)	M	YA (25–30)	LBA/EIA	8.1	2.16	-18.8	8.4	52.93	18.99	3.24
1623	TREL	33	M	YA (25–30)	LBA/EIA	5.1	2.03	-15.1	8.3	34.62	12.36	3.26
1624	KAR	Tum 5, d	M	OA (40–50)	EIA	11	4.05	-17.0	10.1	44.89	16.40	3.19
1625	KAR	Tum 12, c	M	YA (18–30)	EIA	3.5	1.16	-16.4	9.0	35.68	12.83	3.24
1626	KLAD	Tum 1, b1	F	OA (40–50)	EIA	5	1.79	-18.0	8.5	35.35	12.92	3.19
1627	KLAD	Tum 1, d	F	OA (40–50)	EIA	14.6	3.30	-16.2	8.4	36.06	13.31	3.16
1628	KLAD	Tum 1, st	M	OA (40–50)	EIA	6.1	2.14	-17.0	10.1	32.52	11.84	3.20
1629	KLAD	Tum 2, a1	M	YA (18–30)	EIA	4.7	1.78	-16.8	9.5	33.77	11.97	3.29
1630	KLAD	Tum 2, a2	M	OA (40–50)	EIA	13.9	4.41	-17.3	9.8	44.90	15.80	3.31

[1] STAV = Stavroupoli; SPA = Spathes; TREL = Treis Elies; KAR = Karitsa; KLAD = Kladeri.

[2] Y child = young child; YA = young adult; MA = middle adult; OA = older adult; M = male; F = female.

[3] LN = Late Neolithic; LBA = Late Bronze Age; EIA = Early Iron Age; for definitions of the chronological phases, see Table 4.1.

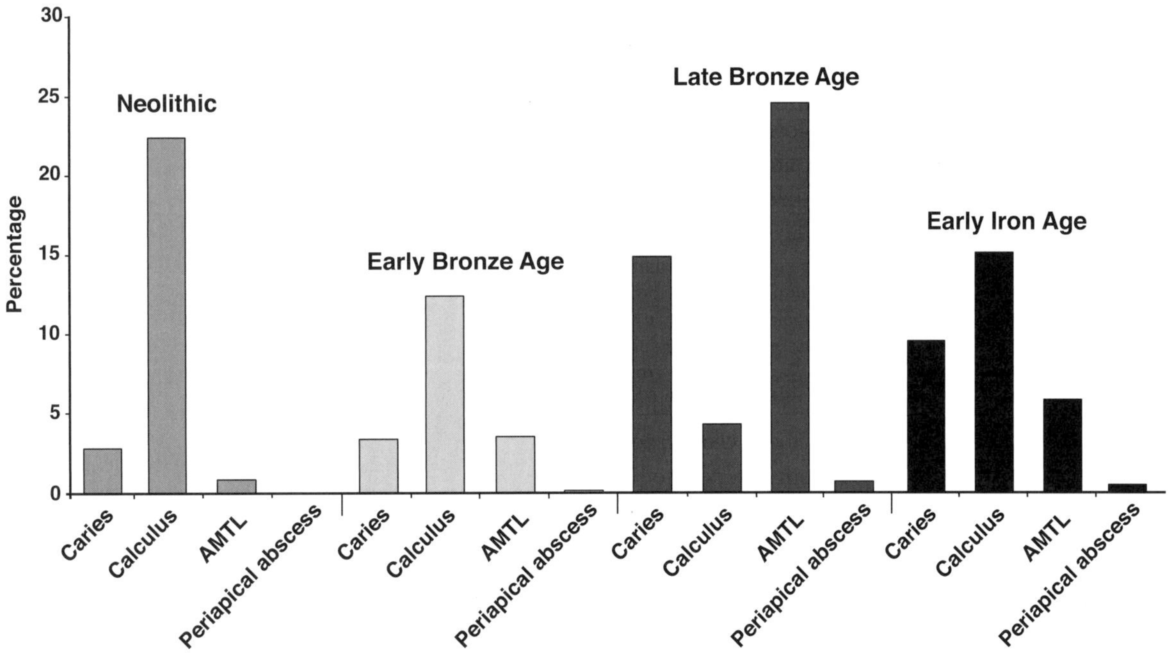

Figure 4.2. Distribution of dental disease in skeletal assemblages of prehistoric Greek Macedonia (tooth count analysis) (overall n - 4,027)

Dental lesions observed include dental caries, dental calculus, and periapical abscesses. Prevalence of tooth loss prior to death (antemortem tooth loss or AMTL) was also recorded.[45] Dental caries and calculus tend to be mutually exclusive due to the mechanisms that produce them in the mouth. Therefore, comparison between the prevalence of the two conditions in a population can provide important information about dietary patterns.[46] The overall distribution of dental conditions through the period spanning the Neolithic/EBA to the LBA/EIA (Fig. 4.2) suggests a shift from a higher reliance on meat consumption, as represented by a high prevalence of calculus versus low rates of dental caries, to a diet rich in carbohydrates such as starchy and processed foodstuffs, as represented by high frequencies of dental caries and AMTL versus a low frequency of calculus deposits.

Macroscopic analysis of dental disease has been conducted on a large sample of 4,027 permanent and deciduous teeth from prehistoric skeletal assemblages in northern Greece.[47] In order to overcome the constraints of differential preservation between skeletal populations,[48] the prevalence of dental lesions was measured by the total prevalence rate, defined as the number of observed lesions against the total number of recovered teeth. In order to calculate the prevalence of AMTL, dental alveoli were also recorded. In addition, the effect of the differential demographic structure of skeletal assemblages has been eliminated by standardizing the crude prevalence of dental conditions. Standardization is a statistical technique commonly applied in epidemiology that balances the average values of crude prevalence rates to allow comparisons between populations with different demographic structures.[49]

45. Triantaphyllou 2001.
46. Hillson 1979.
47. Triantaphyllou 2001.
48. Lukacs 1992; Waldron 1994, pp. 42–60.
49. Coggon, Rose, and Barker 1993; Waldron 1994, pp. 63–67. For application of the method in the case of the present study's skeletal assemblages, see Triantaphyllou 2001, p. 67.

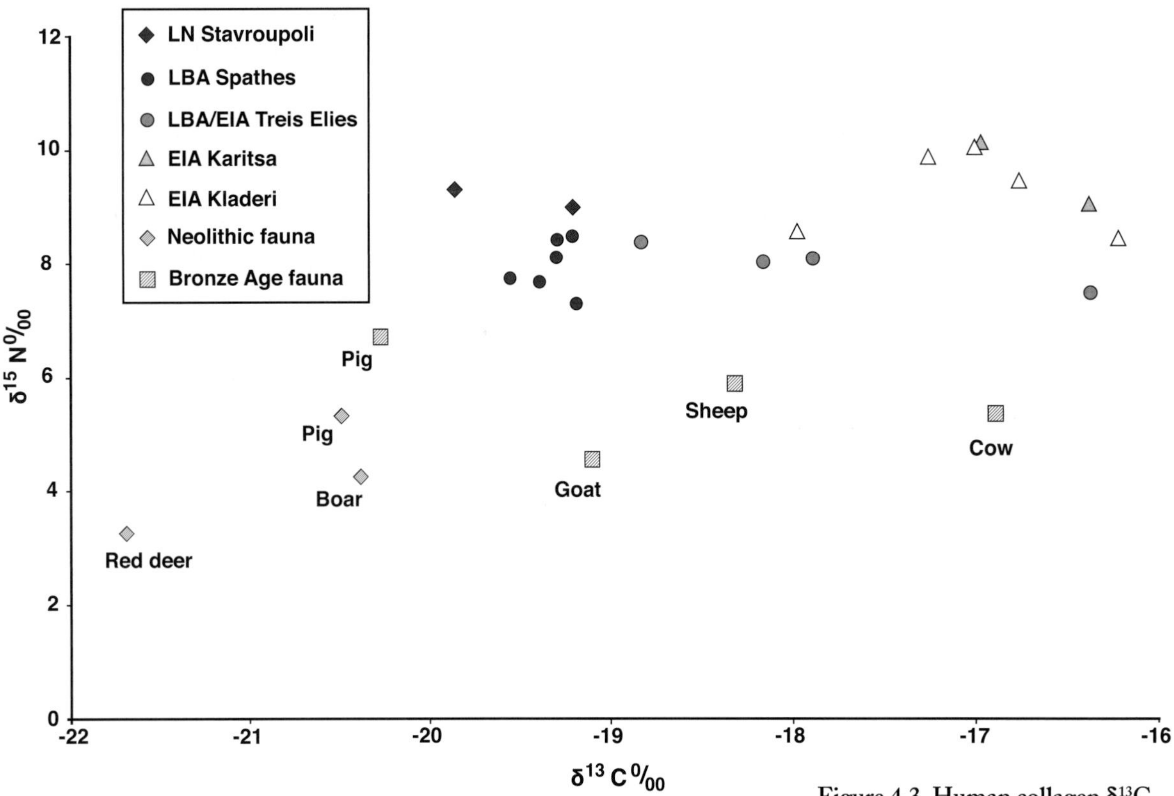

Figure 4.3. Human collagen $\delta^{13}C$ and $\delta^{15}N$ values from prehistoric northern Greek assemblages plotted with faunal isotope values

RESULTS

The carbon and nitrogen stable isotope values from the present study of five prehistoric skeletal assemblages from northern Greece are shown in Table 4.2, with each entry listed by site, skeleton number, sex, age, and date. Also, in Figure 4.3, the human stable isotope values are plotted alongside the isotopic values from Neolithic and Bronze Age animal bones from the same area to provide a basis for the discussion and reconstruction of past diet in prehistoric northern Greece.[50]

The faunal data represent both terrestrial C_3 and C_4 consumers (Table 4.3). It is clear from Table 4.3 that animals such as the herbivorous sheep and cows, due to their overall enriched ^{13}C values indicative of the presence of C_4 plant foods, probably consumed millet or C_4 grass directly in their diet.

The mean $\delta^{13}C$ and $\delta^{15}N$ values of omnivorous pigs *(Sus scrofa domesticus)* are, respectively: -20.4‰ ± 0.2‰ and 5.2 ± 0.8‰ for the Neolithic assemblage (LN II Makriyalos, northern Pieria); and -20.2‰ ± 0.2‰ and 6.7‰ ± 0.9‰ for the Bronze Age assemblage (EBA/MBA Polemistra, western Macedonia). Interestingly, the $\delta^{15}N$ values are different for the Neolithic and the Bronze Age assemblages, and this can be strongly supported statistically—based on Student's *t*-test for independent samples ($t = -2.47$, $p = 0.04$). The more positive $\delta^{15}N$ values of Bronze Age omnivorous pigs suggest a "carnivorous" diet with greater consumption of meat provided perhaps by animal remains from domestic refuse, as opposed to the herbivorous diet of the Neolithic pig.

The mean human bone collagen $\delta^{13}C$ values for the five skeletal assemblages in the present study are presented in Table 4.4. The stable isotope

50. For the Neolithic fauna, see Triantaphyllou 2001, table 7.11; for the Bronze Age fauna, see Triantaphyllou, Richards, and Ziota in prep.

TABLE 4.3. MEAN STABLE ISOTOPE VALUES FOR FAUNA FROM PREHISTORIC NORTHERN GREECE*

Species	*$\delta^{13}C$ (‰ VPDB)*	*SD*	*$\delta^{15}N$ (‰ AIR)*	*SD*
Ovis aries	-18.3	± 0.7	5.9	± 1.0
Capra hircus	-20.5	–	4.4	–
Bos taurus	-16.9	± 1.7	5.3	± 0.7
Sus scrofa	-20.4	± 0.4	4.1	± 0.8
Cervus elaphus	-21.7	± 0.4	3.2	± 1.2

*Combined values for the Neolithic and the Bronze Age from various sources.

TABLE 4.4. MEAN STABLE ISOTOPE VALUES FOR HUMANS FROM PREHISTORIC NORTHERN GREECE

Site	*$\delta^{13}C$ (‰ VPDB)*	*SD*	*$\delta^{15}N$ (‰ AIR)*	*SD*
Stavroupoli	-19.5	± 0.5	9.1	± 0.3
Spathes	-19.3	± 0.1	8.0	± 0.5
Treis Elies	-16.9	± 1.6	8.2	± 0.7
Karitsa tumuli	-16.7	± 0.4	9.6	± 0.7
Kladeri tumuli	-17.0	± 9.3	9.2	± 0.7

values cluster in two primary areas, which represent terrestrial C_3 and C_4-type diets. The higher $\delta^{13}C$ values represented by the transitional LBA/EIA skeletal assemblages of Treis Elies and the EIA skeletal assemblages of the Karitsa and Kladeri tumuli appear to suggest the consumption of millet either directly or indirectly through the consumption of animals fed on millet. This finding is reinforced by the faunal stable isotope values and, in particular, the high $\delta^{13}C$ values of Bronze Age sheep and cattle as shown in Figure 4.3. Indeed, the relatively high $\delta^{15}N$ values of EIA human groups in northern Greece suggest a contribution of animal protein into their diet, perhaps through consumption of sheep and cattle fed on millet or C_4 grass.

Furthermore, it is interesting to note the clustering of the LBA Spathes stable isotope values (range in $\delta^{13}C$: 1.4‰; range in $\delta^{15}N$: 2‰), which strongly suggest a homogenous terrestrial C_3-type diet. Conversely, the EIA groups, and in particular the transitional LBA/EIA skeletal assemblage of Treis Elies, reflect a relatively broad range, especially with regard to the $\delta^{13}C$ values (range: 3.8‰), which suggests the existence of both C_3 and C_4 consumers—probably because, as mentioned above, the use of the cemetery encompasses a broad time span. This relative heterogeneity of diet suggested by the stable isotope values of the LBA/EIA Treis Elies skeletal assemblage is consistent with the relative variability in the site's mortuary evidence (e.g., grave types, burial positions).[51] Poor documentation at LBA Spathes and LBA/EIA Treis Elies, however, sets certain limitations on the correlation of the biological information with the archaeological contexts of the skeletal assemblages from these two sites. Finally, the generally low $\delta^{15}N$ values cannot support the contribution of any marine protein into the dietary regime of LBA and EIA populations in northern Greece, as previously suggested in earlier stable isotope work in the same region.[52]

51. Triantaphyllou 2001, p. 26.
52. Triantaphyllou 2001, pp. 135, 138.

DISCUSSION

The overall distribution of dental conditions suggests higher meat consumption by Neolithic and EBA populations and higher carbohydrate consumption by LBA and EIA populations in northern Greece. The stable isotope values suggest fairly similar animal protein intake throughout these time periods, with the exception of the LN Makriyalos human bone samples, whose less positive $\delta^{15}N$ values suggest heavy reliance on plants. Macroscopic examination of dental pathological lesions and stable isotope analysis address different aspects of diet. Macroscopic investigation indirectly reveals particular types of dietary intake related to the chemical environment of the oral cavity, such as the consumption of meat and carbohydrates. Higher levels of dietary protein result in increased alkalinity of plaque, which predisposes the dentition to the development of calculus deposits.[53] Higher levels of dietary carbohydrates, on the other hand, result in increased acidity, which predisposes the dentition to the development of carious lesions. Stable isotope values reflect patterns of consumption with a special emphasis on the dietary intake of protein. Moreover, stable isotope analysis measures intake of all kinds of animal protein, as reflected in elevated $\delta^{15}N$ values in human bone collagen, and cannot distinguish between meat and dairy products. The association between elevated $\delta^{15}N$ values in human bone collagen and a diet based on animal protein has recently been challenged by Bogaard and collaborators,[54] and it is also discussed in depth for Classical Thebes by Vika.[55] More specifically, recent work on early farming in Europe, based on archaeobotanical evidence and the analysis of stable isotopes of modern cereals, has demonstrated that manuring significantly raises $\delta^{15}N$ values in cereal grain and chaff.[56] When attempting to reconstruct prehistoric diet, therefore, it is necessary to consider the possibility that different manuring levels and frequency may affect the $\delta^{15}N$ values of human collagen. Further research on archaeological material, as well as on manured and non-manured conditions, will allow refinement of the analysis, which may improve the interpretation of stable isotopes.

Figure 4.4 presents the isotopic signals of all the skeletal assemblages that have provided results to date from prehistoric northern Greece (n = 129): LN Makriyalos, on the coast of northern Pieria (n = 21) (Triantaphyllou 2001, p. 136, table 7.10); LN Stavroupoli (n = 2) (Table 4.2); EBA Makriyalos (n = 6) (Triantaphyllou 2001, p. 136, table 7.10); EBA/MBA Koilada, in the upland basin of Kitrini Limni, western Macedonia (n = 54) (Triantaphyllou 2010; Triantaphyllou, Richards, and Ziota in prep.); LBA Korinos, on the coast of southern Pieria (n = 4) (Triantaphyllou 2001, p. 135, table 7.9); LBA Rymnio in Western Macedonia (n = 4) (Triantaphyllou 2001, p. 135, table 7.9); LBA Spathes (n = 11) (Triantaphyllou 2001, p. 135, table 7.9; Table 4.2); LBA/EIA Treis Elies (n = 6) (Table 4.2); EIA Makriyalos (n = 14) (Triantaphyllou 2001, p. 136, table 7.10); EIA Karitsa (n = 2) (Table 4.2); and EIA Kladeri (n = 5) (Table 4.2). Human bone values in Figure 4.4 have been plotted alongside Neolithic and Bronze Age faunal isotopic signals from the same area in order to present a local dietary system as a basis for discussing and reconstructing past diet in prehistoric northern Greece.

53. Hillson 1979; Meiklejohn, Baldwin, and Schentag 1988; Frayer 1989.
54. Bogaard et al. 2007.
55. Vika, Aravantinos, and Richards 2009, p. 1079; Vika 2011, p. 1160; Chap. 3.
56. A. Bogaard (pers. comm.).

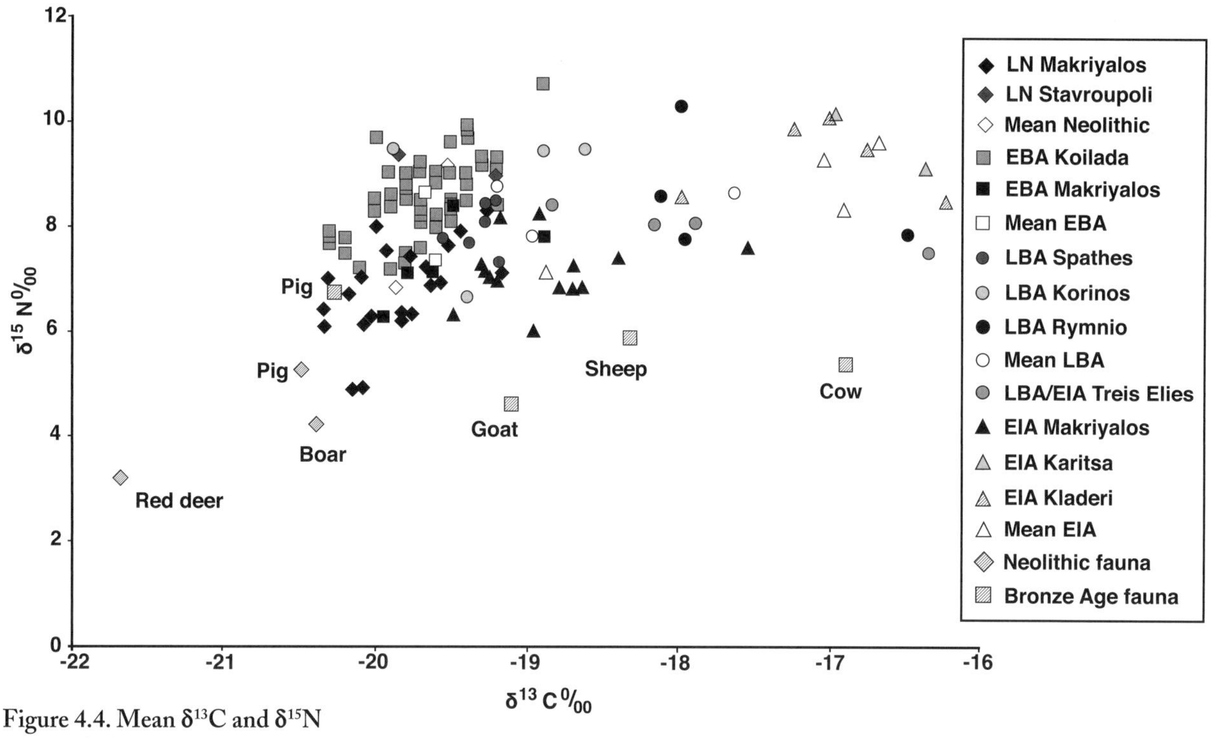

Figure 4.4. Mean $\delta^{13}C$ and $\delta^{15}N$ values of all human skeletal assemblages that have provided stable isotopic signals from prehistoric Greek Macedonia plotted with faunal isotope values

It is clear that isotopic signals from prehistoric populations in northern Greece fall within two main types of foodstuffs indicative of either a terrestrial C_3-type or C_4-type diet. The Neolithic, EBA, and at least some LBA populations (LBA Korinos and Spathes) show a preference for a C_3-type diet. These populations range from heavy plant protein consumers (e.g., LN Makriyalos, with $\delta^{15}N$ values as low as 6.8‰, whose inhabitants may have based their diet primarily on C_3 plants such as wheat, barley, and legumes) to heavy animal protein consumers (e.g., LN Stavroupoli, EBA Koilada, and LBA Korinos). Conversely, most of the EIA populations (LBA/EIA Treis Elies and the EIA Kladeri and Karitsa tumuli) as well as the LBA population at Rymnio clearly show a tendency toward greater inclusion of C_4 dietary items with a high contribution of animal protein, suggesting perhaps that herbivorous animals such as sheep and cattle were consuming C_4 grasses and/or millet.

C_4 Intake and the Contribution of Millet to the Prehistoric Diet

With regard to the occurrence of millet in prehistoric Greece, there is only one mention of the plant in the archaeological literature: at Olynthos in Chalkidike, where it may date as early as the FN period.[57] Millet only appears as an established crop, stored alongside various other cereals, perhaps by the LBA, as has been suggested by the archaeobotanical evidence from LBA Assiros[58] and Kastanas.[59] Furthermore, the adoption of millet in LBA northern Greece is associated with the use of the most common LBA coarse ware form known as *pyraunos*—a jar fixed to a cylindrical stand, pierced for ventilation, and with a space for fuel to be placed underneath.

57. Mylonas 1929.
58. Jones et al. 1986; Jones 1987.
59. Kroll 1983.

This shape is unknown in southern Greece but is commonly found from northern Greece to as far north as Hungary.[60] The significance of human bone stable isotope values from LBA/EIA skeletal assemblages for the early history of millet in Greece is worth discussing here.

According to archaeobotanical evidence, millet may have been introduced to Greek Macedonia via northern routes rather than Anatolian ones.[61] It also seems that millet was known in central and eastern Europe much earlier (5th millennium B.C.: eastern Germany, Czech Republic, Poland, Ukraine, and Romania) than in Greece. The millet discovered at Neolithic Olynthos probably does not represent a locally grown crop.[62] The idea that millet was introduced into Greek Macedonia from Europe may be reinforced by the positive $\delta^{13}C$ values from LBA Rymnio, which is located near Aeani, in western Macedonia—an area that has constituted a main route from north to south since the Neolithic period.[63] The skeletal assemblages from LBA/EIA Treis Elies and the EIA Karitsa and Kladeri tumuli are similarly located on the southern edge of this same north–south route into Greece.

Conversely, EIA Makriyalos appears to divert significantly from the dominant C_4-type diet observed elsewhere among EIA populations. This phenomenon may be consistent with a broader regional influence apparent on the site, which seems to have brought Makriyalos closer to southern cultural practices (i.e., those of Thessaly and the Greek mainland) rather than the northern practices of Upper Macedonia. For example, whereas tumuli were the most common type of group burial within most EIA cemeteries of northern Greece (cf., e.g., Vergina, Olympus mountain), they are not found at EIA Makriyalos. Instead, EIA Makriyalos adopted the type of chamber tomb well-known in the south for the burial of large groups of individuals.[64]

Animal Protein and the Evidence for the Use of Secondary Products

Reliance on animal protein—the predominant consumption of meat and/or dairy products—is the subject of an extensive, ongoing discussion about specialized pastoralism versus mixed farming in prehistoric Greece.[65] On the one hand, high reliance on carbohydrates is indicated by macroscopic investigation of dentition. On the other hand, the rather minimal contribution of animal protein is reflected in the stable isotope values of high mountain populations (e.g., at LBA Spathes and LBA/EIA Treis Elies). These results are in accordance with the model of systematic mixed farming existing even in high mountainous regions.[66]

High $\delta^{15}N$ values from LN Stavroupoli on the Thermaic Gulf coast, however, could suggest relatively high consumption of animal protein, although one out of the two individuals sampled from this site is derived from a 3- to 4-year-old child. It is highly probable therefore that the high nitrogen isotope value of this individual represents a "trophic level effect" from the mother's milk, which may have skewed the results of this extremely small sample. Nevertheless, results from the stable isotope analysis of carbon ($\delta^{13}C$) in organic residues from two pots at Neolithic Stavroupoli—conducted as part of "The Emergence of Dairying in Early Farming Practices

60. Wardle 1993, p. 123.
61. Cf. Valamoti 2007a, pp. 287–288; 2007b, p. 101.
62. Valamoti 2007a, p. 288; 2007b, p. 99.
63. Andreou, Fotiadis, and Kotsakis 1996.
64. Triantaphyllou 2001, p. 27.
65. See also Chap. 11.
66. Halstead 1994, 1996a, 1996b.

of the Fertile Crescent and the Balkans Project"—provide support for the results of the human bone isotopic values,[67] perhaps relating the higher $\delta^{15}N$ human values with the consumption of dairy products. Although this project is ongoing and the results are not yet conclusive, only Stavroupoli out of the three major Neolithic sites investigated to date (Makriyalos, Paliambela, Stavroupoli) provided clear evidence for the presence of dairy products by 5400 B.C.[68]

At LN Makriyalos, by contrast, a combined analysis of organic residues from 11 baking pans and 16 cooking pots, using gas chromatography mass spectrometry and the stable isotopes of carbon, has pointed to: (1) the preparation of boiled or stewed food of both plant and animal origin (with a special preference for pork); (2) the association of specific vessels (i.e., baking pans) with the preparation of plant-based foodstuffs; and (3) the marked absence of evidence related to the preparation or consumption of dairy products.[69]

With regard to LN Stavroupoli, Kotsakis and colleagues conclude that evidence of preparation and consumption of dairy products found in two pots in a single domestic unit suggests that dairy products on this site may have been associated only with certain households and so were not included in the inhabitants' common dietary regime.[70] This conclusion is supported by the mortality curves of domestic animals from Neolithic faunal assemblages, which are compatible with meat production rather than exploitation of secondary products.[71]

Furthermore, in a recent discussion of manipulation of the deceased in Neolithic Greece,[72] it has been argued that evidence for scattered human remains and, therefore, for secondary treatment of the deceased, tends to be extensive. Evidence for primary inhumations, however, is limited. Considering the duration of habitation in certain cases (cf. EN Nea Nikomedeia), primary inhumations could not have been the predominant form of mortuary treatment. If this is also the case at LN Stavroupoli, the different character of the two samples from primary inhumations (out of a total of only five primary inhumations that have been recovered from the site to date) may relate to a different cultural identity of these individuals. This difference in burial custom may help explain the observed difference in the dietary signatures of the two individuals and their possible consumption of dairy products. Further analyses of organic residues from cooking vessels as well as of stable isotopes from primary burials and scattered human bones at LN Stavroupoli will help to define the extent of the role played by dairy products within the population's dietary regime.

Seafood Consumption and Marine Intake

Another noteworthy point concerning the results of stable isotope analysis of prehistoric assemblages from northern Greece is the lack of any clear indication of marine foodstuffs having been consumed at the coastal sites of LN Stavroupoli, located on the Thermaic Gulf, and LN Makriyalos and LBA Korinos, on the northern Pieria coast.[73] This lack of marine resources is also the case at a large number of inland and coastal Neolithic sites in southern Greece and Thessaly,[74] but not at Bronze Age Grave Circle A

67. Evershed et al. 2008.
68. Kotsakis et al. 2008.
69. Urem-Kotsou 2007; Urem-Kotsou and Kotsakis 2007, p. 237.
70. Kotsakis et al. 2008.
71. Cf. Halstead 1996a, 1996b; Isaakidou 2006.
72. Triantaphyllou 2008, p. 149.
73. Triantaphyllou 2001, pp. 135, 138.
74. Papathanasiou 2003.

at Mycenae (1600–1400 B.C.).[75] The isotopic signals from Grave Circle A suggest that the people buried there, who were undoubtedly of very high status, relied on marine foodstuffs.[76] The majority of the stable isotope values derived from prehistoric assemblages, however, lead to the conclusion that seafood was not an important dietary component.

This is not to state that seafood was not consumed at all. The main shellfish species identified from the Gulf of Thessaloniki is *Cerastoderma glaucum*.[77] This species is found in estuarine environments that would be compatible with high input from terrestrial, freshwater systems; therefore, its isotopic signal should indicate low $\delta^{13}C$ values. Faunal isotopic measurements are necessary, however, in order to define the local estuarine ecological system. Additionally, when protein intake from marine resources comprises less than 20% of the dietary regime—representing possibly a seasonal or occasional consumption (e.g., feasts)—marine input probably will not be indicated in the isotopic values of the human samples.[78] Also, experimental research has shown that certain marine resources—especially those of low trophic level species, such as sardines and anchovies, which are often consumed in the traditional Mediterranean diet—might fall below the detection threshold of the analytical methods used.[79] It should be added that studies of marine ecology, as well as accounts of fishing technology in the historical sources, indicate that fish played a limited role in the diet of ancient Greeks.[80]

CONCLUSIONS

The integration of isotopic data with information provided by archaeobotanical, zooarchaeological, and organic residue analyses allows for certain inferences to be made about diet and dietary variation in populations of Neolithic through EIA date in northern Greece. The isotopic signals of these groups fall within two main types of foodstuffs that represent terrestrial C_3- and C_4-type diets. Neolithic, EBA, and at least some of the LBA groups demonstrate a preference for C_3-type diets and ranged from being heavy consumers of plant protein to heavy consumers of animal protein. Conversely, most of the Early Iron Age groups show a clear tendency toward reliance on a C_4-type diet, with a high contribution of animal protein derived in particular from herbivorous animals such as sheep and cattle that had consumed C_4 grasses and/or millet.

ACKNOWLEDGMENTS

I would like to thank Mike Richards, Department of Human Evolution, Max Planck Institute for Evolutionary Anthropology, and especially the Max Planck Society for funding the analysis. Special thanks should go to the archaeologists who are in charge of the human bone material: Efi Poulaki-Pantermali for Spathes, Treis Elies, and the Olympus tumuli; and Dimitris Grammenos, Stavros Kotsos, and Stergios Lioutas for Stavroupoli. I also owe special thanks to Kostas Kotsakis and Duska Urem-Kotsou, who

75. Richards and Hedges 2008.
76. Richards and Hedges 2008.
77. Veropoulidou 2011.
78. For recent discussion of this issue, cf. Milner et al. 2004; Hedges 2004.
79. Garvie-Lok 2001, pp. 326–330; Hedges 2004.
80. Gallant 1985.

gave me access to unpublished material and pointed out to me the preliminary results of their ongoing work on the emergence of dairy products in eastern Europe and the Balkans. Finally, I would like to thank warmly the three anonymous referees and Sherry Fox of the Wiener Laboratory at the American School of Classical Studies at Athens for her efforts to have this volume published.

REFERENCES

Acsádi, G., and J. Nemeskéri. 1970. *History of Human Life Span and Mortality,* Budapest.

Ambrose, S. H. 1993. "Isotopic Analysis of Paleodiets: Methodological and Interpretive Considerations," in *Investigations of Ancient Human Tissue: Chemical Analyses in Anthropology* (Food and Nutrition in History and Anthropology 10), ed. M. K. Sandford, Langhorne, Pa., pp. 59–130.

Andreou, S., M. Fotiadis, and K. Kotsakis. 1996. "Review of Aegean Prehistory V: The Neolithic and Bronze Age of Northern Greece," *AJA* 100, pp. 537–597.

Antikas, L. W., and T. G. Antikas. 2002. "Τζαμάλα Ι: Κοινά και τραυματικά περιστατικά σε σκελετούς της Ύστερης Εποχής του Χαλκού και Πρώιμης Εποχής του Σιδήρου," *Το Αρχαιολογικό Έργο στη Μακεδονία και Θράκη* 16, pp. 509–518.

Becker, C. 1986. *Kastanas: Ausgrabungen in einem Siedlungshügel der Bronze- und Eisenzeit Makedoniens, 1975–1979* 5: *Die Tierknochenfünde* (Prähistorische Archäologie in Südosteuropa 5), Berlin.

Berrizbeitia, E. L. 1989. "Sex Determination with the Head of the Radius," *Journal of Forensic Sciences* 34, pp. 1206–1213.

Bogaard, A., T. H. E. Heaton, P. Poulton, and I. Merbach. 2007. "The Impact of Manuring on Nitrogen Isotope Ratios in Cereals: Archaeological Implications for Reconstruction of Diet and Crop Management Practices," *JAS* 34, pp. 335–343.

Bökönyi, S. 1986. "Animal Remains," in *Excavations at Sitagroi: A Prehistoric Village in Northern Greece* 1 (Monumenta archaeologica 13), ed. C. Renfrew and E. S. Elster, Los Angeles, pp. 63–132.

Buikstra, J. E., and D. H. Ubelaker, eds. 1994. *Standards for Data Collection from Human Skeletal Remains. Proceedings of a Seminar at the Field Museum of Natural History, Organized by Jonathan Haas* (Arkansas Archeological Survey Research Series 44), Fayetteville.

Coggon, D., G. Rose, and D. J. P. Barker. 1993. *Epidemiology for the Uninitiated,* 3rd ed., London.

Evershed, R. P., S. Payne, A. G. Sherratt, M. S. Copley, J. Coolidge, D. Urem-Kotsou, K. Kotsakis, M. Özdoğan, A. Özdoğan, O. Nieuwenhuyse, P. M. M. G. Akkermans, D. Bailey, R.-R. Andeescu, S. Campbell, S. Farid, I. Hodder, N. Yalman, M. Özbaraşan, E. Biçakci, Y. Garfinkel, T. Levy, and M. M. Burton. 2008. "Intensification of Milk Production and Use Related to Increased Cattle Herding by the Prehistoric Farmers of Central/SE Europe and the Near East," *Nature* 455, pp. 528–531.

Frayer, D. W. 1989. "Oral Pathologies in the European Upper Palaeolithic and Mesolithic," in *People and Culture in Change. Proceedings of the Second Symposium* (*BAR-IS* 508), ed. I. Hershkovitz, Oxford, pp. 255–281.

Gallant, T. W. 1985. *A Fisherman's Tale: An Analysis of the Potential Productivity of Fishing in the Ancient World* (Miscellanea graeca 7), Gent.

Garvie-Lok, S. 2001. "Loaves and Fishes: A Stable Isotope Reconstruction of Diet in Medieval Greece" (diss. Univ. of Calgary).

Grammenos, D. B., and S. Triantaphyllou, eds. 2004. *Ανθρωπολογικές μελέτες από τη Βόρεια Ελλάδα* 1

(Δημοσιεύματα του Αρχαιολογικού Ινστιτούτου Μακεδονικών και Θρακικών Σπουδών 5), Thessaloniki.

Halstead, P. 1987. "Man and Other Animals in Later Greek Prehistory," *BSA* 82, pp. 71–83.

———. 1992. "Agriculture in the Bronze Age Aegean: Towards a Model of Palatial Economy," in *Agriculture in Ancient Greece. Proceedings of the Seventh International Symposium at the Swedish Institute at Athens, 16–17 May, 1990* (*ActaAth* 4°, 42), ed. B. Wells, Stockholm, pp. 105–117.

———. 1994. "The North-South Divide: Regional Paths to Complexity in Prehistoric Greece," in *Developments and Decline in the Mediterranean Bronze Age* (Sheffield Archaeological Monographs 8), ed. S. Mathers and S. Stoddart, Sheffield, pp. 195–219.

———. 1996a. "The Development of Agriculture and Pastoralism in Greece: When, How, Who, and What?" in *The Origins and Spread of Agriculture and Pastoralism in Eurasia,* ed. D. R. Harris, Washington, D.C., pp. 296–309.

———. 1996b. "Pastoralism or Household Herding? Problems of Scale and Specialization in Early Greek Animal Husbandry," *WorldArch* 28, pp. 20–42.

Hedges, R. E. M. 2004. "Isotopes and Red Herrings: Comments on Milner et. al. and Lidén et al.," *Antiquity* 78, pp. 34–37.

Hillson, S. 1979. "Diet and Dental Disease," *WorldArch* 11, pp. 147–162.

Holland, T. D. 1991. "Sex Assessment Using the Proximal Tibia," *American Journal of Physical Anthropology* 85, pp. 221–227.

Holman, D. J., and K. A. Bennett. 1991. "Determination of Sex from Arm Bone Measurements," *American Journal of Physical Anthropology* 84, pp. 421–426.

Hubbard, R. N. L. 1979. "Appendix 2: Ancient Agriculture and Ecology at Servia," in M. Wijnen, C. Ridley, K. A. Wardle, R. N. L. Hubbard, J. P. N. Watson and R. E. Jones, "Rescue Excavations at Servia 1971–1973: A Preliminary Report," *BSA* 74 (pp. 185–230), pp. 226–228.

Isaakidou, V. 2006. "Ploughing with Cows: Knossos and 'The Secondary Products Revolution,'" in *Animals in the Neolithic of Britain and Europe* (Neolithic Studies Group Seminar Papers 7), ed. D. Serjeantson and D. Field, Oxford, pp. 95–112.

Jones, G. 1987. "Agricultural Practice in Greek Prehistory," *BSA* 82, pp. 115–123.

Jones, G., K. Wardle, P. Halstead, and D. Wardle. 1986. "Crop Storage at Assiros," *Scientific American* 254, pp. 96–103.

Karliabas, Y., M. Bessios, and S. Triantaphyllou. 2005. "Νεκροταφεία Πρώιμης Εποχής του Σιδήρου στη Βόρεια Πιερία," in *Η Εποχή του Σιδήρου στο Αιγαίο,* ed. A. Giannikouri and N. Stampolidis, Athens, pp. 341–352.

Kelley, M. A. 1978. "Phenice's Visual Sexing Technique for the *Os Pubis:* A Critique," *American Journal of Physical Anthropology* 48, pp. 121–122.

Kotjabopoulou, E., Y. Hamilakis, P. Halstead, C. Gamble, and P. Elefanti. 2003. *Zooarchaeology in Greece: Recent Advances* (*BSA* Studies 9), London.

Kotjamani, G. 2009. "Από τη συλλογή στην καλλιέργεια: Αρχαιοβοτανική διερεύνηση των πρώιμων σταδίων εκμετάλλευσης των φυτών και της αρχής της γεωργίας στον ελλαδικό χώρο (σπήλαιο Θεόπετρας, σπήλαιο Σχιστού, Σιδάρι, Ρεβένια)" (diss. Univ. of Thessaloniki).

Kotsakis, K., D. Urem-Kotsou, R. P. Evershed, A. Sherratt, and M. Copley. 2008. "Έπιναν οι νεολιθικοί γεωργοκτηνοτρόφοι γάλα; Η μαρτυρία για την αρχαιότερη κατανάλωση γάλακτος στην Ελλάδα," in *Η ιστορία του ελληνικού γάλακτος και των προϊόντων του. Γ΄ Τριήμερο Εργασίας, Ξάνθη, 7–9 Οκτωβρίου 2005,* Athens, pp. 83–88.

Krogman, W. M., and M. Y. İşcan. 1986. *The Human Skeleton in Forensic Medicine,* 2nd ed., Springfield, Ill.

Kroll, H. 1983. *Kastanas: Ausgrabungen in einem Siedlungshügel der Bronze- und Eisenzeit Makedoniens, 1975–1979* 1: *Die Pflanzenfunde* (Prähistorische Archäologie in Südosteuropa 2), Berlin.

Larsen, C. S. 1997. *Bioarchaeology: Interpreting Behavior from the Human*

Skeleton (Cambridge Studies in Biological Anthropology 21), Cambridge.

Lovell, N. C. 1989. "Test of Phenice's Technique for Determining Sex from the *Os Pubis*," *American Journal of Physical Anthropology* 79, pp. 117–120.

Lukacs, J. R. 1992. "Dental Palaeopathology and Agricultural Intensification in South Asia: New Evidence from Bronze Age Harappa," *American Journal of Physical Anthropology* 87, pp. 133–150.

MacLaughlin, S. M., and M. F. Bruce. 1986. "The Sciatic Notch/Acetabular Index as a Discriminator of Sex in European Skeletal Samples," *Journal of Forensic Sciences* 31, pp. 1380–1390.

Margaritis, E. 2002. "Αρχαιοβοτανικά δεδομένα από το νεολιθικό οικισμό της Σταυρούπολης: Μια πρώτη προσέγγιση," in *Σωστικές ανασκαφές στο νεολιθικό οικισμό Σταυρούπολης Θεσσαλονίκης*, ed. D. Grammenos and S. Kotsos, Thessaloniki, pp. 804–825.

———. 2004. "Γεωργικές πρακτικές στο στερεοσκόπιο: Η ανάλυση των φυτικών καταλοίπων από το νεολιθικό οικισμό της Σταυρούπολης. Οι ανασκαφικές περίοδοι 1998–2003," in *Σωστικές ανασκαφές στο νεολιθικό οικισμό Σταυρούπολης Θεσσαλονίκης, Μέρος II (1998–2003)*, ed. D. Grammenos and S. Kotsos, Thessaloniki, pp. 605–611.

Meiklejohn, C., J. H. Baldwin, and C. T. Schentag. 1988. "Caries as a Probable Dietary Marker in the Western European Mesolithic," in *Diet and Subsistence: Current Archaeological Perspectives. Proceedings of the Nineteenth Annual Conference of the Archaeological Association of the University of Calgary*, ed. B. V. Kennedy and G. M. LeMoine, Calgary, pp. 273–279.

Milner, N., O. E. Craig, G. N. Bailey, K. Pedersen, and S. H. Andersen. 2004. "Something Fishy in the Neolithic? A Re-evaluation of Stable Isotope Analysis of Mesolithic and Neolithic Coastal Populations," *Antiquity* 78, pp. 9–22.

Mylonas, G. E. 1929. *Excavations at Olynthus* 1: *The Neolithic Settlement*, Baltimore.

Nikolaidou, D. 2010. "Ανάλυση και μελέτη ζωοαρχαιολογικού υλικού από την Τούμπα Θεσσαλονίκης (τομή 761, κτήριο Ζ, δρόμος Χ1)" (diss. Univ. of Thessaloniki).

Papathanasiou, A. 2003. "Stable Isotope Analysis in Neolithic Greece and Possible Implications on Human Health," *International Journal of Osteoarchaeology* 13, pp. 314–324.

Pappa, M., P. Halstead, K. Kotsakis, A. Bogaard, R. Fraser, V. Isaakidou, I. Mainland, D. Mylona, D. Skourtopoulou, S. Triantaphyllou, C. Tsoraki, D. Urem-Kotsou, S.-M. Valamoti, and R. Veropoulidou. 2013. "The Neolithic Site of Makriyalos, Northern Greece: Reconstruction of Social and Economic Structure of the Settlement through Comparative Study of the Finds," in *Diet, Economy, and Society in the Ancient Greek World: Towards a Better Integration of Archaeology and Science. Proceedings of the International Conference Held at the Netherlands Institute at Athens on 22–24 March 2010* (*Pharos* Suppl. 1), ed. S. Voutsaki and S. M. Valamoti, Leuven, pp. 77–88.

Phenice, T. W. 1969. "A Newly Developed Visual Method of Sexing the *Os Pubis*," *American Journal of Physical Anthropology* 30, pp. 298–301.

Poulaki-Pantermali, E. 1985. "Οι πιερικοί πρόποδες," *ArchDelt* 40, Β′, pp. 240–243.

———. 1986a. "Νεκροταφείο Αγίου Δημητρίου," *ArchDelt* 41, Β′, pp. 140–141.

———. 1986b. "Όλυμπος 2. Μακεδονικόν όρος, μετεορώτατον," in *Αμητός: Τιμητικός τόμος για τον καθηγητή Μανόλη Ανδρόνικο*, Thessaloniki, pp. 705–706.

———. 1987a. "Ανασκαφή Αγίου Δημητρίου, Ολύμπου," *Το Αρχαιολογικό Έργο στη Μακεδονία και Θράκη* 1, pp. 201–208.

———. 1987b. "Σπάθες Αγίου Δημητρίου," *ArchDelt* 42, Β′2, pp. 363–364.

———. 1988. "Τρεις Ελιές," *ArchDelt* 43, Β′2, p. 366.

———. 1989. "Τρεις Ελιές," *ArchDelt* 44, Β′2, p. 324.

———. 1990. "Όλυμπος 3: Μια παλιά Ιστορία," in *Οι αρχαιολόγοι μιλούν για την Πιερία, Δίον 1986*, pp. 84–95.

———. 1991. "Ανασκαφή Αγίου Δημητρίου, Ολύμπου," *Το Αρχαιολογικό Έργο στη Μακεδονία και Θράκη* 2, pp. 173–180.

———. 1992. "Τα περιβόλια των θεών," *Το Αρχαιολογικό Έργο στη Μακεδονία και Θράκη* 3, pp. 149–154.

Renfrew, C. 1971. "Sitagroi, Radiocarbon, and the Prehistory of South-East Europe," *Antiquity* 45, pp. 275–282.

Richards, M. P., and R. E. M. Hedges. 2008. "Stable Isotope Results from the Sites of Gerani, Armenoi, and Mycenae," in *Archaeology Meets Science: Biomolecular Investigations in Bronze Age Greece. The Primary Scientific Evidence, 1997–2003*, ed. Y. Tzedakis, H. Martlew, and M. K. Jones, Oxford, pp. 220–230.

Richards, M. P., R. E. M. Hedges, T. I. Molleson, and J. C. Vogel. 1998. "Stable Isotope Analysis Reveals Variations in Human Diet at the Poundbury Camp Cemetery Site," *JAS* 25, pp. 1247–1252.

Richards, M. P., J. A. Pearson, T. I. Molleson, N. Rusell, and L. Martin. 2003. "Stable Isotope Evidence of Diet at Neolithic Çatal Höyük, Turkey," *JAS* 30, pp. 67–76.

Scheuer, L., and S. Black. 2000. *Developmental Juvenile Osteology*, London.

Sutherland, L. D., and J. M. Suchey. 1991. "Use of the Ventral Arc in Pubic Sex Determination," *Journal of Forensic Sciences* 36, pp. 501–511.

Theodoropoulou, T. 2007. "L'exploitation des ressources aquatiques en Égée Septentrionale aux périodes Pré- et Protohistoriques" (diss. Université Paris I Panthéon-Sorbonne).

Triantaphyllou, S. 1998a. "An Early Iron Age Cemetery in Ancient Pydna, Pieria: What Do the Bones Tell Us?" *BSA* 93, pp. 353–364.

———. 1998b. "Prehistoric Populations from Northern Greece: A Breath of Life for the Skeletal Remains," in *Cemetery and Society in the Aegean Bronze Age* (Sheffield Studies in Aegean Archaeology 1), ed. K. Branigan, Sheffield, pp. 150–164.

———. 2001. *A Bioarchaeological Approach to Prehistoric Cemetery*

Populations from Central and Western Greek Macedonia (*BAR-IS* 976), Oxford.

———. 2002. "Πρώτα αποτελέσματα της οστεολογικής εξέτασης του σκελετικού υλικού της νεολιθικής θέσης Σταυρούπολης Θεσσαλονίκης," in *Σωστικές ανασκαφές στο νεολιθικό οικισμό Σταυρούπολης Θεσσαλονίκης*, ed. D. Grammenos and S. Kotsos, Thessaloniki, pp. 829–846.

———. 2003. "A Bioarchaeological Approach to Bronze Age Cemetery Populations from Western and Central Greek Macedonia," in *Metron: Measuring the Aegean Bronze Age. Proceedings of the 9th International Aegean Conference* (*Aegeum* 24), ed. K. P. Foster and R. Laffineur, Liège, pp. 217–222.

———. 2004. "Αποτελέσματα της οστεολογικής εξέτασης του σκελετικού υλικού της νεολιθικής θέσης Σταυρούπολης Θεσσαλονίκης κατά την ανασκαφική χρονιά 2002–2003," in *Σωστικές ανασκαφές στον νεολιθικό οικισμό Σταυρούπολης Θεσσαλονίκης, Μέρος II (1998–2003)*, ed. D. Grammenos and S. Kotsos, Thessaloniki, pp. 613–624.

———. 2008. "Living with the Dead: A Re-Consideration of Mortuary Practices in the Greek Neolithic," in *Escaping the Labyrinth: The Cretan Neolithic in Context* (Sheffield Studies in Aegean Archaeology 8), ed. V. Isaakidou and P. D. Tomkins, Oxford, pp. 139–157.

———. 2010. "Aspects of Life Histories from the Bronze Age Cemetery at Xeropigado Koiladas, Western Macedonia," in *Mesohelladika: The Greek Mainland in the Middle Bronze Age. Actes du colloque international organisé par l'École française d'Athènes, en collaboration avec l'American School of Classical Studies at Athens et le Netherlands Institute in Athens, Athènes, 8–12 mars 2006* (*BCH* Suppl. 52), ed. A. Philippa-Touchais, G. Touchais, S. Voutsaki, and J. Wright, Athens, pp. 969–974.

Triantaphyllou, S., M. P. Richards, and C. Ziota. In prep. "Stable Isotope Analysis of Human Bone from EBA/LBA Xeropigado Koiladas Cemetery, W. Macedonia."

Tsoraki, C. 2007. "Unravelling Ground Stone Life Histories: The Spatial Organization of Stone Tools and Human Activities at LN Makriyalos, Greece," *Documenta Praehistorica* 34, pp. 289–297.

———. 2008. "Neolithic Society in Northern Greece: The Evidence of Ground Stone Artefacts" (diss. Univ. of Sheffield).

Urem-Kotsou, D. 2007. "Νεολιθική κεραμική του Μακρυγιάλου. Διατροφικές συνήθειες και οι κοινωνικές διαστάσεις της κεραμικής" (diss. Univ. of Thessaloniki).

Urem-Kotsou, D., and K. Kotsakis. 2007. "Pottery, Cuisine, and Community in the Neolithic of North Greece," in *Cooking Up the Past: Food and Culinary Practices in the Neolithic and Bronze Age Aegean,* ed. C. Mee and J. Renard, Oxford, pp. 225–246.

Valamoti, S.-M. 2003. "Neolithic and Early Bronze Age 'Food' from Northern Greece: The Archaeobotanical Evidence," in *Food, Culture, and Identity in the Neolithic and Early Bronze Age* (*BAR-IS* 1117), ed. M. Parker Pearson, Oxford, pp. 97–111.

———. 2004. *Plants and People in Late Neolithic and Early Bronze Age Northern Greece: An Archaeobotanical Investigation* (*BAR-IS* 1258), Oxford.

———. 2007a. "Food across Borders: A Consideration of the Neolithic and Bronze Age Archaeobotanical Evidence from Northern Greece," in *Between the Aegean and Baltic Seas: Prehistory across Borders. Proceedings of the International Conference Bronze and Early Iron Age Interconnections and Contemporary Developments between the Aegean and the Regions of the Balkan Peninsula, Central and Northern Europe, University of Zagreb, 11–14 April 2005* (*Aegaeum* 27), ed. I. Galanaki, H. Tomas, Y. Galanakis, and R. Laffineur, Liège, pp. 282–291.

———. 2007b. "Traditional Foods and Culinary Novelties in Neolithic and Bronze Age Northern Greece: An Overview of the Archaeobotanical Evidence," in *Cooking Up the Past:*

Food and Culinary Practices in the Neolithic and Bronze Age Aegean, ed. C. Mee and J. Renard, Oxford, pp. 89–108.

Valamoti, S.-M., M. Mangafa, C. Koukouli-Chrysanthaki, and D. Malamidou. 2007. "Grape-Pressings from Northern Greece: The Earliest Wine in the Aegean?" *Antiquity* 81, pp. 54–61.

van Zeist, W., and S. Bottema. 1971. "Plant Husbandry in Early Neolithic Nea Nikomedeia, Greece," *Acta botanica Neerlandica* 20, pp. 524–538.

Vaughan, S. J., and W. D. E. Coulson, eds. 2000. *Palaeodiet in the Aegean. Papers from a Colloquium Held at the 1993 Meeting of the Archaeological Institute of America in Washington D. C.* (Wiener Laboratory Monograph 1), Oxford.

Veropoulidou, C.-E. 2011. "Όστρεα από τους οικισμούς του Θερμαϊκού Κόλπου: Ανασυνθέτοντας την κατανάλωση των μαλακίων στη Νεολιθική και την Εποχή του Χαλκού" (diss. Univ. of Thessaloniki).

Vika, E. 2011. "Diachronic Dietary Reconstructions in Ancient Thebes, Greece: Results from Stable Isotope Analyses," *JAS* 38, pp. 1157–1163.

Vika, E., V. Aravantinos, and M. P. Richards. 2009. "Aristophanes and Stable Isotopes: A Taste for Freshwater Fish in Classical Thebes (Greece)?" *Antiquity* 83, pp. 1076–1083.

Waldron, T. 1994. *Counting the Dead: The Epidemiology of Skeletal Populations*, Chichester.

Wardle, K. 1993. "Mycenaean Trade and Influence in Northern Greece," in *Wace and Blegen: Pottery as Evidence for Trade in the Aegean Bronze Age, 1939–1989. Proceedings of the International Conference Held at the American School of Classical Studies at Athens, December 2–3 1989*, ed. C. Zerner, P. Zerner, and J. Winder, Amsterdam, pp. 117–141.

Wood, J. W., G. R. Milner, H. C. Harpending, and K. M. Weiss. 1992. "The Osteological Paradox: Problems of Inferring Prehistoric Health from Skeletal Samples," *CurrAnth* 33, pp. 343–358.

Ziota, C. 1995. "Κίτρινη Λίμνη 1995: Νέες ερευνητικές δραστηριότητες," *Το Αρχαιολογικό Έργο στη Μακεδονία και Θράκη* 9, pp. 47–58.

———. 1996. "Κίτρινη Λίμνη: Φυσικό περιβάλλον και ανθρώπινη δραστηριότητα κατά την προϊστορία," *Δυτικομακεδονικά Γράμματα* Z, pp. 286–303.

———. 1998a. "Κίτρινη Λίμνη 1998. Νέες ερευνητικές δραστηριότητες," *Το Αρχαιολογικό Έργο στη Μακεδονία και Θράκη* 12, pp. 503–516.

———. 1998b. "Προϊστορικό νεκροταφείο στην Κοιλάδα Κοζάνης: Μία πρώτη αναλυτική παρουσίαση της ανασκαφικής έρευνας (1995–1996)," *Τόμος στη Μνήμη Μαίρης Σιγανίδου*, ed. M. Lilimpaki-Akamati and K. Tsakalou-Tzanavari, Thessaloniki, pp. 81–102.

———. 2007. "Ταφικές πρακτικές και κοινωνίες της Εποχής του Χαλκού στη Δυτική Μακεδονία" (diss. Univ. of Thessaloniki).

Ziota, C., and S. Triantaphyllou. 2004. "Early Bronze Age Burial Practices and Skeletal Populations: A Case Study from Western Macedonia," in *Spatial Analysis of Funerary Areas*, ed. L. Šmejda and J. Turek, Plzeň, pp. 38–47.

CHAPTER 5

Variations in Diet in Prehistoric Thebes: The Case of the Bronze Age Mass Burial

by Efrossini Vika

The Bronze Age in Greece is represented by a rich archaeological record that has been extensively studied for more than a century. Bioarchaeological approaches, however, such as human osteological analyses and stable isotope analyses, only began to be applied extensively to this period within approximately the last couple of decades. Much of this research has focused on the LH period, characterized by well-known sites such as Mycenae, while little attention has been paid to the EH period. This paper reports the results of stablc isotopc palacodictary analysis of human bonc samplcs from prehistoric burials located in and around the city of Thebes, Greece, in order to test the hypothesis that location and time act as formative variables in dietary choices.

Material from three sites and two time periods (EH and MH)[1] was examined. Twelve of the individuals represent an unusual prehistoric mass burial discovered under a mudbrick mound that covered EH II architectural remains. Results of analyses on these collectively buried humans indicate that they had a relatively homogeneous diet primarily based on terrestrial protein. Analysis of 12 humans of MH date interred at the same site showed that they had very similar dietary patterns, although results from the study of a different group of individuals buried 2 km away demonstrated statistically significant differences in dietary patterns. This study represents the first reconstruction of diet through isotope analysis for prehistoric Thebes.

THE REGION: BOIOTIA AND THEBES

Thebes is located in the region of Boiotia, in mainland central Greece (see Fig. 1.1). The region consists of two extensive plains crossed by rivers and surrounded by mountains. Large freshwater lakes and many natural springs were present in the area. The ancient natural geomorphology and climate did not differ considerably from that of today. The alluvial soils were fertile, suitable for crops and livestock, and therefore the area was beneficial for human settlement. Several important Bronze Age settlements were located in Boiotia: Thebes, Orchomenos, and Gla.[2] Thebes was constructed on a plateau, dominating the surrounding flat plain.

1. For the chronology of these two periods, see Table 1.1.
2. Bintliff and Snodgrass 1985; Fossey 1998.

Thebes and its inhabitants feature prominently in ancient Greek legends. The acropolis of the city was discovered at the beginning of the 20th century, but major excavations took place during the 1960s, following modern urban development.[3]

Early Helladic Thebes

The first phase of the Bronze Age in mainland Greece is the EH period. The architectural finds of this period in Thebes indicate to some researchers that there was urban organization in the city as early as the 3rd millennium B.C.[4] Excavations have revealed elaborate buildings, probably of public, administrative, religious, or economic character, which coexisted with small buildings made of poor, perishable materials—a combination likely indicating social stratification.[5] In EH Thebes, the existence of a centralized authority was unlikely. Instead, the system appears to have been organized in household units, with leading groups sharing the responsibility of public works.[6] Recent studies demonstrate that the EH Theban economy was based on surplus production and storage at the household level.[7] A regular burial ground attributed to the EH settlement has not yet been discovered.

THE SITE

The primary focus of this study is a mass burial discovered atop the ruins of an EH apsidal building.[8] The destruction level was dated to the end of EH II. Subsequent to the collapse of the building, the ground was levelled, a burial was placed above the ruins, and mudbrick slabs were used to construct a mound over the entire site. The mound, covering an area of 400 m^2, must have been clearly visible on the landscape. After its construction, the site was abandoned for almost 200 years. Evidence for human presence at the site appears next during the MH period, when the upper levels were destroyed by the construction of at least 25 tombs.

A cemetery of Bronze Age date has not been discovered in or around the city of Thebes, so it is not possible to place the mass burial within a specific comparative context. All of the individuals from the burial died within a very short time span and were hastily interred. This may have been related to the cause of death (if it was considered dangerous) or may have been affected by reasons pertaining to the surviving community.[9] Construction of the tumulus can indicate a need to cover a "source of evil"—a point not unknown in archaeological interpretation. Yet the tumulus covers the entire site, not just the burial. It is, therefore, set apart from the burial tumuli seen in this time period in other regions of Greece and the Balkans.[10] The tumulus of Thebes has been compared to Lerna III in the

3. Symeonoglou 1985.
4. Hägg and Konsola 1996.
5. Konsola 1981, 1988.
6. Pullen 1985.
7. Psaraki in press.
8. Aravantinos 2004.
9. Wiencke 1989; Chapman 2000.
10. See, e.g., Hammond 1974.

Argolid; there are strong differences between the sites, however: at Thebes, there is a placement of a burial over the ruins, whereas at Lerna, the end of phase III is marked by a fire.[11] Modern urban development in Thebes does not allow for extensive archaeological excavations. As a result, much fragmentary archaeological information needs to be integrated before the significance of this site can be explained.

THE BURIAL

The burial, covering an area of ca. 2.7 × 4 m exactly over the ruins of the building, contained a minimum of 12 individuals. The skeletons were overlapping, indicating deposition without organization, but recovery of the bones in correct anatomical position also indicates interment in a fully articulated state. Seven vessels, found intact, had been placed around the bodies.[12]

The presence of an on-site osteologist is not required by the Ephorates of Antiquities in Greece. As a result, and due to the nature of the burial, the skeletons were not removed and stored as discrete individuals. Although the tumulus placed over the burial protected the assemblage from post-depositional damage, erosion due to acidity and humidity in the soil had altered surface features of the bones, resulting in the loss of potentially diagnostic information. All skeletal remains were inventoried, and methods for recording commingled remains were applied during the analysis.[13] Fragmentation was recorded using the "zonation method":[14] 1,300 skeletal elements were counted, which corresponded to a minimum number of 12 individuals. Age was estimated and sex determined as initial steps in the analysis, since these data provide the baseline for subsequent observations.

GENERAL OSTEOLOGICAL OBSERVATIONS

Estimation of age and determination of sex followed established methods as defined by Todd, Suchey-Brooks, Lovejoy et al., Phenice, and Buikstra and Ubelaker.[15] Cranial suture closure[16] and dental wear[17] were used when pelvic morphology was not conclusive or when the pelvis was absent. For younger individuals, age was estimated based on either epiphyseal fusion or dental and bone development.[18] In most cases, poor preservation only allowed assignment of individuals to very broad age categories.

The burial consisted of two adult females, a female between 20–25 years of age at death, an adult male, an adult of indeterminate sex, and four adolescents (<15–18 years old). Also present in the assemblage were the bones of two children, with ages at death estimated to be approximately 5 years and 6–8 years, respectively, and bones of an infant between 4–12 months. The adult:subadult ratio is 5:7, while 80% of the elements that retained landmarks for age estimation indicate an approximate age of <25 years. A number of bones remain unidentified and unassigned to individual skeletons; therefore, the size of each age and sex category could increase.[19]

11. Caskey and Blackburn 1977.
12. For a detailed discussion of the typology and significance of the pottery, see Psaraki 2004.
13. Cf. Buikstra and Ubelaker 1994.
14. Knüsel and Outram 2004.
15. Todd 1920, 1921; Phenice 1969; Lovejoy et al. 1985; Brooks and Suchey 1990; Buikstra and Ubelaker 1994.
16. After Todd and Lyon 1924; Meindl and Lovejoy 1985.
17. Brothwell 1989.
18. Scheuer and Black 2004.
19. Gordon and Buikstra 1981; Walker, Johnson, and Lambert 1988; Bello et al. 2006.

PALAEODIET: STABLE ISOTOPE ANALYSIS FOR DIETARY RECONSTRUCTION

Stable isotope analysis for dietary reconstruction is the only direct way for archaeologists to reconstruct past human diets.[20] Samples were prepared following protocols based on Brown et al.,[21] modified by Richards and Hedges.[22] It was necessary to increase the weight of the samples (start mass) to 300–500 mg due to poor skeletal preservation. Whole bone pieces were cleaned with air-abrasion and then demineralized in 0.5 M HCl at 4°C for several days. They were then rinsed to neutrality with de-ionized water, placed in sealed glass tubes in a pH 3 solution, and gelatinized at 72°C for 48 hours. Subsequently, 5–8μm Ezee® filters (Elkay Laboratory Products) were used to remove reflux-insoluble residues, and Ultrafree®-4 Centrifugal Filter Units with Biomax-30 membranes (Millipore) were employed to concentrate the remaining solution. The obtained collagen (>30 kDa) was freeze-dried for 48 hours. Samples were weighed out for duplicate runs, combusted in a Europa Scientific Roboprep–CN, and measured in a Europa 20/20 continuous flow isotope ratio mass spectrometer. Measurement error was ± 0.2‰ or less.

RESULTS

Of the 15 samples that were prepared for analysis from the skeletons of the EH mass burial, one had a C/N ratio outside the acceptable range, and four did not yield any collagen.[23] Therefore, 10 human skeletal samples were suitable for isotopic analysis (Table 5.1, Fig. 5.1). The $\delta^{13}C$ values for these 10 samples range between -20.5‰ and -19.4‰, with a mean of -19.8‰ ± 0.28 (1σ). The corresponding $\delta^{15}N$ values range between 8.8‰ and 10.9‰, with a mean of 9.9‰ ±0.65 (1σ). These values demonstrate a homogeneous diet for the group, with C_3 plant protein and contributions from terrestrial animal protein. It is unlikely that fish protein would have been a frequent part of the diet of these people. Attempts to calculate percentage inputs for each foodstuff cannot be adequately supported without knowing the end members[24] for the population under discussion.[25] It is clear, however, that animal protein was consumed regularly.

DISCUSSION

The values presented here are similar to those observed at other sites in temperate, prehistoric Europe and other sites in Greece,[26] in recording protein in the diet deriving mainly from C_3 cereals and occasionally terrestrial animals.

20. See Chap. 2.
21. Brown et al. 1988.
22. Richards and Hedges 1999.
23. See DeNiro 1985 for quality criteria.
24. The term "end members" refers to the values that would be exhibited if a diet consisted 100% of terrestrial or 100% of marine protein.
25. Newsome et al. 2004.
26. Triantaphyllou 2001; Papathanasiou 2003; Richards and Vika 2006; Lagia, Petroutsa, and Manolis 2007; Vika, Aravantinos, and Richards 2009; Vika 2011; and see Chaps. 3, 4, 6.

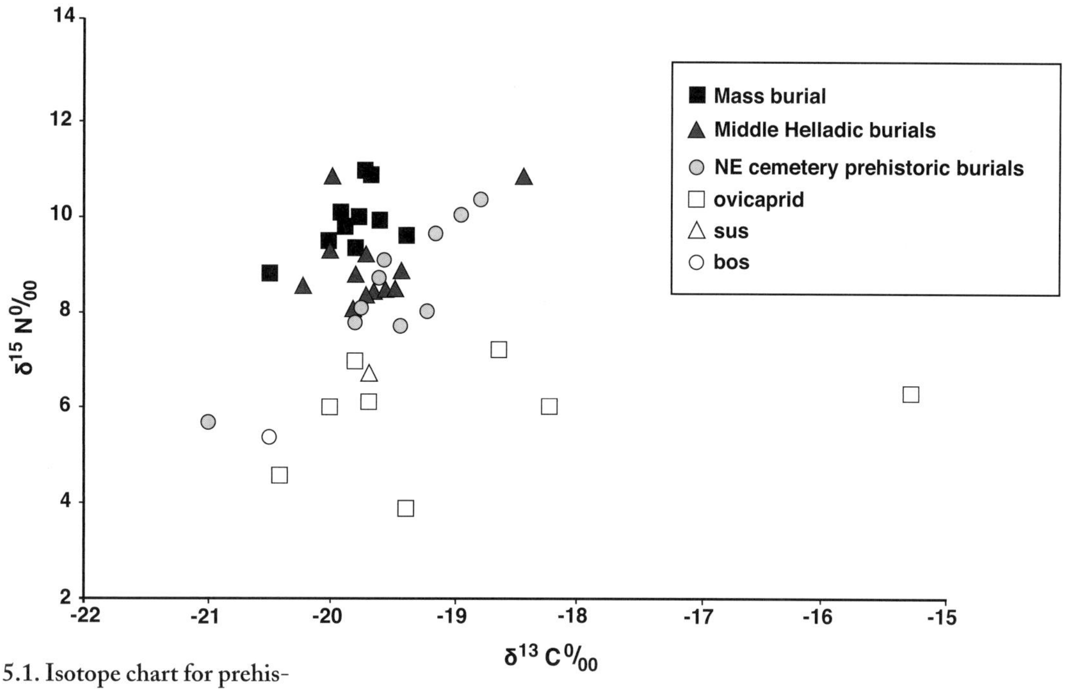

Figure 5.1. Isotope chart for prehistoric human and faunal data in Thebes

Ethnology demonstrates that prerogatives of high social status include differential access to food, with certain types of food being exclusively available to rulers or to the ruling lineage.[27] Iconographic and textual evidence in ancient Egypt suggests that animal products were ranked and that food choices were to some extent related to status.[28] The integration of textual and archaeological evidence would suggest, for example, that in Egypt beef was consumed by the upper class and fish by the workers.[29] Gender divisions in food consumption can also be seen. In classical Greek antiquity, for example, men and women participated equally in family celebrations or sacrifices, but formed different groups and often kept different eating hours. Normally, the women would eat after the men, consuming any part of the animal that was left.[30]

In order to investigate whether sociocultural parameters acted as criteria in dietary differentiation within the group, male and female values were examined separately using non-parametric statistics (Kolmogorov-Smirnov or KS tests). There were four females in the sample from the mass burial, with an average $\delta^{13}C$ value of -19.8‰ and an average $\delta^{15}N$ of 9.4‰. No difference is apparent when comparing the female average values with the values from the single male in the sample, but the sample is far too small for further statistical analyses.

Ten animal samples were analyzed to provide a reference baseline for the human samples. All of the animal bones come from stratigraphic levels associated with the mass burial. There are eight sheep/goats, one pig, and one cattle sample. The $\delta^{13}C$ and $\delta^{15}N$ values are as follows—ovicaprids: $\delta^{13}C$ -20‰ to -15.3‰, $\delta^{15}N$ 4.6‰ to 7.2‰; pig: $\delta^{13}C$ -19.7‰, $\delta^{15}N$ 6.7‰; cattle: $\delta^{13}C$ -20.5‰, $\delta^{15}N$ 5.4‰. The values are characteristic of herbivorous animals that fed on a C_3 environment.

27. Hatch and Geidel 1985.
28. Rossel 2004.
29. Rossel 2004, p. 201.
30. Dalby 1996; for Bronze Age Greece, see also Chap. 3.

TABLE 5.1. STABLE ISOTOPE VALUES FOR HUMANS AND FAUNA FROM PREHISTORIC THEBES

Name	*Location**	*Details***	*Sample Weight (g)*	*δ¹⁵N (‰ AIR)*	*% N*	*δ¹³C (‰ VPDB)*	*% C*	*C:N*	*% Coll.*
OSE1	NE	UI	0.89	8.1	15.46	-19.7	43.85	3.3	1.5
OSE10	NE	UI	0.93	9.1	14.72	-19.6	43.02	3.4	2.2
OSE12	NE	UI	1.12	8.8	9.38	-19.6	27.15	3.4	2.7
OSE13	NE	UI	0.87	10.1	8.31	-19.4	23.32	3.3	2.4
OSE14	NE	Ad M	0.92	9.7	14.92	-19.2	43.18	3.4	1.3
OSE15	NE	UI	0.52	5.7	9.73	-21.0	28.60	3.4	2.5
OSE2	NE	Ad M	0.99	7.8	15.18	-19.5	43.13	3.3	2.0
OSE3	NE	Ad M	1.16	7.7	11.09	-19.4	31.55	3.3	1.4
OSE5	NE	Ad M	0.95	8	12.05	-19.2	34.21	3.3	4.2
OSE6	NE	UI	0.94	10.4	13.29	-18.8	37.24	3.3	2.2
OSE7	NE	Ad M	0.8	10	10.02	-18.9	28.76	3.4	2.6
OSE8	NE	UI	0.63	7.9	8.64	-19.8	24.81	3.3	4.0
THA1	Mass	Pig	0.95	6.7	15.26	-19.7	44.27	3.4	0.8
THA4	Mass	Cattle	0.93	5.4	15.72	-20.5	44.54	3.3	0.8
THA5O	Mass	Ovicaprid	0.51	6	11.4	-18.2	31.22	3.2	2.8
THA6	Mass	Ovicaprid	1.05	5.9	15.52	-20.0	44.44	3.3	0.7
THA61	Mass	Ovicaprid	1.11	4.6	14.90	-19.4	42.38	3.3	1.0
THA62	Mass	Ovicaprid	1.12	6.1	13.63	-19.7	38.82	3.3	0.8
THA63	Mass	Ovicaprid	1.1	6.2	14.87	-15.3	42.85	3.4	3.9
THA7	Mass	Ovicaprid	1.05	6.8	16.01	-20.0	45.67	3.3	1.5
THA8O	Mass	Ovicaprid	0.64	7.2	13.4	-18.6	36.91	3.2	2.2
THA9O	Mass	Ovicaprid	0.55	4.5	13.4	-20.0	36.36	3.2	2.7
THOP1	Mass	Ad F	1.13	9.6	16.42	-19.4	46.94	3.3	2.4
THOP3	Mass	Ad M	0.96	9.4	15.84	-19.8	45.42	3.3	1.3
THOP4	Mass	UI	1.07	11	15.89	-19.7	45.89	3.4	2.5
THOP5	Mass	SA–adoles	1.04	10.9	16.24	-19.7	46.41	3.3	2.0
THOP9	Mass	SA–adoles	1.02	10	15.53	-19.8	44.53	3.3	1.4
THOP14	Mass	Ad F	1.07	8.8	15.57	-20.5	44.08	3.3	4.2
THOP15	Mass	Ad F	0.94	9.7	8.16	-19.6	23.46	3.4	2.2
THOP24	Mass	SA–child	0.94	9.8	15.37	-19.9	44.09	3.3	2.6
THOP25	Mass	SA–adoles	1.08	10.1	14.74	-19.9	43.19	3.4	4
THOP28	Mass	Ad F	1.2	9.5	15.50	-20.0	43.03	3.4	0.5
THOP40	MH	Ad F	0.84	9.3	13.99	-20.0	41.81	3.5	0.5
THOP41	MH	Ad F	1.06	8.5	15.07	-19.6	43.48	3.4	2.8
THOP42	MH	UI	1.16	8.5	13.41	-20.2	39.36	3.4	0.7
THOP43	MH	UI	1.3	8.3	15.15	-19.7	43.65	3.4	1.0
THOP44	MH	UI	1.07	8.3	14.49	-19.7	41.25	3.3	0.8
THOP45	MH	Ad F	1.2	8.4	14.49	-19.7	44.27	3.3	3.9
THOP46	MH	SA–adoles	0.89	8.4	15.18	-19.5	42.74	3.3	2.3
THOP47	MH	Ad F	0.98	8.7	14.80	-19.8	42.17	3.3	2.1
THOP48	MH	SA–child	1.04	10.8	14.63	-18.5	41.48	3.3	1.0
THOP49	MH	Ad M	1.03	10.8	14.43	-20.0	41.62	3.4	0.9
THOP50	MH	SA–adoles	1.13	8.8	16.16	-19.4	46.04	3.3	0.9
THOP51	MH	UI	0.99	9.2	14.96	-19.7	42.47	3.3	2.7

*NE = NE cemetery; Mass = mass burial and mass burial associated context; MH = Middle Helladic burials.

**UI = unidentifiable skeleton; Ad = adult; SA = subadult; adoles = adolescent; M = male; F = female.

Interestingly, one of the sheep samples exhibits a $\delta^{13}C$ value of -15.3‰. The sample was taken from a clear EH context and is associated with a terrestrial $\delta^{15}N$ value. This result most likely indicates a diet based significantly on C_4 protein for this animal. This result can aid in the investigation of the cultivation and use of C_4 plants in the Greek prehistoric landscape. Archaeobotanical evidence demonstrates the presence of varieties of millet in MBA and LBA northern Greece (Ayios Mamas, Assiros, Kastanas),[31] but isotopic analyses on human and animal populations have not yet clearly demonstrated consumption of this type of protein, since a slight enrichment in ^{13}C may be the result of different climatic conditions and not necessarily the input of C_4 protein. Serious evidence of C_4 in diets is still lacking in prehistoric Greece.[32]

Alternatively, enrichment of ^{13}C could result from consumption of seaweed by domesticated animals as demonstrated by Balasse.[33] The effect of lighter $\delta^{13}C$ values in seagrasses is reported in the literature.[34] Greek farmers today bring their animals to the shore in the summer months to drink salted water, but sheep are known to refuse to drink if the water is not clean and holds algae. Moreover, eating seaweed is a specific adaptation and induces changes in animal physiology; at this point, it is unknown whether the particular animal would have been adapted to this practice. In the sole consideration of diet, it remains more plausible that this particular animal would have been browsing in a different region, as consistent consumption of a lighter carbon source would have been necessary for the overall dietary value to reflect it.

Although a recent study on Aegean fish isotopes[35] clearly changes the previous interpretations of fish consumption in Greek antiquity, here it appears that fish (marine or freshwater) were not consumed regularly by this group, as might have been expected considering the geographic location of the area and availability of freshwater resources.[36] Although only 30 km from the sea, Thebes is surrounded by low mountains that may have hindered access to the coast. It is also worth noting that fishing in the Bronze Age was probably neither a cost-effective activity nor one capable of providing food for the entire community on a systematic, continuous basis.[37] The land around Thebes has reportedly always been fertile, apparently enough to cover the basic needs of the inhabitants. No fish bones were recovered from Thebes, but this is probably due to lack of systematic sieving and not to an absence of Bronze Age fish. Therefore, fish cannot be fully excluded as a potential dietary item, although they may have existed in smaller overall quantities than terrestrial protein.[38]

Variation through Time

In order to explore dietary variation through time at Thebes, a further 12 human skeletal samples were analyzed; they were recovered from burials dug into the tumulus during the MH period (Table 5.1, Fig. 5.1). These remains represent inhumations in single tombs constructed in the area that destroyed the upper parts of the tumulus. The hypothesis examined was that the lapse of time would signify changes in culture or resources and consequently change the distribution of procured foodstuffs. Therefore, time would be a variable for the study of dietary patterns.

31. See Renfrew 1973; Kroll 1984; Halstead 1994.
32. See also Chaps. 3, 4, 6, 11.
33. Balasse et al. 2005.
34. E.g., Cooper and DeNiro 1989.
35. Vika and Theodoropoulou 2012.
36. Dufour, Bocherens, and Mariotti 1999.
37. Purcell 1995.
38. See also Chap. 11.

The mean $\delta^{13}C$ value for the twelve samples is -19.6 ± 0.5‰ (1σ), while the respective mean $\delta^{15}N$ value is 9.0 ± 0.9‰ (1σ). These values indicate continuity in dietary habits for the area. There is no statistically significant difference in the $\delta^{13}C$ (KS test, D = 0.2, p = 0.7) between the two time periods. The difference in the $\delta^{15}N$ values between the two time periods is highly statistically significant (KS test, D = 0.7, p = 0.002), but this is due to an outlier: THOP48. When the outlier is excluded, the values are statistically the same.

It should be noted that qualitative or other differences in the provenance of observed protein cannot be detected through the present analyses. Although there is no evidence from other disciplines that the environment changed drastically in Thebes over these 200 years, certain changes would not be directly measurable with stable isotope analysis. When discussing dietary variations, one should always keep in mind the resolution of each methodology and enhance this with other data when available. For example, within the 200-year time lapse, certain species, faunal or botanical, may have become less abundant. If they were replaced by species of similar dietary signatures or photosynthetic pathways, however, the difference will not be observed isotopically. The general continuation of patterns is, nonetheless, an important observation, considering that the EH III marks the introduction of innovations in material culture that have been attributed to either new populations or to internal societal reorganization.[39]

Within the MH group, sample THOP48 is higher in both isotopes (with nitrogen of 10.8‰ and carbon of -18.4‰), causing a false statistical difference between the two time periods. Although this skeleton is poorly preserved, it is recorded as belonging to a child. One possible explanation for the high values of both $\delta^{13}C$ and $\delta^{15}N$ is that the child was still being breastfed at the time of death.[40] Exclusive breastfeeding would place the infant one trophic level above the mother. A fuller investigation of children's values is necessary, however, to confirm this possibility. It should be noted that children entering the burial record probably had endured severely compromised health, and this could potentially have affected their isotopic signatures.

Variation through Space

In order to investigate whether spatial location can act as a variable in the formation of dietary patterns, samples were also collected from an area of the city's Northeastern Cemetery—located approximately 2 km from the ancient acropolis toward the direction of the sea. Prehistoric pit burials and the remains of prehistoric habitation are also found there. It is hypothesized that individuals living in rural areas would follow a less "urbanized" way of life, which would be reflected in their diets. At the same time, proximity to the sea could offer readily available seafood for these individuals.

Twelve samples contained collagen in amounts sufficient for analysis (Table 5.1: samples THOP40–THOP51; Fig. 5.1). The $\delta^{13}C$ values range between -21.0‰ and -18.8‰, with an average of -19.5‰ ± 0.6 (1σ). The corresponding $\delta^{15}N$ values range between 5.7‰ and 10.4‰, with an average of 8.6‰ ± 1.4 (1σ). The much wider variation in $\delta^{15}N$ observed

39. See, e.g., Caskey and Blackburn 1977; Buck 1979.

40. Fuller et al. 2006.

for this group is caused by the outlier with a value of 5.7‰. The values follow a linear distribution.

One sample stands out with a very low $\delta^{15}N$ value (5.7‰). Three separate collagen extractions were conducted for this sample. The low value is indicative of a diet with minimal contributions from animal protein. Although this finding is unique among the present samples analyzed, it is paralleled elsewhere among Greek samples previously studied: for example, at Neolithic Makriyalos,[41] as well as at Theopetra Cave and Alepotrypa Cave.[42] If this result is explained with diet alone, it identifies individuals consuming very little animal protein and a high proportion of legumes. Ancient literature (e.g., Empedocles) describes a practice of vegetarianism, but there is no indication supportive of such notions during prehistory. Pathological reasons or resource availability should also be proposed in the interpretation of these values.[43]

In absolute terms, the nitrogen values from the prehistoric burials within this cemetery are lower than those from the mass burial in the tumulus. Although these values lie within the same trophic level, and although it remains impossible to calculate percentage contributions without accurate determination of dietary end points for this population, it can be suggested that individuals from the mass burial consumed greater amounts of animal protein, including milk, milk products, and meat from terrestrial animals.

The same can be said for the values of the MH burials from the tumulus. It is possible that spatial location or distance could account for the slight difference observed in the isotopic values. Individuals from the tumulus burials lived inside the city, where perhaps a greater abundance and variation of natural resources existed. It is certain that a variety of livestock would have reached the palace as offerings or sacrificial animals.[44] The values from the Northeastern Cemetery follow a linear distribution characteristic of marine protein consumption. Proximity to the sea may account for this pattern, but the values are not so significantly high that full confidence cannot be placed in this interpretation. Nonetheless, consumption of fish cannot be ruled out altogether.

CONCLUSIONS

Individuals buried *en masse* over the ruins of an EH apsidal building at Thebes are the focus of this paper. The mass burial contained a minimum of 12 individuals of both sexes, primarily young adults and adolescents. Results of dietary reconstruction for these individuals through stable isotope analysis reveals a diet based primarily on terrestrial and C_3 protein. No clear marine or freshwater protein signal is present, despite proximity to the sea and nearby lakes. No significant differences in isotopic values exist between individuals. These results conform to the general picture of ancient diet in Greece, as previously reconstructed through similar studies.[45] Two cases in the present study that demonstrate consumption of C_4 resources are particularly noteworthy and may contribute to a more systematic understanding of C_4 plant cultivation strategies in Bronze Age Greece. Additional analyses performed on two groups of human samples

41. Triantaphyllou 2001.
42. Papathanasiou 2003.
43. As discussed more recently in Schwarcz, Dupras, and Fairgrieve 1999; Sponheimer et al. 2003; Fuller et al. 2004, 2005.
44. Piteros, Olivier, and Melena 1990.
45. See also Chap. 11.

of later date than the mass burial in the tumulus, and from a more remote location, reveal that although no significant dietary differences exist within sites through time, differences in dietary practices do appear to exist between sites based on location.

ACKNOWLEDGMENTS

I feel the need to thank Anastasia Papathanasiou, Sherry Fox, and Mike Richards for inviting me to participate in the session they organized so successfully in Santorini and consequently to participate in this volume. Funds for this study were granted by the Basile and Elise Goulandris Foundation, the Max Planck Society, and Bradford University. I would also like to thank Vassilis Aravantinos, Ephor emeritus in Thebes, for his enthusiastic support throughout the years of my work there, and Kyriaki Psaraki, for the useful discussions we had on this site. Reviewers' comments were greatly appreciated.

REFERENCES

Aravantinos, V. 2004. "New Evidence about the EH II Period in Thebes: A New Architectural Complex and a Group Burial within the Kadmeia," in *Die ägäische Frühzeit: Forschungsbericht 1975–2002* II.2. *Die Frühbronzezeit in Griechenland, mit Ausnahme von Kreta* (Veröffentlichungen der mykenischen Kommission 21), ed. E. Alram-Stern, Vienna, pp. 1255–1259.

Balasse, M., A. Tresset, K. Dobney, and S. H. Ambrose. 2005. "The Use of Isotope Ratios to Test for Seaweed Eating in Sheep," *Journal of Zoology* 266, pp. 283–291.

Bello, S. M., A. Thomann, M. Signoli, O. Dutour, and P. Andrews. 2006. "Age and Sex Biases in the Reconstruction of Past Population Structures," *American Journal of Physical Anthropology* 129, pp. 24–38.

Bintliff, J. L, and A. M. Snodgrass. 1985. "The Cambridge/Bradford Boeotian Expedition: The First Four Years," *JFA* 12, pp. 123–161.

Brooks, S., and J. M. Suchey. 1990. "Skeletal Age Determination Based on the *Os Pubis*: A Comparison of the Acsádi-Nemeskéri and Suchey-Brooks Methods," *Human Evolution* 5, pp. 227–238.

Brothwell, D. R. 1989. "The Relationship of Tooth Wear to Aging," in *Age Markers in the Human Skeleton*, ed. M. Y. İşcan, Springfield, pp. 303–318.

Brown, T. A., D. E. Nelson, J. S. Vogel, and J. R. Southon. 1988. "Improved Collagen Extraction by Modified Longin Method," *Radiocarbon* 30, pp. 171–177.

Buck, R. J. 1979. *A History of Boeotia*, Edmonton.

Buikstra, J. E., and D. H. Ubelaker, eds. 1994. *Standards for Data Collection from Human Skeletal Remains. Proceedings of a Seminar at the Field Museum of Natural History, Organized by Jonathan Haas* (Arkansas Archeological Survey Research Series 44), Fayetteville.

Caskey, J. L., and E. T. Blackburn. 1977. *Lerna in the Argolid: A Short Guide*, Athens.

Chapman, J. 2000. *Fragmentation in Archaeology: People, Places, and Broken Objects in the Prehistory of South Eastern Europe*, London.

Cooper, L. W., and M. J. DeNiro. 1989. "Stable Carbon Isotope Variability in the Seagrass *Posidonia Oceanica:* Evidence for Light Intensity Effects," *Marine Ecology Series* 50, pp. 225–229.

Dalby, A. 1996. *Siren Feasts: A History of Food and Gastronomy in Greece*, London.

DeNiro, M. J. 1985. "Post-Mortem Preservation and Alteration of *in vivo* Bone Collagen Isotope Ratios in Relation to Paleodietary Reconstruction," *Nature* 317, pp. 806–809.

Dufour, E., H. Bocherens, and A. Mariotti. 1999. "Palaeodietary Implications of Isotopic Variability in Eurasian Lacustrine Fish," *JAS* 26, pp. 617–627.

Fossey, J. M. 1998. *Topography and Population of Ancient Boiotia,* Chicago.

Fuller, B. T., J. L. Fuller, D. A. Harris, and R. E. M. Hedges. 2006. "Detection of Breastfeeding and Weaning in Modern Human Infants with Carbon and Nitrogen Stable Isotope Ratios," *American Journal of Physical Anthropology* 129, pp. 279–293.

Fuller, B. T., J. L. Fuller, N. E. Sage, D. A. Harris, T. C. O'Connell, and R. E. M. Hedges. 2004. "Nitrogen Balance and δ^{15}N: Why You're Not What You Eat during Pregnancy," *Rapid Communications in Mass Spectrometry* 18, pp. 2889–2896.

———. 2005. "Nitrogen Balance and δ^{15}N: Why You're Not What You Eat during Nutritional Stress," *Rapid Communications in Mass Spectrometry* 19, pp. 2497–2506.

Gordon, C. C., and J. E. Buikstra. 1981. "Soil pH, Bone Preservation and Sampling Bias at Mortuary Sites," *AmerAnt* 46, pp. 566–571.

Hägg, R., and D. Konsola, eds. 1996. *Early Helladic Architecture and Urbanization. Proceedings of a Seminar Held at the Swedish Institute in Athens, June 8, 1985* (*SIMA* 76), Göteborg.

Halstead, P. 1994. "The North–South Divide: Regional Paths to Complexity in Prehistoric Greece," in *Development and Decline in the Mediterranean Bronze Age* (Sheffield Archaeological Monographs 8), ed. C. Mathers and S. Stoddart, pp. 195–219.

Hammond, N. G. L. 1974. "The Tumulus-Burials of Leukas and Their Connections in the Balkans and Northern Greece," *BSA* 69, pp. 129–144.

Hatch, J. W., and R. A. Geidel. 1985. "Status-Specific Dietary Variation in Two World Cultures," *Journal of Human Evolution* 14, pp. 469–476.

Knüsel, C. J., and A. K. Outram. 2004. "Fragmentation: The Zonation Method Applied to Fragmented Human Remains from Archaeological and Forensic Contexts," *Environmental Archaeology* 9, pp. 85–97.

Konsola, D. 1981. *Προμυκηναϊκή Θήβα: Χωροταξική και οικιστική διάρθρωση,* Athens.

———. 1988. "Θήβα, Πρώιμες φάσεις οικιστικής ανάπτυξης," *Επετηρίς της Εταιρείας Βοιωτικών Μελετών,* Τόμος Α΄, Τεύχος Α, Thebes.

Kroll, H. 1984. "Bronze Age and Iron Age Agriculture in Kastanas, Macedonia," in *Plants and Ancient Man: Studies in Palaeoethnobiology. Proceedings of the Sixth Symposium of the International Work Group for Palaeoethnobotany, Groningen, 30 May–3 June 1983,* ed. W. Van Zeist and W.A. Casparie, Rotterdam, pp. 243–246.

Lagia, A., E. Petroutsa, and S. Manolis. 2007. "Health and Diet during the Middle Bronze Age in the Peloponnese: The Site of Kouphovouno," in *Cooking Up the Past: Food and Culinary Practices in the Neolithic and Bronze Age Aegean,* ed. C. Mee and J. Renard, Oxford, pp. 313–328.

Lovejoy, C. O., R. S. Meindl, T. R. Pryzbeck, and R. P. Mensforth. 1985. "Chronological Metamorphosis of the Auricular Surface of the Ilium: A New Method for the Determination of Age at Death," *American Journal of Physical Anthropology* 68, pp. 15–28.

Meindl, R. S., and C. O. Lovejoy. 1985. "Ectocranial Suture Closure: A Revised Method for the Determination of Skeletal Age at Death Based on the Lateral-Anterior Sutures," *American Journal of Physical Anthropology* 68, pp. 57–66.

Newsome, S. D., D. L. Phillips, B. J. Culleton, T. P. Guilderson, and P. L. Koch. 2004. "Dietary Reconstruction of an Early to Middle Holocene Human Population from the Central California Coast: Insights from Advanced Stable Isotope Mixing Models," *JAS* 31, pp. 1101–1115.

Papathanasiou, A. 2003. "Stable Isotope Analysis in Neolithic Greece and Possible Implications on Human Health," *International Journal of Osteoarchaeology* 13, pp. 314–324.

Phenice, T. W. 1969. "A Newly Developed Visual Method of Sexing the *Os Pubis,*" *American Journal of Physical Anthropology* 30, pp. 297–302.

Piteros, C., J. P. Olivier, and J. L. Melena. 1990. "Les inscriptions de linéaire B des nodules de Thèbes (1982): La fouille, les documents, les possibilités d'interpretation," *BCH* 114, pp. 103–184.

Psaraki, K. 2004. "A New EH II Pottery Assemblage from Thebes," in *Die ägäische Frühzeit: Forschungsbericht 1975–2002* II.2. *Die Frühbronzezeit in Griechenland, mit Ausnahme von Kreta* (Veröffentlichungen der mykenischen Kommission 21), ed. E. Alram-Stern, Vienna, pp. 1259–1265.

———. In press. "Πρακτικές αποθήκευσης και οικιακή οικονομία στη Θήβα στο τέλος της ΠΕ ΙΙ," *5th International Congress of Boeotian Studies, Thebes September 2005.*

Pullen, D. J. 1985. "Social Organization in Early Bronze Age Greece: A Multi-Dimensional Approach" (diss. Indiana Univ.).

Purcell, N. 1995. "Eating Fish. The Paradoxes of Seafood," in *Food in Antiquity,* ed. J. Wilkins, D. Harvey, and M. Dobson, Exeter, pp. 132–149.

Renfrew, J. M. 1973. *Palaeoethnobotany: The Prehistoric Food Plants of the Near East and Europe,* New York.

Richards, M. P., and R. E. M. Hedges. 1999. "Stable Isotope Evidence for Similarities in the Types of Marine Foods Used by Late Mesolithic Humans at Sites along the Atlantic Coast of Europe," *JAS* 26, pp. 717–722.

Richards, M. P., and E. Vika. 2006. "Stable Isotope Results from New Sites in the Peleponnese: Sykia, Kalamaki, and Spaliareka," in *Archaeology Meets Science: Biomolecular Investigations in Bronze Age Greece. The Primary Scientific Evidence, 1997–2003,* ed. Y. Tzedakis, H. Martlew, and M. K. Jones, Oxford, pp. 231–235.

Rossel, S. 2004. "Food for the Dead, the Priest, and the Mayor: Looking for Status and Identity in the Middle Kingdom Settlement at South Abydos, Egypt," in *Behaviour*

Behind Bones: The Zooarchaeology of Ritual, Religion, Status, and Identity. Proceedings of the 9th Conference of the International Council of Archaeozoology, Durham, August 2002, ed. S. J. O'Day, W. Van Neer, and A. Ervynck, Oxford, pp. 198–202.

Scheuer, L., and S. Black. 2004. *The Juvenile Skeleton,* New York.

Schwarcz, H. P., T. L. Dupras, and S. I. Fairgrieve. 1999. "^{15}N Enrichment in the Sahara: In Search of a Global Relationship," *JAS* 26, pp. 629–636.

Sponheimer, M., T. Robinson, L. Ayliffe, B. Roeder, J. Hammer, B. Passey, A. West, T. Cerling, D. Dearing, and J. Ehleringer. 2003. "Nitrogen Isotopes in Mammalian Herbivores: Hair δ^{15}N Values from a Controlled Feeding Study," *International Journal of Osteoarchaeology* 13, pp. 80–87.

Symeonoglou, S. 1985. *The Topography of Thebes: From the Bronze Age to Modern Times,* Princeton.

Todd, T. W. 1920. "Age Changes in the Pubic Bone: I. The Male White Pubis," *American Journal of Physical Anthropology* 3, pp. 285–334.

———. 1921. "Age Changes in the Pubic Bone: II. The Pubis of the Male Negro-White Hybrid; III: The Pubis of the White Female; IV: The Pubis of the Female Negro-White Hybrid," *American Journal of Physical Anthropology* 4, pp. 1–70.

Todd, T. W., and D. W. Lyon. 1924. "Endocranial Suture Closure, Its Progress, and Age Relationship: Part I. Adult Males of White Stock," *American Journal of Physical Anthropology* 7, pp. 325–384.

Triantaphyllou, S. 2001. *A Bioarchaeological Approach to Prehistoric Cemetery Populations from Central and Western Greek Macedonia* (*BAR-IS* 976), Oxford.

Vika, E. 2011. "Diachronic Dietary Reconstructions in Ancient Thebes, Greece: Results from Stable Isotope Analyses," *JAS* 38, pp. 1157–1163.

Vika, E., V. Aravantinos, and M. P. Richards. 2009. "Aristophanes and Stable Isotopes: A Taste for Freshwater Fish in Classical Thebes, (Greece)?" *Antiquity* 83, pp. 1076–1083.

Vika, E., and T. Theodoropoulou. 2012. "Re-Investigating Fish Consumption in Greek Antiquity: Results from δ^{13}C and δ^{15}N Analysis from Fish Bone Collagen," *JAS* 39, pp. 1618–1627.

Walker, P. L., J. R. Johnson, and P. M. Lambert. 1988. "Age and Sex Biases in the Preservation of Human Skeletal Remains," *American Journal of Physical Anthropology* 76, pp. 183–188.

Wiencke, M. H. 1989. "Change in Early Helladic II," *AJA* 93, pp. 495–509.

CHAPTER 6

Existence and Subsistence in Mycenaean-Era East Lokris: The Isotopic Evidence

by Carina A. Iezzi

The Mycenaean LBA was a time of great social and political change. Evidence from burials, archaeological finds, and Linear B tablets indicate that Mycenaean society was hierarchical and highly structured. Until relatively recently, studies of LBA civilization on the Greek mainland have primarily focused on the large, wealthy, impressive settlements in this prehistoric, pre-*polis* society. The small numbers of LBA skeletons that have been excavated and studied have primarily come from palatial centers[1] and thus represent the uppermost levels of the social structure. Since socioeconomic status is correlated with nutrition and disease levels,[2] individuals from the highest socioeconomic strata would have had the best access to high-quality nutrition and less exposure to certain diseases, and they would therefore have had better overall health than those from lower socioeconomic strata. According to skeletal evidence from Mycenae[3] and, more cautiously, Homeric written sources, elite individuals from palatial sites feasted on high-quality, nutrient-dense foods and had a good level of health compared to those from both earlier time periods and from lower socioeconomic statuses. The study of major centers, while adding greatly to our knowledge of the biological manifestations of culture and environment, has also served to skew perceptions about what life was like during the LBA by offering information from only one level of society.

Given the archaeological evidence, East Lokris (Fig. 6.1) seems to have been on the periphery of the Mycenaean world; the closest major site, approximately 75 km to the southeast, was Thebes. By the LBA, Thebes had developed into a strong political and intellectual center on par with Mycenae.[4] Based on Linear B evidence, leaders of the center controlled agriculture and developed trade relations with groups throughout the Aegean and the East.[5] Therefore, the skeletal remains assessed in the present study can provide information about non-elite populations as opposed to those from immediately around major centers. In this way, the East Lokrian materials will serve to broaden our perspective on how social and environmental factors may have affected the biology of the non-elite socioeconomic stratum, therefore providing a more holistic portrait of the Mycenaean world.

1. Angel 1971, 1973.
2. Angel 1971; Bogin 1988.
3. Angel 1971.
4. Piteros, Olivier, and Melena 1990.
5. Vika 2011.

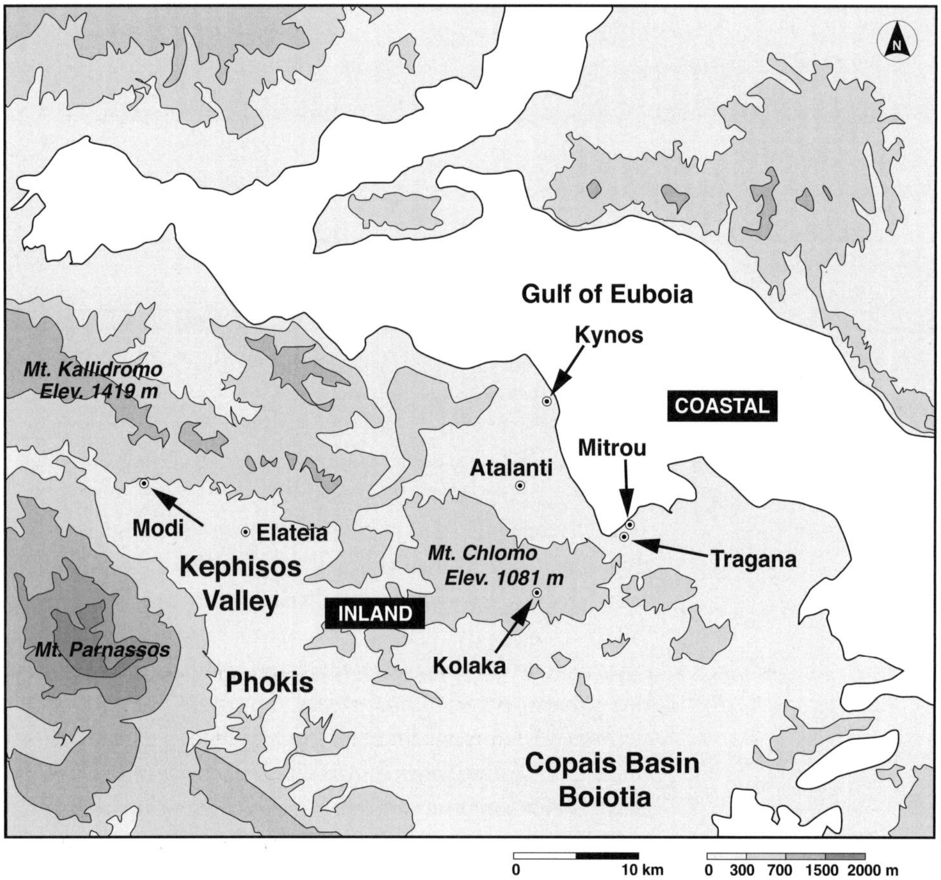

Figure 6.1. Map of Lokris area showing sites referenced in the text

The focus of this study is the reconstruction of the LBA Mycenaean diet in the East Lokris region of central Greece. To this end, $^{13}C/^{12}C$ and $^{15}N/^{14}N$ stable isotopes of LBA human bone samples from coastal and inland regions will be assessed with the goal of deciphering potential dietary differences between the two environmental zones. These data will paint a clearer portrait of how LBA populations utilized their local ecological settings.

STABLE ISOTOPE STUDIES

Isotopic analysis of bone affords a direct and precise means by which the basic composition of ancient diets can be determined.[6] These data can complement the morphological study of archaeological human bone to reconstruct past human foodways. The phrase "you are what you eat" forms the basic principle underlying the field of isotope ecology and palaeodietary reconstruction. In bones and other body tissues, certain foods leave characteristic isotopic records when they are consumed. Stable isotopes are thus used to assess the structure and dynamics of ecosystems since isotope ratios vary according to environmental components such as soil composition, altitude, and temperature. These factors affect the types, quantity, and quality of foods that grow in a region.

6. DeNiro and Schoeninger 1983; Schoeninger, DeNiro, and Tauber 1983; Schoeninger 1989; Ambrose and Norr 1992; Norr 1995.

PREHISTORIC SUBSISTENCE IN GREECE

According to osteological studies, throughout prehistory, the acquisition of sufficient nutrients and calories was problematic and the Mycenaean diet was somewhat deficient in protein, calories, and iron.[7] Traditionally, palaeodietary reconstructions have relied on indirect means of study, including zooarchaeology, palaeobotany, artifact assessment, and Linear B tablets.[8] Based on palynological evidence, fruits—including figs, pears, strawberries, olives, and grapes—and nuts—walnuts, chestnuts, and shelled acorns—were included in the LBA diet. The Linear B tablets and palaeobotanical studies have revealed that millet, wheat, emmer, barley, peas, chickpeas, and lentils were also available.[9] Linear B tablets from Knossos and Pylos indicate that pig, sheep, goat, and cattle were raised at those sites[10] and that wheat, barley, olives, and figs were eaten by individuals there.[11] Although the Linear B tablets emphasize the society of the palace, they do provide a context into which additional information from other social strata can be placed to produce a more balanced understanding of LBA society.

Relatively recently, stable isotope studies have increased because they provide direct evidence of palaeodiet composition. Over the past decade or so, a small number of stable isotope investigations of LBA dietary patterns have been undertaken. Based on these studies, an image has emerged of the typical LBA diet. From what can be discerned at this point, it was based primarily on C_3 plants, such as temperate grasses, and included some terrestrial animal protein. Interestingly, however, it did not habitually include marine food resources,[12] even among sites located close to the sea.

Generally, the pattern from Neolithic through Classical times periodically fluctuated from higher plant consumption to higher animal protein consumption.[13] In terms of morphological study of skeletal remains, Proskynas, a site spanning the FN through Mycenaean periods (4500 to 1300 B.C.) and located near the present study area, revealed indications of anemia, most likely acquired from chronic dietary deficiency or inadequacy and/or high pathogen loads.[14] Similarly, in northern Greece, skeletal studies indicate a pattern of declining health from the EBA to the EIA.[15] In this same population, a dietary isotope study revealed that between the EBA and the LBA, there was a significant reduction in plant consumption, while from the LBA to the EIA, samples point to a significant increase in animal protein intake. Individuals from the LBA coastal site in this region consumed more animal protein than did those from inland, who relied more heavily on carbohydrates. Several indications from the LBA and EIA samples also revealed indications of regular consumption of C_4 plants such as millet.[16]

7. Bisel and Angel 1985.
8. Chadwick 1976.
9. Halstead 1995.
10. Chadwick 1976; Vermeule 1983; Treuil et al. 1996.
11. Killen 2004.
12. Triantaphyllou 1999; Petroutsa et al. 2009; Petroutsa and Manolis 2010; Vika 2011.
13. Triantaphyllou 1999.
14. Papathanasiou, Zachou, and Richards 2009.
15. Triantaphyllou 1999.
16. Triantaphyllou 1999.

QUESTIONS TO BE ADDRESSED BY THE EAST LOKRIS MATERIALS

• Was there a heavier dietary reliance on C_3 or C_4 plants?

C_3 plants include temperate grasses, all trees and shrubs, all fruits and nuts, cultivated roots, and tubers. C_4 plants include tropical grasses, millet, domesticated maize, sorghum, and sugar cane. The ratio of ^{13}C to ^{12}C can elucidate these dietary differences. C_4 plants have less negative $\delta^{13}C$ values than do C_3 plants.[17] For example, a diet consisting of only C_4 plants would exhibit $\delta^{13}C$ values in the -9‰ to -14‰ range, while a diet of only C_3 plants would exhibit values in the -27‰ to -19‰ range.[18]

• How much animal protein was consumed?

Animal protein includes meat and/or secondary products such as dairy. Nitrogen ($\delta^{15}N$) values indicate the level of dietary reliance on animal protein. The $\delta^{15}N$ value of atmospheric nitrogen is about 0‰, while soils have a $\delta^{15}N$ value of approximately 10‰. Thus, nitrogen-fixing plants, including legumes, have lower $\delta^{15}N$ values than non-nitrogen-fixing plants.[19] Very low $\delta^{15}N$ values (0‰–6‰) indicate fairly heavy consumption of legumes.[20]

• How important were marine food resources as opposed to terrestrial foods?

Marine plants and animals rely on a carbon source that has a $\delta^{13}C$ value of about 0‰. Conversely, terrestrial plants and animals rely on CO_2 in the atmosphere, which has a $\delta^{13}C$ value of -7‰. Marine species tend to have higher $\delta^{15}N$ values than terrestrial species and can range to more than 20‰.[21] A marine-based diet is indicated by more positive $\delta^{13}C$ and $\delta^{15}N$ values.

THE BURIALS

Relatively little is currently known about the LBA bioarchaeology of East Lokris. The detailed examination of skeletal material previously carried out by the author,[22] and from which the sample for the present isotope analysis pilot study was selected, represents the first analysis of a large-scale collection of human remains from this time period and geographic area. All of the burials described here are tentatively dated to the LH IIIB–IIIC phases (1325 to 1100 B.C.) and were recovered by the Greek Archaeological Service during rescue excavations (pers. comm., P. Dakoronia). The skeletal remains were from secondary, commingled burials and were fragmentary and incomplete. Burials from a total of 27 chamber tombs were studied and included a minimum of 186 individuals. Since burial structure can be associated with socioeconomic level[23] and since all burials were derived from chamber tombs, all individuals likely came from the same general socioeconomic level. In this way, tomb type serves as a general control for potential major socioeconomic differences. The present isotopic analysis pilot study focuses on a small sample from within this collection.

17. Katzenberg 1992, p. 106.
18. See Chap. 2.
19. Katzenberg 1992, p. 107.
20. Keegan 1989, p. 226.
21. Katzenberg 1992, p. 112.
22. Iezzi 2005.
23. Graziadio 1991.

TOPOGRAPHY OF EAST LOKRIS

By virtue of its geologic history, Greece's topography is rugged and there is a sharp juxtaposition of environmental zones and ecological niches.[24] Greece's major tectonic movements over the millennia have generated wide lines of agricultural and fishing zones, upland plateaus, and lowland basins.[25] The southeastern part of the Greek mainland, or "Aegean Belt," is notorious for lacking consistent distributions of fertile soil, adequate degrees of moisture, light, and shelter. Rather, its resources are present in pockets of high quality land that serve as oases for settlements.[26]

In East Lokris, the coastal areas are divided from those of the inland by Mount Kallidromo, which divides the Kephissos valley from the gulf of Euboia. The tombs themselves were located in four main areas of East Lokris, which was thought to be on the periphery of the Mycenaean world. In modern times, two of the areas under study—Atalanti and Tragana—are very fertile and can support the largest population density in the area. Both are currently heavily cultivated with corn, tobacco, and garden produce. Other sites in the region are primarily involved in either agriculture or a mixture of herding and cultivation. The third area studied, Kolaka, is located in the middle of an upland plateau on Mount Chlomo at an elevation of 480 m. The northern slopes of the Kolaka plateau are relatively steep and drop to a torrent bed.[27] Modi, the final study area, is located on the northern side of the Kephissos valley in the foothills of Mount Kallidromo, where the land is quite fertile.[28] Both the coastal and inland zones have their own ecological nuances and would have provided unique adaptive challenges in terms of food and terrain to populations who lived among them.

METHODS

Age and Sex Assessment

Age estimates for adults were made by assessing morphological changes in the sternal end of ribs[29] and pelves,[30] as well as the degree of cranial suture closure.[31] Epiphyseal fusion rates[32] and hand and wrist ossification[33] were also used for young adults. Average age at death for adults (18 years and older) in each regional group was calculated using only individuals for whom a numeric age range could be assessed using the methods described above. Cranial, pelvic, and postcranial morphology[34] were used to determine sex, as were long bone (femur, tibia, fibula, radius, ulna) measurements.[35]

24. Anastassiades 1949.
25. Bintliff 1977, p. 8.
26. Bintliff 1977, p. 8.
27. Fossey 1990.
28. McInerney 1999.
29. McKern and Stewart 1957; İşcan, Loth, and Wright 1984a, 1984b, 1985.
30. Todd 1920; McKern and Stewart 1957; Gilbert and McKern 1973; Lovejoy et al. 1985; Webb and Suchey 1985; Suchey et al. 1986.
31. McKern and Stewart 1957; Baker 1984; Meindl and Lovejoy 1985; Krogman and İşcan 1986.
32. McKern and Stewart 1957; Suchey et al. 1984; Webb and Suchey 1985.
33. Greulich and Pyle 1959.
34. Phenice 1969; İşcan and Derrick 1984; Krogman and İşcan 1986; Bass 1987.
35. İşcan and Miller-Schaivitz 1984a, 1984b; Bass 1987.

Dental Data

Historically, caries prevalence in a population can indicate changes in food production and diet with very low incidences of decay before the advent of agriculture and large increases after the FN period. The higher caries prevalence in the Coastal group (as defined below) indicates that there was a difference.

Isotopic Analysis

Samples were selected based on five criteria that take into account the commingled, incomplete, and fragmentary nature of the skeletal remains. Although most of the skeletal elements had broken over time in the burial context, the internal and external bone surfaces (cortex, marrow cavity, trabecular bone) themselves were well preserved. The sampling criteria were as follows:

1. Only adults were selected because subadults may have eaten a somewhat different diet.
2. Only bones of individuals whose sex could be determined were selected so that dietary differences between the sexes could be studied.
3. The femur was selected as the sampling bone of choice both because of its exceptional sturdiness and because sex is more reliably assessed from it than from other long bones.
4. All bones sampled were free from expressions of skeletal pathology or anomaly since this might have an unknown effect on isotopic results.
5. Only bones that were not taphonomically altered were selected for isotopic analysis.

One-gram bone samples were taken from 16 individuals, who consisted of four adults (two males and two females) from each of the four sites. Since sample sizes were small, those from Tragana and Atalanti were combined to form a Coastal group (n = 8), while those from Kolaka and Modi were combined to form an Inland group (n = 8). The samples were subjected to mass spectrometry using a modified version of the Longin method[36] for collagen extraction by GeoChron Laboratories, Cambridge, Massachusetts.

In this study, ratios of both carbon ($^{13}C/^{12}C$) and nitrogen ($^{15}N/^{14}N$) isotopes enmeshed in the collagen of bone samples were assessed. Carbon and nitrogen in collagen are derived from ingested protein. Isotopic analysis results are presented as delta (δ) values of carbon ($^{13}C/^{12}C$) and nitrogen ($^{15}N/^{14}N$) in per mil notation (‰) as compared to international standards. The international standards used for isotopic analysis are Vienna Pee Dee Belemnite (VPDB), a marine limestone, for carbon, and atmospheric nitrogen (Ambient Inhalable Reservoir or AIR) for nitrogen.[37] Student's *t*-tests for independent samples were used to discern any statistically significant differences between sub-sample means.

36. Longin 1971.
37. Craig 1957; Schoeninger and Moore 1992.

TABLE 6.1. PALAEODEMOGRAPHY OF EAST LOKRIS (MNI = 186)*

	Subadult	*Males*	*Females*	*Adult*
Coastal	25	35	38	88
Inland	18	27	23	55
Total	43	62	61	143

*Due to commingled, incomplete, and fragmentary condition of skeletal remains, population MNI (minimum number of individuals) was determined per tomb using skeletal elements most prevalent per age and sex. MNI for isotopic samples was based on male and female adult femora.

RESULTS

Age and Sex

Of the 186 individuals present (Table 6.1), the average age of adults was 37 years (n = 50).[38] Sixty-two adults were identified as male (average age of 36 years), and 61 were identified as female (average age of 38 years). The average age of Coastal adults was 39 years (n = 88). Average ages of males and females in this group were 32 (n = 35) and 44 years (n = 38), respectively. The average age of Inland adults was 34 years (n = 55). Average ages of males and females in this region were 39 (n = 27) and 31 years (n = 23), respectively.

Stable Isotopes

The $\delta^{13}C$ and $\delta^{15}N$ stable isotope values for the Coastal and Inland samples of the East Lokris population are presented in Table 6.2 and Figure 6.2. In the Coastal group, the mean averages are $\delta^{13}C$ -19.9‰ ± 1.2‰ and $\delta^{15}N$ 8.5‰ ± 1.2‰. The Inland group had means of $\delta^{13}C$ -18.8‰ ± 1.2‰ and $\delta^{15}N$ 7.6‰ ± 1.4‰. The mean human bone collagen $\delta^{13}C$ and $\delta^{15}N$ values for both groups combined are -19.3‰ ± 1.3‰ and 8.0‰ ± 1.4‰, respectively.

Descriptive statistical analyses of the data revealed that among the East Lokris, Coastal, Inland, male, and female groups, there were wide ranges in both $\delta^{13}C$ and $\delta^{15}N$ values. Overall, males had means of $\delta^{13}C$ -19.2‰ ± 1.4‰ and $\delta^{15}N$ 8.4‰ ± 1.5‰, while females had means of $\delta^{13}C$ -19.5‰ ± 1.2‰ and $\delta^{15}N$ 7.7‰ ± 1.2‰. Within the Coastal group, males had means of $\delta^{13}C$ -20.0‰ ± 0.3‰ and $\delta^{15}N$ 8.9‰ ± 1.1‰, and females had $\delta^{13}C$ -19.8‰ ± 1.8‰ and $\delta^{15}N$ 8.1‰ ± 1.2‰. Inland, males averaged $\delta^{13}C$ -18.3‰ ± 1.6‰ and $\delta^{15}N$ 7.9‰ ± 1.8‰, while females averaged $\delta^{13}C$ -19.2‰ ± 0.5‰ and $\delta^{15}N$ 7.3‰ ± 1.2‰.

Student's *t*-tests comparing the means of Coastal versus Inland groups as well as males versus females found no significant differences in $\delta^{13}C$ or $\delta^{15}N$ values. One exception to these results was when the sexes between regions were tested.

Figure 6.3 and Table 6.3 present a comparison of $\delta^{13}C$ and $\delta^{15}N$ means between East Lokris and samples from the LBA[39] and other prehistoric

38. Iezzi 2005.

39. Triantaphyllou 1999; Petroutsa et al. 2009, pp. 237–244; Petroutsa and Manolis 2010, pp. 614–620; Vika 2011, pp. 1157–1163.

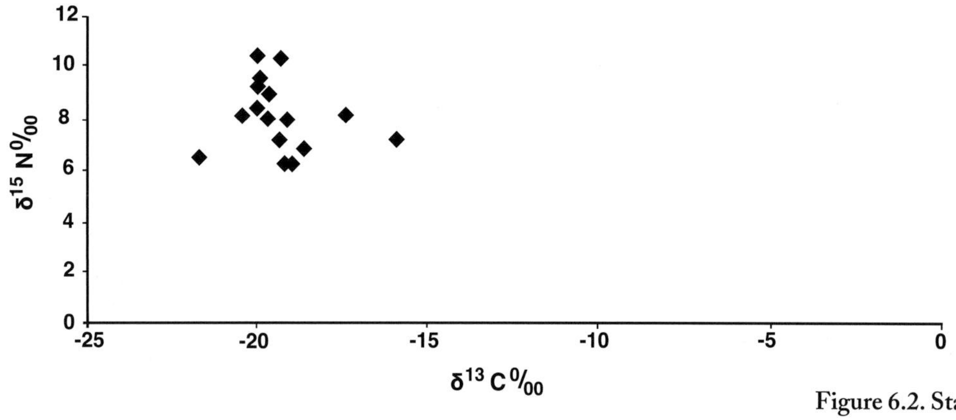

Figure 6.2. Stable carbon and nitrogen isotope values from LBA East Lokris

TABLE 6.2. STABLE ISOTOPE VALUES FROM EAST LOKRIS*

Sample	*Site*	*Region*	*Sex*	*$\delta^{13}C$ (‰ VPDB)*	*$\delta^{15}N$ (‰ AIR)*
Tr1	Tragana	Coast	M	-20.0	9.2
Tr2	Tragana	Coast	F	-19.9	9.5
Tr3	Tragana	Coast	M	-20.0	10.4
Tr4	Tragana	Coast	F	-17.4	8.1
At1	Atalanti	Coast	M	-19.7	8.0
At2	Atalanti	Coast	F	-21.7	6.5
At3	Atalanti	Coast	M	-20.4	8.1
At4	Atalanti	Coast	F	-20.0	8.3
Ko1	Kolaka	Inland	M	-15.9	7.1
Ko2	Kolaka	Inland	F	-19.3	7.1
Ko3	Kolaka	Inland	M	-19.3	10.3
Ko4	Kolaka	Inland	F	-18.6	6.8
Mo1	Modi	Inland	M	-19.0	6.2
Mo2	Modi	Inland	F	-19.7	8.9
Mo3	Modi	Inland	M	-19.1	7.9
Mo4	Modi	Inland	F	-19.2	6.2

*Geochron Laboratories does not include C:N ratio, an indicator of sample quality, in their results; therefore they are not presented here.

TABLE 6.3. MEAN STABLE ISOTOPE VALUES FROM PREHISTORIC GREECE

Site and Author	*$\delta^{13}C \pm 1\sigma$ (‰ VPDB)*	*$\delta^{15}N \pm 1\sigma$ (‰ AIR)*
East Lokris (Iezzi, this chapter)	-19.3 ± 1.2	8.0 ± 1.3
Thebes (Vika 2011)	-19.7 ± 1.0	8.4 ± 1.7
Mainland Greece (Petroutsa and Manolis 2010)	-19.7 ± 0.6	7.9 ± 1.0
Northern Greece (Triantaphyllou 1999)	-18.8 ± 0.9	8.0 ± 1.1

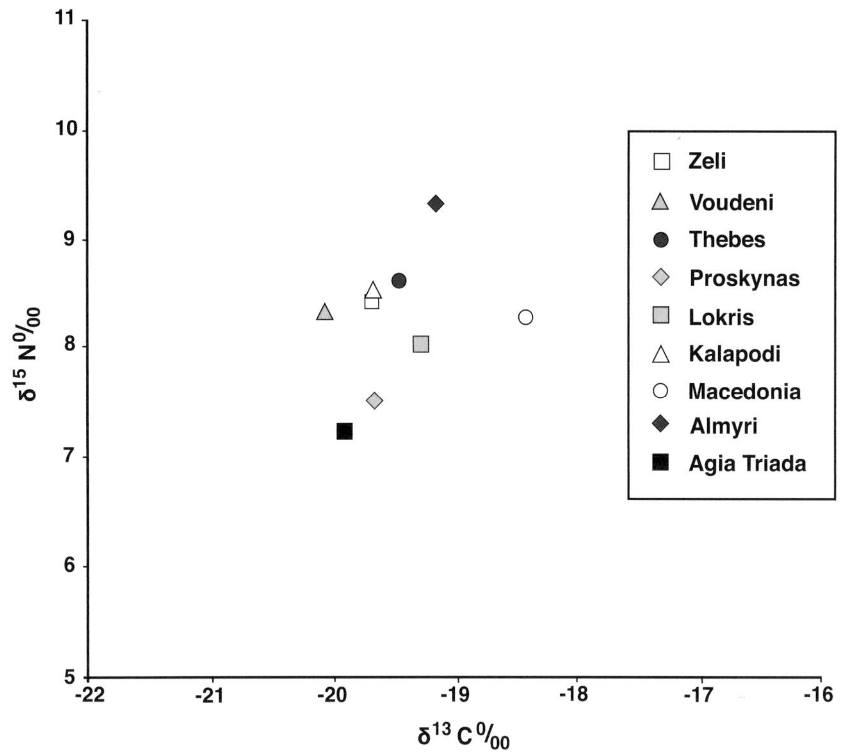

Figure 6.3. East Lokris stable carbon and nitrogcn isotopc mcans com pared to means of LBA and prehistoric samples

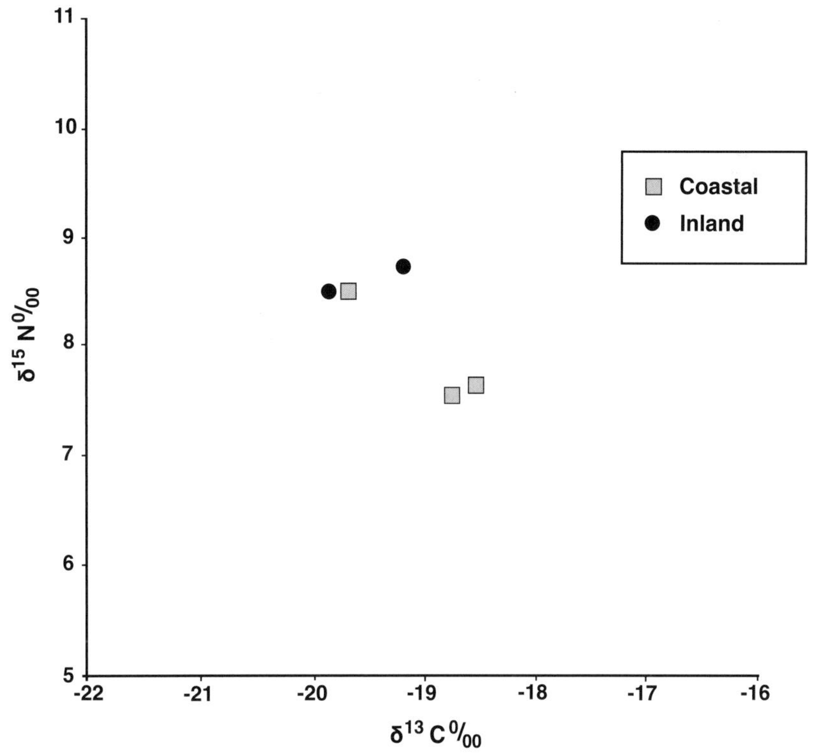

Figure 6.4. LBA Coastal and Inland stable carbon and nitrogen means from East Lokris

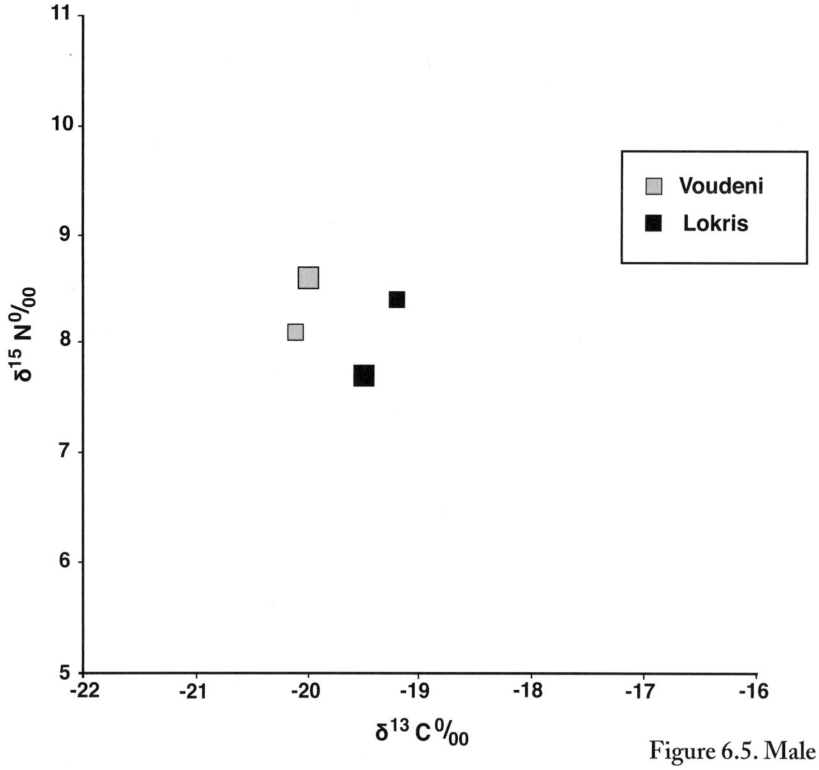

Figure 6.5. Male and female stable carbon and nitrogen means from EBA East Lokris

sites in Greece.[40] Figure 6.4 depicts δ^{13}C and δ^{15}N values for coastal and inland sites including East Lokris and two other LBA sites.[41] Figure 6.5 shows a comparison between δ^{13}C and δ^{15}N means between males and females from East Lokris and one other LBA site.[42]

DISCUSSION

Stable Isotopes

The East Lokris samples fall within the range of means (Table 6.3) and values observed in other published isotopic studies of LBA samples. East Lokris values had the following ranges: δ^{13}C -21.7‰ to -15.9‰, and δ^{15}N 6.2‰ to 10.4‰. Comparable LBA samples had the following ranges: δ^{13}C -21.7‰[43] to -16.5‰,[44] and δ^{15}N 5.9‰[45] to 10.4‰.[46] The results from both the Coastal and Inland groups generally indicate a diet similar to those of other Greek Bronze Age individuals[47] including a diet centered on cereals with some animal protein included.

C_3 versus C_4 Plants

Means of δ^{13}C for East Lokris as a whole (Table 6.3, Fig. 6.3), and for all males and all females (Fig. 6.5), are relatively more negative and reflect a primarily C_3 plant-based diet. Similar results have consistently been reported in other published reconstructions of prehistoric Greek paleodiets.[48] Inland, δ^{13}C means were less negative (-18.8‰) than they were coastally

40. Papathanasiou, Zachou, and Richards 2009, pp. 223–236.
41. Triantaphyllou 1999; Petroutsa and Manolis 2010.
42. Petroutsa et al. 2009.
43. Vika 2011.
44. Triantaphyllou 1999.
45. Vika 2011.
46. Petroutsa and Manolis 2010; Vika 2011.
47. Triantaphyllou 1999; Petroutsa et al. 2009; Petroutsa and Manolis 2010; Vika 2011.
48. Triantaphyllou 1999; Triantaphyllou et al. 2006, 2008; Ingvarsson-Sundström, Richards, and Voutsaki 2009; Petroutsa et al. 2009; Petroutsa and Manolis 2010; Vika 2011.

(-19.9‰) and reflect a possible incorporation of C_4 plants into the Inland diet either directly (by consumption of C_4 plants) or indirectly (by consumption of animals feeding on C_4 plants) (Fig. 6.4). This was true for both males ($\delta^{13}C$ -18.3‰) and females ($\delta^{13}C$ -19.2‰) (Fig. 6.5). Additionally, the wide ranging of $\delta^{13}C$ values further supports a more heterogeneous diet among Inland males. One Inland male in particular (sample Ko1) had the least negative $\delta^{13}C$ value (-15.9‰), suggesting he consumed more C_4 resources than any other individual from the sample. One Coastal female (sample Tr4) also had a relatively less negative $\delta^{13}C$ value (-17.4‰), suggesting that she, too, incorporated a substantial amount of C_4 protein into her diet. To date, these appear to be among the earliest stable isotope indications of possible C_4 consumption in Greece.[49]

In her study of LBA and EIA samples from northern Greece, Triantaphyllou also found less negative $\delta^{13}C$ values (-16.48‰ to -18.7‰) indicative of a relatively high consumption of C_4 resources,[50] and she suggests that millet would have been the most probable candidate among the relevant plant species. According to another study of LBA samples from four combined regions of mainland Greece,[51] again, the general picture for the majority of the sample is a C_3-based diet, with, however, sporadic indications of individuals incorporating some C_4 protein in their diets, particularly in the Peloponnese. In the same study it was also suggested that millet would have been the C_4 plant of choice.[52] C_4 plants could have been used either as animal fodder, entering human diet through the consumption of these animals, and/or as a supplementary or occasional human food source, therefore entering the human diet directly.[53] Either pathway can explain their sporadic and inconsistent appearance in the stable isotope record.

Animal Protein

The $\delta^{15}N$ mean for East Lokris (8.0‰) is comparable to other published LBA samples (Table 6.3, Fig. 6.3). These results indicate that a small to moderate amount of animal protein was regularly consumed by the East Lokris individuals. More positive $\delta^{15}N$ means in the Coastal group (8.5‰) revealed that they consumed more animal protein than did Inland-group individuals ($\delta^{15}N$ 7.6‰). While a comparison of males and females within Coastal and Inland groups did not reveal statistically significant results, males in both groups had more positive $\delta^{15}N$ values (Coastal-group male mean 8.9‰; Inland-group male mean 7.9‰) than did females (Coastal-group female mean 8.1‰; Inland-group female mean 7.3‰). This indicates that males in both groups consumed more animal protein than did females from the respective groups. The individual with the least positive $\delta^{15}N$ value (sample At2; $\delta^{15}N$ 6.5‰) in the Coastal group was female. Three Inland-group individuals had less positive $\delta^{15}N$ values as well; of these, two were female (samples Ko4, 6.8‰; Mo4, 6.2‰) and the third was male (sample Mo1, 6.2‰). Less positive $\delta^{15}N$ values likely indicate a heavier dietary consumption of legumes.

A morphological study of dental carious lesions among LBA Greek teeth found that females generally consumed more cariogenic foods, while males consumed more meat, which resulted in females exhibiting more

49. Cf. Chaps. 3–5, 7, 10.
50. Triantaphyllou 1999.
51. Petroutsa and Manolis 2010.
52. Larsen 1997; Richards and Hedges 2003; Keenleyside, Schwarcz, and Panayotova 2006.
53. Petroutsa and Manolis 2010.

carious lesions than males.[54] These data complement the results from the present isotopic study that found females consumed less animal protein and, by default, would have eaten more carbohydrate-containing plant foods than did males. Similarly, from the author's study comparing dental carious lesions between the Coastal and Inland groups,[55] the Coastal group demonstrated more carious lesions, which may reflect a greater reliance on agricultural foodstuffs when compared to the Inland group, which may have depended more on a mixed economy.

A palaeodietary reconstruction by Petroutsa and Manolis investigated isotope ratios of samples from four regions of mainland Greece,[56] including two near the present study area in central Greece and two from the Peloponnese in southern Greece. Results indicated that the diet of one of the Peloponnesian samples was fairly homogeneous and based more on plant proteins than were samples from Kalapodi, a site located inland from Atalanti (Fig. 6.1). At both Kalapodi and Zeli, an additional site located inland from Atalanti, $\delta^{15}N$ values were very similar to one another ($\delta^{15}N$ 8.5‰ and $\delta^{15}N$ 8.4‰, respectively) and more positive than those from the Peloponnesian site ($\delta^{15}N$ 7.2‰), indicating higher levels of animal protein consumption. Values of $\delta^{15}N$ from all four sites studied by Petroutsa and Manolis indicated that individuals consumed a significant amount (50% to 70%) of terrestrial animal protein.[57] It should be noted that the LBA Kalapodi samples were derived from tholos tombs that contained wealthy grave goods, including golden artifacts, pots, jewels, and weapons, and this may have skewed the results toward more positive $\delta^{15}N$ values.[58]

Tomb type can provide a reliable indicator of social status.[59] Tholos tombs are commonly thought to have been used by elite/high status groups while chamber tombs were used by sub-elite/lower status groups.[60] Bone samples used in the present study were derived from chamber tomb burials. The Zeli comparative samples were obtained from LBA pit burials containing few artifacts.[61] The fact that individuals from Kalapodi consumed more animal protein than those from one of the present study's Inland sites (Modi) may be linked to differential social status. However, the pit burials from which the Zeli samples were derived cannot be explained in this way; the similarities therefore may have more to do with the closer proximity of Kalapodi and Zeli to the East Lokris Coastal sites of Atalanti and Tragana (Fig. 6.1). While samples from the East Lokris sites were derived from chamber tombs, $\delta^{15}N$ results from the Coastal group indicated that individuals there consumed amounts of animal protein on par with samples from Zeli and Kolaka.

In terms of protein, the more positive average $\delta^{15}N$ value in the Coastal group indicates a greater reliance on animal protein than that of the Inlanders. This protein would have likely more often been in the form of secondary products, such as dairy, rather than meat, although isotopically the exact source cannot be distinguished. During the Bronze Age, people came to rely more on secondary animal products.[62] Herded animals were probably not often slaughtered for meat since they would have represented wealth and would have been exploited for their secondary products. However, hunted game could have provided some meat as well.

54. Triantaphyllou 1999; Schepartz, Miller-Antonio, and Murphy 2009.
55. Iezzi 2005.
56. Petroutsa and Manolis 2010.
57. Petroutsa and Manolis 2010.
58. Petroutsa and Manolis 2010.
59. Graziadio 1991.
60. Dickinson 1994, p. 223.
61. Petroutsa and Manolis 2010.
62. Sherratt 1981.

Marine Resources

None of the East Lokris samples exhibited the relatively less negative $\delta^{13}C$ and more positive $\delta^{15}N$ values indicative of regular consumption of marine food resources (Table 6.2, Fig. 6.2). These results are similar to the findings of other isotopic investigations of prehistoric Greek marine resource utilization.[63] So far, the only evidence of Mycenaean era marine food consumption has come from the aristocracy at MBA Mycenae.[64]

Based on archaeological evidence throughout the mainland, the importance of seafood is unclear in the Mycenaean diet. According to Vickery,[65] some portions of the Odyssey suggest that Homeric Greeks disliked and were reluctant to eat fish. Published studies have indicated that it is normal to find that prehistoric Greeks did not utilize many or any marine food resources, even populations living near coastal areas.[66] While people may have occasionally eaten seafood, it did not seem to play an important part in the normal diet. As Papathanasiou points out, this may have been due to an ever-increasing dependence on agriculture as a reliable source of sustenance.[67]

CONCLUSION

In central Greece, a close relationship developed between the physical environment and human geography,[68] with populations adapting to a series of distinct topographic zones, micro-regions, and niches.[69] The natural division of land here had an important effect on prehistoric culture by having an influence on the size and location of settlements, favoring self-sufficiency and encouraging autonomy.[70] Communities inland remained independent of one another largely due to the fact that each possessed sufficient land to maintain itself.[71] Alternately, along the coast, larger pockets of arable land, coupled with access to the sea, facilitated the development of larger populations and ports as hubs of trade and travel.

Despite several limitations to this study—small sample size, lack of sample-quality assessment, and no faunal comparative materials—the results compare well with the small but growing body of direct evidence about the LBA diet. The present study revealed distinct dietary variations between the two regional groups investigated. While both relied on a C_3 plant-based diet largely comprised of cereals and grains, populations near the coast consumed higher amounts of both this type of plant as well as more animal protein. Males in both Coastal and Inland groups consumed

63. Triantaphyllou 1999; Ingvarsson-Sundström, Richards, and Voutsaki 2009; Papathanasiou, Zachou, and Richards 2009; Petroutsa et al. 2009; Petroutsa and Manolis 2010; Vika 2011.
64. Hedges and Richards 1999a, 1999b.
65. Vickery 1936.
66. Triantaphyllou 1999; Papathanasiou 2003; Papathanasiou, Zachou, and Richards 2009; Petroutsa et al. 2009; Petroutsa and Manolis 2010; Vika 2011.
67. Papathanasiou 1999, p. 140.
68. McInerney 1999, p. 40.
69. McInerney 1999, p. 40.
70. McInerney 1999, p. 54.
71. McInerney 1999, p. 54.

more animal protein than did the respective females from these areas. Coastal males consumed the most animal protein, while Inland females consumed the least. Neither regional group relied on marine food resources to any great extent.

Inland populations incorporated greater amounts of C_4 resources—most likely in the form of millet—into their regular diet. Overall, the palaeodiet profile of the Inland group appears to fit that of a mixed economy where C_4 plants regularly supplement a largely C_3 plant-based diet and also include small amounts of animal protein. The Coastal group isotopic profile paints a picture of individuals who subsisted on a C_3 plant diet and moderate amounts of animal protein. Coupled with archaeological finds from Kynos, the LBA port for the site of Atalanti in the Coastal group, evidence in support of an agricultural subsistence base in the Coastal region is provided. At Kynos, one or more complexes of storerooms and workshops were found. Several of these had clay bins in them for storing agricultural products.[72] Further, Linear B data from Thebes indicate that it had trade relations with various areas throughout the Aegean. Thus, it seems plausible that the coastal populations of East Lokris were involved in supplying agricultural and perhaps animal products to Thebes. Based on a comparison of the East Lokris isotopic results with a previously published study of LBA samples derived from tholos and pit burials in central Greece, regardless of social status, populations closer to the coast may have had a more plentiful food supply associated with raising grain, and perhaps animals, for trade with Thebes.

The goal of this project has been to determine what the isotopic evidence can tell us about dietary patterns among East Lokrian Mycenaeans and what these differences indicate. Further stable isotope and other direct-method studies in this region should prove even more enlightening.

ACKNOWLEDGMENTS

Special thanks to Anastasia Papathanasiou, Ephorate of Paleoanthropology and Speleology; Michael Richards, Max Planck Institute for Evolutionary Anthropology; Sherry Fox, Wiener Laboratory, American School of Classical Studies at Athens (ASCSA); Phanouria Dakoronia, 14th Ephorate of Prehistoric and Classical Antiquities; and GeoChron Laboratories, Cambridge, Massachusetts. Funding was provided by the J. L. Angel Fellowship, a Research Associateship from the Wiener Laboratory at the ASCSA, and the Doreen Spitzer fellowship (ASCSA). Thanks are due, also, to three anonymous reviewers for providing constructive comments.

72. Dakoronia 1993.

REFERENCES

Ambrose, S. H., and L. Norr. 1992. "On Stable Isotopic Data and Prehistoric Subsistence in the Soconusco Region," *CurrAnthr* 33, pp. 401–404.

Anastassiades, P. A. 1949. "General Features of the Soils of Greece," *Soil Science* 67, pp. 347–362.

Angel, J. L. 1971. "Disease and Culture in the Ancient East Mediterranean," in *Anthropological Congress Dedicated to Aleš Hrdlička. Proceedings, 30th August–5th September 1969,* ed. V. V. Novotný, Prague, pp. 503–508.

———. 1973. "Human Skeletons from Grave Circles at Mycenae," in *Ο ταφικός κύκλος Β των Μυκηνών,* G. E. Mylonas, Athens, pp. 379–397.

Baker, R. K. 1984. "The Relationship of Cranial Suture Closure and Age Analyzed in a Modern Multi-Racial Sample of Males and Females" (diss. California State University, Fullerton).

Bass, W. M. 1987. *Human Osteology: A Laboratory and Field Manual,* 3rd ed. (Missouri Archaeological Society Special Publication 2), Columbia, Mo.

Bintliff, J. L. 1977. *Natural Environment and Human Settlement in Prehistoric Greece: Based on Original Fieldwork* 1 (*BAR* Suppl. 28), Oxford.

Bisel, S. C., and J. L. Angel. 1985. "Health and Nutrition in Mycenaean Greece: A Study in Human Skeletal Remains," in *Contributions to Aegean Archaeology. Studies in Honor of William A. McDonald* (Publications in Ancient Studies 1), ed. N. C. Wilkie and W. D. E. Coulson, Dubuque, pp. 197–210.

Bogin, B. 1988. *Patterns of Human Growth* (Cambridge Studies in Biological Anthropology 3), Cambridge.

Chadwick, J. 1976. *The Mycenaean World,* Cambridge.

Craig, H. 1957. "Isotopic Standards for Carbon and Oxygen and Correction Factors for Mass-Spectrometric Analysis of Carbon Dioxide," *Geochimica et cosmochimica acta* 12, pp. 133–149.

Dakoronia, P. 1993. "Homeric Towns in East Lokris: Problems of Identification," *Hesperia* 62, pp. 115–127.

DeNiro, M. J., and M. J. Schoeninger. 1983. "Stable Carbon and Nitrogen Isotope Ratios of Bone Collagen: Variations within Individuals, between Sexes, and within Populations Raised on Monotonous Diets," *JAS* 10, pp. 199–203.

Dickinson, O. T. P. K. 1994. *The Aegean Bronze Age,* New York.

Fossey, J. M. 1990. *The Ancient Topography of Opountian Lokris,* Amsterdam.

Gilbert, B. M., and T. W. McKern. 1973. "A Method for Aging the Female *Os Pubis,*" *American Journal of Physical Anthropology* 38, pp. 31–38.

Graziadio, G. 1991. "The Process of Social Stratification at Mycenae in the Shaft Grave Period: A Comparative Examination of the Evidence," *AJA* 95, pp. 403–440.

Greulich, W. W., and S. I. Pyle. 1959. *Radiographic Atlas of Skeletal Development of the Hand and Wrist,* 2nd ed., Stanford.

Halstead, P. 1995. "Late Bronze Age Grain Crops and Linear B Ideograms *65, *120, and *121," *BSA* 90, pp. 229–234.

Hedges, R. E. M., and M. P. Richards. 1999a. "Grave Circle A Mycenae," in *Minoans and Mycenaeans: Flavours of Their Time. National Archaeological Museum, 12 July–27 November 1999,* ed. Y. Tzedakis and H. Martlew, Athens, pp. 227–228.

———. 1999b. "Grave Circle B Mycenae," in *Minoans and Mycenaeans: Flavours of Their Time. National Archaeological Museum, 12 July–27 November 1999,* ed. Y. Tzedakis and H. Martlew, Athens, pp. 223–224.

Iezzi, C. A. 2005. "Regional Differences in the Health Status of Late Bronze Age Populations from East Lokris, Greece" (diss. State Univ. of New York at Buffalo).

Ingvarsson-Sundström, A., M. P. Richards, and S. Voutsaki. 2009. "Stable Isotope Analysis of the Middle Helladic Population from the Two Cemeteries at Asine: Barbouna and the East Cemetery," *Mediterranean Archaeology and Archaeometry* 9, pp. 1–14.

İşcan, M. Y., and K. Derrick. 1984. "Determination of Sex from the Sacroiliac Joint: A Visual Assessment Technique," *Florida Scientist* 47, pp. 94–98.

İşcan, M. Y., S. R. Loth, and R. K. Wright. 1984a. "Age Estimation from the Ribs by Phase Analysis: White Males," *Journal of Forensic Sciences* 29, pp. 1094–1104.

———. 1984b. "Metamorphosis at the Sternal Rib End: A New Method to Estimate Age at Death in White Males," *American Journal of Physical Anthropology* 65, pp. 147–156.

———.1985. "Age Estimation from the Rib by Phase Analysis: White Females," *Journal of Forensic Sciences* 30, pp. 853–863.

İşcan, M. Y., and P. Miller-Shaivitz. 1984a. "Determination of Sex from the Femur in Blacks and Whites," *Collegium Antropologicum* 8, pp. 169–177.

———. 1984b. "Determination of Sex from the Tibia," *American Journal of Physical Anthropology* 64, pp. 53–57.

Katzenberg, M. A. 1992. "Advances in Stable Isotope Analysis of Prehistoric Bones," in *Skeletal Biology of Past Peoples: Research Methods,* ed. S. R. Saunders and M. A. Katzenberg, New York, pp. 105–120.

Keegan, W. F. 1989. "Stable Isotope Analysis of Prehistoric Diet," in *Reconstruction of Life from the Skeleton,* ed. M. Y. İşcan and K. A. R. Kennedy, New York, pp. 223–236.

Keenleyside, A., H. Schwarcz, and K. Panayotova. 2006. "Stable Isotopic Evidence of Diet in a Greek Colonial Population from the Black Sea," *JAS* 33, pp. 1205–1215.

Killen, J. 2004. "Wheat, Barley, Flour, Olives, and Figs on Linear B Tablets," in *Food, Cuisine, and Society in Prehistoric Greece* (Sheffield Studies in Aegean Archaeology 5), ed. P. Halstead and J. C. Barrett, Oxford, pp. 155–173.

Krogman, W. M., and M. Y. İşcan. 1986. *The Human Skeleton in Forensic Medicine,* 2nd ed., Springfield, Ill.

Larsen, C. S. 1997. *Bioarchaeology: Interpreting Behavior from the Human Skeleton* (Cambridge Studies in

Biological Anthropology 21), Cambridge.

Longin, R. 1971. "New Method of Collagen Extraction for Radiocarbon Dating," *Nature* 230, pp. 241–242.

Lovejoy, C. O., R. S. Meindl, T. R. Pryzbeck, and R. P. Mensforth. 1985. "Chronological Metamorphosis of the Auricular Surface of the Ilium: A New Method for the Determination of Adult Skeletal Age at Death," *American Journal of Physical Anthropology* 68, pp. 15–28.

McInerney, J. 1999. *The Folds of Parnassos: Land and Ethnicity in Ancient Phokis,* Austin.

McKern, T. W., and T. D. Stewart. 1957. *Skeletal Age Changes in Young American Males: Analysis from the Standpoint of Age Identification* (Quartermaster Research and Development Center [U.S.], Environmental Protection Research Division, Technical Report, EP-45), Natick, Mass.

Meindl, R. S., and C. O. Lovejoy. 1985. "Ectocranial Suture Closure: A Revised Method for the Determination of Skeletal Age at Death Based on the Lateral-Anterior Sutures," *American Journal of Physical Anthropology* 68, pp. 57–66.

Norr, L. 1995. "Interpreting Dietary Maize from Bone Stable Isotopes in the American Tropics: The State of the Art," in *Archaeology in the Lowland American Tropics: Current Analytical Methods and Applications,* ed. P. W. Stahl, Cambridge, pp. 198–223.

Papathanasiou, A. 1999. "A Bioarchaeological Analysis of Health, Subsistence, and Funerary Behavior in the Eastern Mediterranean Basin: A Case Study from Alepotrypa Cave, Diros, Greece" (diss. Univ. of Iowa).

———. 2003. "Stable Isotope Analysis in Neolithic Greece and Possible Implications on Human Health," *International Journal of Osteoarchaeology* 13, pp. 314–324.

Papathanasiou, A., E. Zachou, and M. P. Richards. 2009. "Bioarchaeological Analysis of the Human Osteological Material from Proskynas, Lokris," in Schepartz, Fox, and Bourbou 2009, pp. 223–235.

Petroutsa, E. I., and S. K. Manolis. 2010. "Reconstructing Late Bronze Age Diet in Mainland Greece Using Stable Isotope Analysis," *JAS* 37, pp. 614–620.

Petroutsa, E. I., M. P. Richards, L. Kolonas, and S. K. Manolis. 2009. "Isotope Paleodietary Analysis of Humans and Fauna from the Late Bronze Age Site of Voudeni," in Schepartz, Fox, and Bourbou 2009, pp. 237–243.

Phenice, T. W. 1969. "A Newly Developed Visual Method of Sexing the *Os Pubis,*" *American Journal of Physical Anthropology* 30, pp. 297–301.

Piteros, C., J.-P. Olivier, and J. L. Melena. 1990. "Les inscriptions de linéaire B des nodules de Thèbes (1982): La fouille, les documents, les possibilities d'interpretation," *BCH* 114, pp. 103–184.

Richards, M. P., and R. E. M. Hedges. 2003. "Variations in Bone Collagen δ^{13}C and δ^{15}N Values of Fauna from Northwest Europe over the Last 40,000 Years," *Palaeogeography, Palaeoclimatology, Palaeoecology* 193, pp. 261–267.

Schepartz, L. A., S. C. Fox, and C. Bourbou, eds. 2009. *New Directions in the Skeletal Biology of Greece* (*Hesperia* Suppl. 43, OWLS 1), Princeton.

Schepartz, L. A., S. Miller-Antonio, and J. M. A. Murphy. 2009. "Differential Health among the Mycenaeans of Messenia: Status, Sex, and Dental Health at Pylos," in Schepartz, Fox, and Bourbou 2009, pp. 155–174.

Schoeninger, M. J. 1989. "Reconstructing Prehistoric Human Diet," in *The Chemistry of Prehistoric Human Bone,* ed. T. D. Price, Cambridge, pp. 38–67.

Schoeninger, M. J., M. J. DeNiro, and H. Tauber. 1983. "Stable Nitrogen Isotope Ratios of Bone Collagen Reflect Marine and Terrestrial Components of Prehistoric Human Diet," *Science* 220, pp. 1381–1383.

Schoeninger, M. J., and K. Moore. 1992. "Bone Stable Isotope Studies in Archaeology," *Journal of World Prehistory* 6, pp. 247–296.

Sherratt, A. G. 1981. "Plough and Pastoralism: Aspects of the Secondary Products Revolution," in *Pattern of the Past. Studies in Honour of David Clarke,* ed. I. Hodder, G. Isaac, and N. Hammond, Cambridge, pp. 261–305.

Suchey, J. M., S. Brooks, D. V. Wisely, and D. Katz. 1986. "Evaluation of the Todd and McKern-Stewart Methods for Aging the Male *Os Pubis,*" in *Forensic Osteology: Advances in the Identification of Human Remains,* ed. K. J. Reichs, Springfield, Ill., pp. 33–67.

Suchey, J. M., P. A. Owens, D. V. Wisely, and T. T. Noguchi. 1984. "Skeletal Aging of Unidentified Persons," in *Human Identification: Case Studies in Forensic Anthropology,* ed. T. A. Rathbun and J. E. Buikstra, Springfield, Ill., pp. 278–297.

Todd, T. W. 1920. "Age Changes in the Pubic Bone: I. The Male White Pubis," *American Journal of Physical Anthropology* 3, pp. 285–334.

Treuil, R., P. Darque, J.-C. Poursat and G. Touchais. 1996. *Les civilisations égéennes du Neolithique et de l'Âge du Bronze,* Athens.

Triantaphyllou, S. 1999. "A Bioarchaeological Approach to Prehistoric Cemetery Populations from Western and Central Greek Macedonia" (diss. Univ. of Sheffield).

Triantaphyllou, S., M. P. Richards, G. Touchais, A. Philippa-Touchais, and S. Voutsaki. 2006. "Analyses of Middle Helladic Skeletal Material from Aspis, Argos, 2: Stable Isotope Analysis of Human Remains," *BCH* 130, pp. 627–637.

Triantaphyllou, S., M. P. Richards, C. Zerner, and S. Voutsaki. 2008. "Isotopic Dietary Reconstruction of Humans from Middle Bronze Age Lerna, Argolid, Greece," *JAS* 35, pp. 3028–3034.

Vermeule, E. 1983. *Greece in the Bronze Age,* Athens.

Vickery, K. F. 1936. *Food in Early Greece* (Illinois Studies in the Social Sciences 20.3), Urbana, Ill.

Vika, E. 2011. "Diachronic Dietary Reconstructions in Ancient Thebes, Greece: Results from Stable Isotope Analyses," *JAS* 38, pp. 1157–1163.

Webb, P. A. O., and J. M. Suchey. 1985. "Epiphyseal Union of the Anterior Iliac Crest and Medial Clavicle in a Modern Multiracial Sample of American Males and Females," *American Journal of Physical Anthropology* 68, pp. 457–466.

CHAPTER 7

Dietary Reconstruction at the Geometric-Period Burial Site of Ayios Dimitrios

by Eleni Panagiotopoulou and Anastasia Papathanasiou

The Protogeometric and Geometric periods (ca. 1100 to 700 B.C.) comprise the EIA in Greece, which is a formative period in Greek history when major transformations took place, many of which served as foundations for later Greek society and the rise of the ideology of the *polis*. After the fall of the Bronze Age Mycenaean palatial system, social hierarchy broke down and populations were more scattered and mobile and less culturally unified than before, but at the same time, the primary Greek institutions developed as well as the alphabet, trade, colonies, and a new social structure.[1] Nevertheless, this period is understudied archaeologically relative to the periods before and after, and it is almost unknown bioarchaeologically. This period has been known for decades as the "Dark Ages," a name echoing precisely the lack of archaeological and other research.

Evidence from this time period increases as more research focuses on it; new data, however, come not from settlement sites but from cemeteries. This study focuses on a Sub-Protogeometric III to Late Geometric cemetery (850–740 B.C.) that provides the opportunity to study a sizable population from a necropolis that was active for more than 100 years. The corresponding settlement has not yet been located. The nearest known ancient settlement (about one kilometer away) is Thronion, which is referred to in ancient literature[2] and flourished during the Classical period.[3] The Geometric cemetery of Ayios Dimitrios is archaeologically very significant, because it bridges the temporal gap for habitation in this area between the Mycenaean and the Classical periods and provides key information about cemetery structure and organization, burial practices, and pottery typology for this time period. Thus, this site fills the void in Coldstream's map for this area, which Desborough[4] characterized as "no man's land" due to the lack of any archaeological evidence.[5]

MATERIAL

The site of Ayios Dimitrios is located in Phthiotis, about 40 km southeast of Lamia, at the base of Mt. Knimis, and less than a kilometer from the present coast (which was significantly closer in antiquity). Rescue excavations conducted by the 14th Ephorate of Prehistoric and Classical Antiquities,

1. Snodgrass 1971; Desborough 1972; Coldstream 1977; Hornblower and Spawforth 2003.
2. Homer (*Il.* 2.533); Thuc. 2.26.2; Strabo 9.4.4 C426.
3. Papakonstantinou and Sipsi 2009.
4. Desborough 1972.
5. Papakonstantinou and Sipsi 2009.

during the expansion of the E-65 highway, revealed an extended necropolis underneath a Byzantine building, possibly with 70 graves belonging to both children and adults, reflecting a settlement of similar magnitude. The graves are organized in groups, which may correspond to burial mounds or burial circles, as is implied by their circular arrangement and the remnants of stone pilings and a partial enclosure.[6]

The osteological series of Ayios Dimitrios consists mainly of undisturbed but fragmentary skeletal remains derived from, in most cases, single primary burials, and from four cremations. The analysis of the material followed the standard procedures of Buikstra and Ubelaker for complete skeletons and commingled remains.[7] In order to determine the health status of the Ayios Dimitrios population, a number of health/stress indicators were observed on the osteological assemblage. The basic demographic parameters of the population—minimum number of individuals (MNI), age, sex, and stature—were assessed. As the sample was fragmentary and incomplete, a number of different methods were used to estimate age-at-death and determine sex. The detailed methodology is presented in Papathanasiou et al.[8] Pathological conditions were identified following Ortner and Putschar,[9] Resnick,[10] and Buikstra and Ubelaker.[11] Observations were made under normal light conditions without the aid of microscopy. The pathological conditions were recorded by presence-absence, and the percentages reflect the observed over the observable, which is the number of elements or individuals that display the characteristic over the total number of the same elements or individuals where the condition was observable.

A summary of the results of the osteological analysis[12] is presented here. Based on counts of skeletal elements, there is a minimum of 51 individuals, consisting of 28 adults and 23 subadults, including 6 males or probable males, and 11 females or probable females. The mean adult age-at-death is 36.3 years. The ages range from newborn to over 55 years. The mortality curve for the total excavated group exhibits one very significant peak at ages 0–5 years that is related to infant mortality. Eighteen out of the 51 total individuals of the group were under the age of 2 years at the time of death, with 11 being newborns or preterm births (including two 7-month-old fetuses). The average stature for males is 165.3 ± 2.6 cm (n = 3), and for females it is 153.8 ± 4.9 cm (n = 4). The most frequently observed pathological conditions in this group are cribra orbitalia, with a prevalence of 27.8% or 5 out of 18 individuals that exhibit lesions in either a mild or severe form. Also, another 4 out of 23 individuals (17.4%) exhibit porotic hyperostosis. Both cribra orbitalia and porotic hyperostosis are related to anemic conditions of genetic or acquired origin. Furthermore, two cases of hemorrhage and hematomas were observed in two infants (6 months and 8–9 months), implying possible scurvy or other systemic malnutrition. Dental pathology at Ayios Dimitrios is represented primarily by a high prevalence of antemortem tooth loss of 46.5% (93 out of 200 alveolar spaces), a high prevalence of dental caries of 9.4% (33 out of 352 teeth), and by enamel hypoplasias with a very low prevalence of 2.3% (7 out of 352 teeth).

6. Papakonstantinou and Sipsi 2009.
7. Buikstra and Ubelaker 1994.
8. Papathanasiou et al. 2013, pp. 2925–2926.
9. Ortner and Putschar 1985.
10. Resnick 1995.
11. Buikstra and Ubelaker 1994.
12. Papathanasiou et al. 2013, pp. 2926–2928.

The main focus of this chapter is the dietary reconstruction of the excavated population using stable carbon and nitrogen isotope analysis of human bone collagen. This reconstruction provides direct information about the population's diet and possible weaning age and helps to assess the dietary habits, subsistence strategy, and economy of this community. More specifically, attention is given to intra-population differences related to gender and/or age. The following questions are addressed: Could diet be regulated by social means such as gender or age? What do diet and certain pathological conditions imply for the subsistence strategy adopted by this population? Is there the same level of social diversification in the Geometric period as in the LBA? In addition to the human bones, a number of faunal bones from a nearby context have also been analyzed with stable isotopes in order to establish a baseline for the food-web on which humans relied.[13]

METHODS

A total of 36 human and 10 faunal samples were analyzed. The analysis of stable carbon and nitrogen isotopes for dietary reconstruction took place at the Max Planck Institute for Evolutionary Anthropology, Department of Human Evolution, under the supervision of Michael P. Richards. Samples for isotopic analysis were taken from 36 discrete individuals. Since there was no faunal osteological material from the site of Ayios Dimitrios, 10 faunal samples were obtained from the adjacent and almost contemporary site of Kynos, which presumably belonged to the same broad environment and ecosystem. The faunal samples that were obtained come from cattle, goat/sheep, turtle shell, and pig.

For the bone collagen extraction, the Richards and Hedges method was followed.[14] The demineralization started with the immersion of the samples in weak solution of hydrochloric acid (0.5 M HCl), and then they were heated at 70°C for 48 hours in order to be converted to solution. These solutions were Ezee filtered and Ultra filtered in order to remove molecular contaminants from the samples. The last step of this procedure is the lyofilisation–freeze drying of the resultant residue, a process whereby the collagen in solution becomes solid. From the solid collagen, approximately 0.5 mg aliquots were weighed into tin capsules for the measurement of the stable carbon and nitrogen isotopes on the continuous flow GC-IRMS (Thermo Finnigan Delta XP spec with a Flash 1112 elemental analyser) at the laboratory at the Max Planck Institute.

RESULTS

The percentage collagen, carbon, and nitrogen yield, as well as the C:N ratio, were used as criteria to distinguish the diagenetically altered samples from the ones that maintained unaltered collagen (see Tables 7.1, 7.2). Eleven human samples were excluded from this study as they were outside the range of acceptable collagen. The rejected samples have undergone postmortem diagenesis, and they would yield incorrect δ values.

13. Hedges and Reynard 2007.
14. Richards and Hedges 1999a, p. 722; Müldner and Richards 2005; Richards, Fuller, and Molleson 2006; Chap. 2.

TABLE 7.1. STABLE ISOTOPE VALUES FOR HUMANS FROM AYIOS DIMITRIOS

S-EVA	*Ref. No.*	*Age/Sex**	*C:N*	*$\delta^{13}C$ (‰ VPDB)*	*Mean $\delta^{13}C$*	*$\delta^{15}N$ (‰ AIR)*	*Mean $\delta^{15}N$*	*% C*	*% N*	*% Coll.*
12064	T49	Ad	No collagen							0.19
12065a	T57	7 mo fetus	3.29	-18.64	-18.84	10.70	10.75	43.82	15.53	4.79
12065b			3.23	-19.03		10.79		42.75	15.42	
12066a	T59	7 mo fetus	3.33	-18.97	-18.83	8.54	8.33	43.48	15.25	1.47
12066b			–	-18.68		8.12		–	–	
12067a	T15	0 mo	3.36	-19.59	-19.72	10.00	9.93	41.82	14.50	1.85
12067b			–	-19.85		9.86				
12068a	T32	0 mo	3.41	-19.85	-19.71	9.33	9.19	39.15	13.39	1.13
12068b			–	-19.56		9.04		–	–	
12069a	T47	0 mo preterm	3.64	–	–	–	–	–	–	0.28
12069b			–	–		–		–	–	
12070a	T52	0 mo	3.40	-19.54	-19.83	9.75	9.64	43.05	14.76	0.76
12070b			3.34	-20.11		9.52		41.98	14.65	
12071a	T55	0 mo	3.48	-19.52	-19.28	8.99	8.53	42.70	14.31	0.50
12071b			–	-19.03		8.07		–	–	
12072a	T29	6 mo	3.29	-18.71	-19.02	8.84	8.88	43.98	15.57	2.44
12072b			3.21	-19.32		8.91		42.80	15.53	
12073a	T19	6 mo	3.40	-19.47	-19.83	11.36	11.13	37.13	12.76	1.84
12073b			3.45	-20.18		10.90		35.30	11.95	
12074a	T31	8 mo	3.43	-19.27	-19.73	10.40	10.38	35.23	11.97	0.47
12074b			3.41	-20.18		10.35		34.36	11.76	
12075a	T28	8–9 mo	3.28	-18.88	-18.72	10.47	10.32	41.17	14.66	3.93
12075b			–	-18.56		10.17		–	–	
12076a	T22	9 mo	3.26	-19.22	-19.39	10.71	10.69	43.62	15.63	3.08
12076b			3.22	-19.55		10.67		43.07	15.61	
12077a	T30	1 y	3.26	-19.03	-19.10	11.60	11.53	42.15	15.07	8.56
12077b			3.18	-19.17		11.45		41.72	15.33	
12078a	T44	2+ y	3.55	-20.03	-20.25	8.41	8.25	41.73	13.70	0.40
12078b			3.49	-20.46		8.08		40.03	13.39	
12079	T50	5 y	No collagen							0.00
12080a	T35	7 y	4.12	–	–	–	–	–	–	0.25
12080b			–	–		–		–	–	
12081	T51	10–15 y	No collagen							0.00
12082a	T27	16 y	3.27	-19.46	-19.57	8.13	8.11	42.73	15.25	1.69
12082b			3.23	-19.67		8.08		42.07	15.19	
12083	T26	YA F	No collagen							0.11
12084a	T40b	ca. 20 y F	3.37	-19.94	-20.12	6.91	6.79	40.34	13.97	1.08
12084b			3.32	-20.29		6.66		39.49	13.89	
12085a	T37	20–25 y F	3.46	-19.61	-19.89	9.02	8.83	33.39	11.25	1.12
12085b			3.42	-20.16		8.64		35.13	11.98	

TABLE 7.1—*Continued*

S-EVA	*Ref. No.*	*Age/Sex**	*C:N*	*δ¹³C (‰ VPDB)*	*Mean δ¹³C*	*δ¹⁵N (‰ AIR)*	*Mean δ¹⁵N*	*% C*	*% N*	*% Coll.*
12086	T48	20–30 y				No collagen				0.05
12087a	T36	Ad F	–	–	-19.23		8.66	–	–	2.83
12087b			3.21	-19.23		8.66		37.58	13.65	
12088a	T39	Ad F	3.40	-18.69	-18.98	8.87	8.77	43.01	14.77	0.95
12088b			3.35	-19.27		8.66		41.78	14.54	
12089a	T40a	Ad F	3.27	-19.84	-20.23	6.38	6.25	32.32	11.52	2.11
12089b			3.25	-20.61		6.12		32.51	11.68	
12090a	T45	Ad	3.48	-19.73	-20.07	7.80	7.62	28.88	9.69	0.83
12090b			3.39	-20.41		7.43		27.07	9.31	
12091a	T46	Ad	–	–	–	–	–	–	–	0.15
12091b			–	–		–		–	–	
12092a	T53	Ad	3.38	-19.35	-19.50	9.25	9.18	43.24	14.94	1.22
12092b			3.30	-19.64		9.11		42.82	15.13	
12093a	T42	30+ y	3.36	-19.31	-19.62	9.07	9.11	42.30	14.70	1.98
12093b			3.28	-19.93		9.15		41.33	14.72	
12094	T38	30–35 y M				No collagen				0.00
12095a	T21	35–40 y M	3.50	-19.45	-19.64	8.65	8.61	42.91	14.29	0.48
12095b			3.43	-19.82		8.57		42.47	14.44	
12096a	T20	40 y F	3.38	-19.46	-19.84	8.73	8.71	43.91	15.14	1.02
12096b			3.29	-20.21		8.68		42.46	15.05	
12097a	T33	ca. 40 y F	3.45	-19.41	-19.86	8.74	8.64	83.81	28.35	0.42
12097b			3.46	-20.31		8.54		41.97	14.15	
12098a	T43	45+ y M	3.43	-19.54	-19.80	8.41	8.39	43.43	14.77	0.83
12098b			3.36	-20.06		8.36		42.24	14.69	
12099	T54	45+ y	–	–	–	–	–	–	–	0.00

*Ad = adult; YA = young adult; mo = month; y = years; M = male; F = female.

Faunal Samples

The $\delta^{13}C$ and $\delta^{15}N$ faunal values vary between -18.2‰ ≥ $\delta^{13}C$ ≥ -22.6‰ and 7.7‰ ≥ $\delta^{15}N$ ≥ 4.6‰, with mean values of $\delta^{13}C$ = -19.9‰ ± 1.4‰ and $\delta^{15}N$ = 6.0‰ ± 1.0‰ (see Table 7.2, Fig. 7.1).

Human Samples

The human values vary between -18.7‰ ≥ $\delta^{13}C$ ≥ -20.3‰ and 11.5‰ ≥ $\delta^{15}N$ ≥ 6.3‰ (see Table 7.1, Fig. 7.2). The human samples can be separated into two groups: the infants (13 individuals: two preterm fetuses in the 7th month of gestation, four newborns, two infants of 6 months, three infants of 8–9 months, one infant of 1 year, and one infant of 2+ years); and the adults (12 individuals: 16–45 years).

Accordingly, the δ values form two groups reflecting the two age categories (infants and adults). Thus, the δ values of the infants are -18.7‰ ≥

TABLE 7.2. STABLE ISOTOPE VALUES FOR FAUNA FROM AYIOS DIMITRIOS

S-EVA	*Ref. No*	*Species*	*C:N*	*$\delta^{13}C$ (‰ VPDB)*	*Mean $\delta^{13}C$*	*$\delta^{15}N$ (‰ AIR)*	*Mean $\delta^{15}N$*	*% C*	*% N*	*% Coll.*
12124a	AB1	Turtle shell	3.50	-22.50	-22.56	5.35	5.30	43.15	14.38	0.53
12124b			–	-22.61		5.24		–	–	
12125a	AB2	Sheep/goat radius	3.27	-20.64	-20.70	6.24	6.15	41.40	14.78	3.36
12125b				-20.76		6.06		–	–	
12126a	AB3	Pig maxilla	3.35	-20.98	-21.01	7.46	7.37	31.02	10.79	1.70
12126b			–	-21.04		7.28		–	–	
12127a	AB4	Cattle shaft fragments	3.25	-19.21	-19.36	7.71	7.65	34.96	12.56	5.80
12127b			–	-19.51		7.58		–	–	
12128a	AB5	Sheep/goat mandible	3.24	-18.18	-18.38	5.48	5.50	43.79	15.76	3.57
12128b			–	-18.58		5.52		–	–	
12129a	AB6	Sheep humerus	3.29	-19.01	-19.07	4.64	4.61	42.82	15.17	5.49
12129b			–	-19.13		4.57		–	–	
12130a	AB7	Pig maxillary tooth	3.31	-21.02	-21.24	4.86	4.83	43.15	15.21	1.83
12130b			–	-21.45		4.80		–	–	
12131a	AB8	Sheep mandible	3.33	-18.16	-18.23	5.66	5.86	42.95	15.05	4.06
12131b			3.39	-18.29		6.06		43.60	14.98	
12132a	AB9	Sheep metacarpal	3.25	-18.51	-18.77	6.58	6.45	43.37	15.59	4.20
12132b			–	-19.03		6.32		–	–	
12133a	AB10	Cattle 2nd phalanx	3.24	-19.28	-19.20	6.32	6.29	43.66	15.70	4.39
12133b			3.25	-19.11		6.25		43.73	15.69	

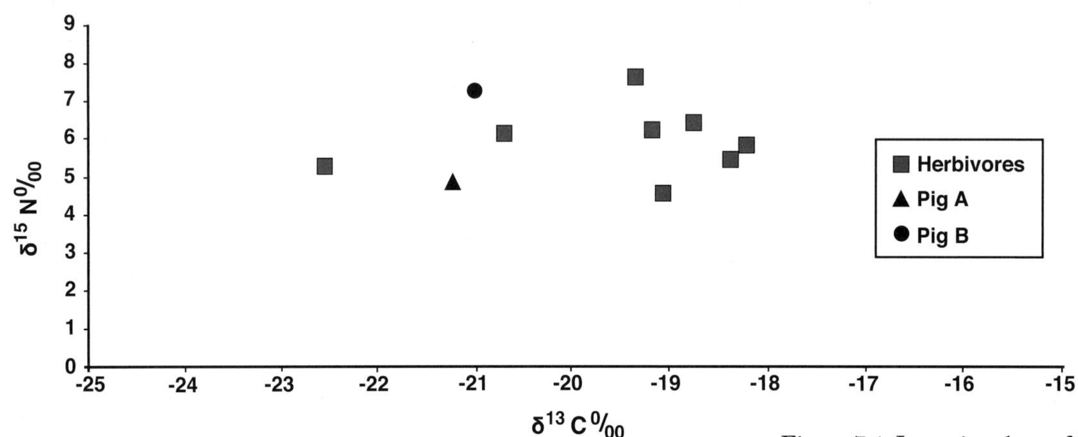

Figure 7.1. Isotopic values of faunal samples from Ayios Dimitrios

$\delta^{13}C \geq -20.3‰$ and $11.5‰ \geq \delta^{15}N \geq 8.3‰$, with a mean value of $\delta^{13}C = -19.4‰ \pm 0.5‰$ and $\delta^{15}N = 9.8‰ \pm 1.1‰$. The adults' values vary between $-19.0‰ \geq \delta^{13}C \geq -20.2‰$ and $9.2‰ \geq \delta^{15}N \geq 6.3‰$, with mean values of $\delta^{13}C = -19.7‰ \pm 0.4‰$ and $\delta^{15}N = 8.3‰ \pm 0.9‰$ (see Table 7.1).

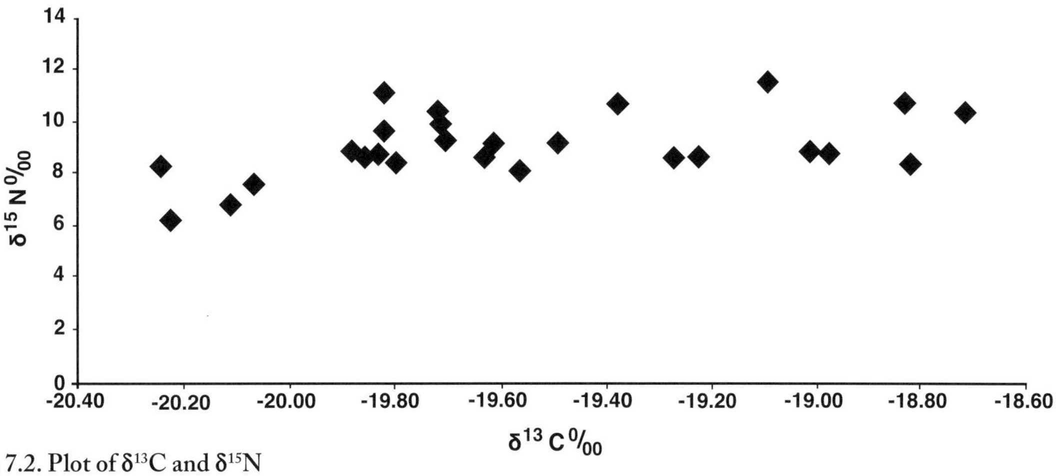

Figure 7.2. Plot of $\delta^{13}C$ and $\delta^{15}N$ values of human samples from Ayios Dimitrios

DISCUSSION

This section will focus first on the results from the stable isotope analysis of the faunal samples, which provide the baseline reference. The interpretation of the results from the human samples will follow, culminating with human health and socioeconomic implications.

Faunal/Human Samples

The isotope values of all the animals that were analyzed are plotted in Figure 7.1. The herbivore values, as presented in the results section, indicate C_3 plant protein consumption, as was expected. The two pigs have δ values that are similar to herbivores, which may imply that the pigs were free-ranging and not feeding on human refuse.[15]

Assuming that the aforementioned animal isotopic values represent the values of the food-web upon which humans were feeding, two observations can be made: (1) in terms of $\delta^{13}C$, the human isotopic values are in the same range as a C_3 diet, like the faunal values; and (2) in terms of $\delta^{15}N$ values, the two groups (adults vs. faunal) exhibit a trophic level difference (Fig. 7.3), which means that animal meat protein and/or dairy products were consumed. It is not possible, however, to distinguish between the two sources of protein, because they exhibit the same isotopic values.

Human Diet

A C_3 diet is indicated by the range of the human and faunal isotopic values of carbon and nitrogen (see Table 7.1). On the chart (Fig. 7.4) of the isotopic composition of foodstuffs,[16] the gray area corresponds to the human values, and the dashed-line area represents the faunal ones. According to the four extreme types of diet determined by Richards and Hedges (Fig. 7.5),[17] the human isotopic values indicate the consumption of mostly plant and some animal protein. Given that the $\delta^{15}N$ values are relatively high, it can be inferred that leguminous plants formed a minor part of the

15. See also Chap. 3.
16. Vika and Richards 2003, p. 84.
17. Richards and Hedges 1999b, p. 719.

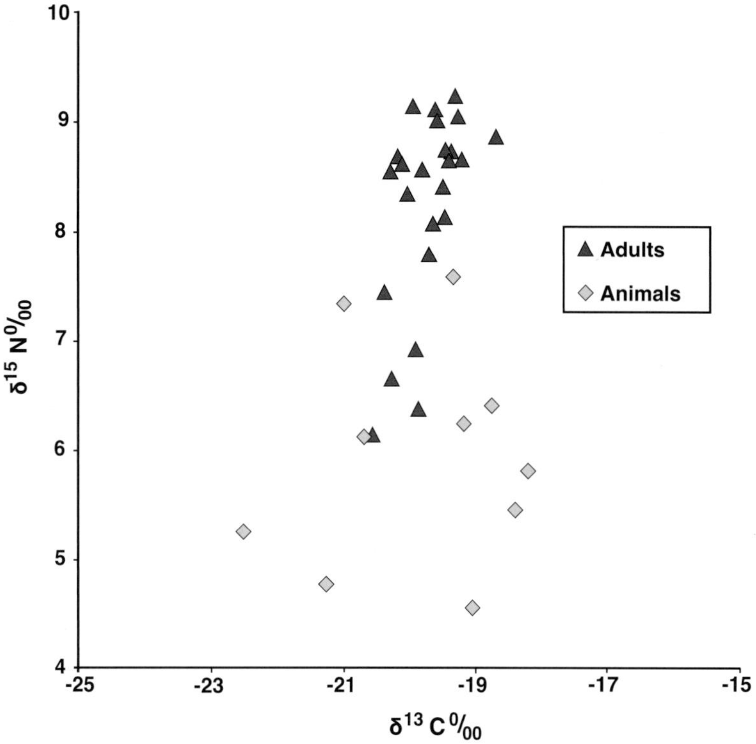

Figure 7.3. Comparative depiction of the isotopic values of human and faunal samples from Ayios Dimitrios

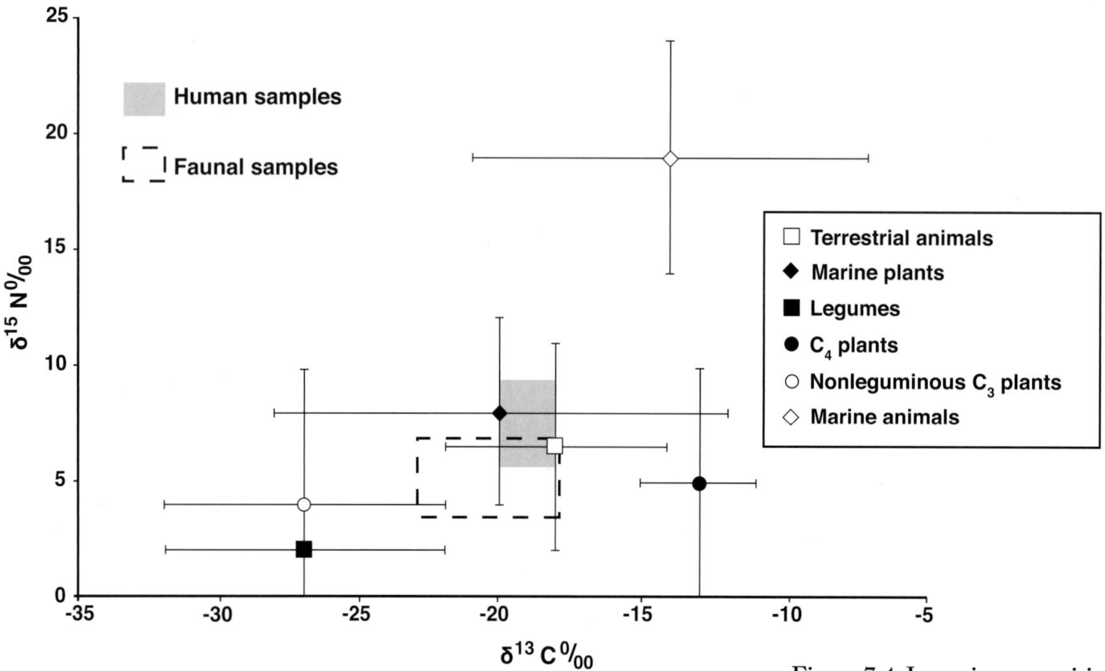

Figure 7.4. Isotopic compositions.
After Vika and Richards 2003, p. 84, fig. 1

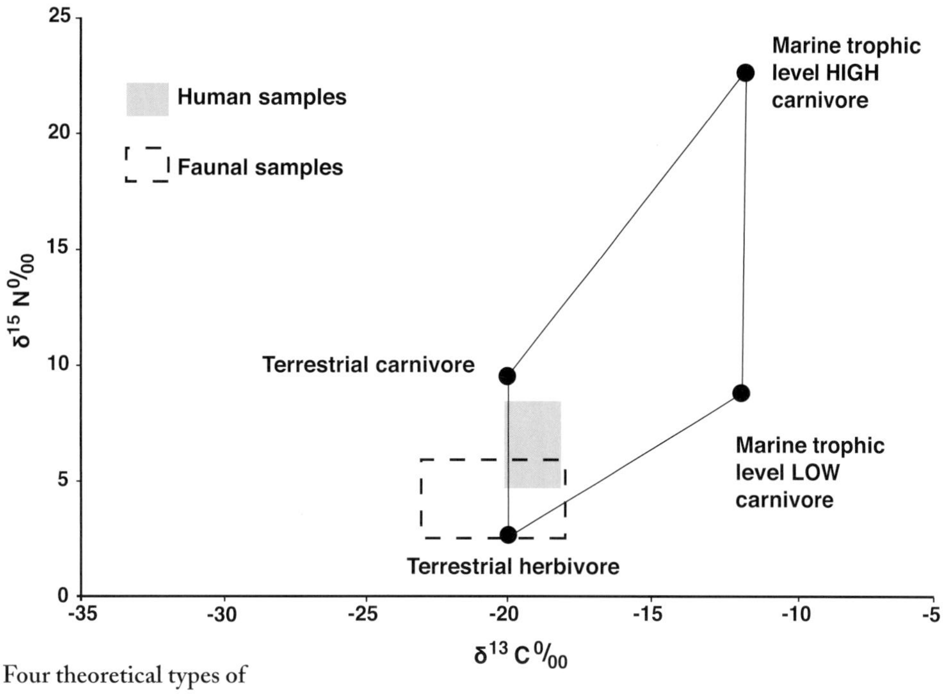

Figure 7.5. Four theoretical types of human diet and their isotopic values. After Richards and Hedges 1999a, p. 719, fig. 2

diet, which was mainly focused on cereals, fruits, and vegetables. Seafood was in non-detectable quantities, if any, and therefore did not constitute a regular part of the daily diet.

The isotopic signal for $\delta^{13}C$ indicates minimal amounts of C_4 protein if the cut-off point for C_4 consumption is set at -19.00‰. At Ayios Dimitrios, the human $\delta^{13}C$ samples range between -20.30‰ and -18.70‰. Compared with EIA sites in northern Greece that have exhibited a C_4 signal, such as Korinos (-19.88‰ to -18.62‰), Spathes (-19.30‰ to -18.19‰), and Rymnio (-18.13‰ to -16.48‰), there is little overlapping.[18] Therefore, the consumption of C_4 resources does not exhibit a systematic pattern of cultivation and consumption but rather appears to be sporadic in some individuals who may feed on animals that range and feed on larger biomasses and randomly include C_4 plants in their diets.

Ayios Dimitrios clearly represents a predominantly terrestrial diet consistent with an emphasis on the consumption of terrestrial plant protein with a minor animal protein component, which cannot be distinguished as to whether it is derived from meat or dairy products. Marine-derived protein does not seem to have made any major contribution to the diet of this population even though Ayios Dimitrios is a coastal site and the sea was a known trade route during the EIA.[19] Obviously, this population chose to focus on agricultural products for their subsistence, which comprise the bulk of their diet and, apparently, their economy, while marine resources, even if they were used, would have been in very small quantities and therefore untraceable by stable isotope analysis.[20]

18. Triantaphyllou 2001.
19. Kotsonas 2009.
20. See also Chap. 11.

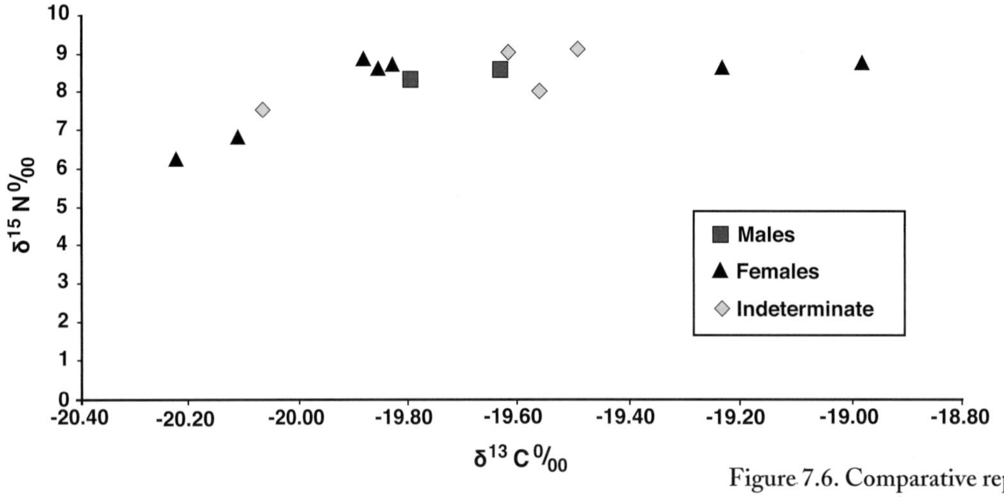

Figure 7.6. Comparative representation of the isotopic values of carbon and nitrogen between males and females from Ayios Dimitrios

Sex Differences

One of the main purposes of this study was to assess if there is the same level of social diversification as documented for the LBA. The mortuary data suggest homogeneity and an inclusiveness with respect to social, gender, and age categories. Additionally, the pathological conditions show no differences between males and females.[21] The diet was a third axis that could be explored. In the Bronze Age, although there is generally homogeneity in the diet, more variability was detected at Mycenae as individuals in the royal tombs (Grave Circles A and B) demonstrate significantly higher levels of animal and fish protein in their diets than individuals buried in chamber tomb cemeteries.[22] Schepartz and Papathanasiou arrived at similar conclusions in a study at Pylos,[23] where there is a statistically significant difference in the amount of animal protein consumed, with females exhibiting less animal protein in their diet. The immediate conclusions are that the life experiences were different for Mycenaean males and females and that diet reflects social diversification.[24]

For Ayios Dimitrios, the isotopic values of males and females are plotted in Figure 7.6. The samples belong to two males, seven females, and four unidentified individuals. Statistical analysis of the sample means revealed that there are no statistically significant differences (Student's t-test, $t = 1.5$, $p>0.37$) when males and females were compared, indicating similar access to all food categories. This finding may suggest more homogeneity within the community with no apparent exclusions, but the sample sizes are admittedly small.

Comparing Infants to Adults

Very few isotopic studies have dealt with weaning age in antiquity.[25] The Ayios Dimitrios assemblage consists of a sizable group of infants and children. Although there are 12 individuals below the age of 1, there is only one individual between the ages of 2 and 5 (the weaning and post-weaning age), so it is difficult to determine weaning practices with confidence. A trophic level effect occurs among the two human subgroups (infants and adults) (Fig. 7.7). The difference in the mean values of nitrogen between the two groups is about 1.5‰, and the differences between the maximum and

21. Papathanasiou et al. 2013, p. 2931.

22. Richards and Hedges 1999a.

23. Papathanasiou, Schepartz, and Richards 2006.

24. Schepartz, Miller-Antonio, and Murphy 2009.

25. Dupras, Schwarcz, and Fairgrieve 2001; Prowse 2001; Fuller et al. 2006; Keenleyside, Schwarcz, and Panayotova 2006; Triantaphyllou et al. 2008; Chap. 9.

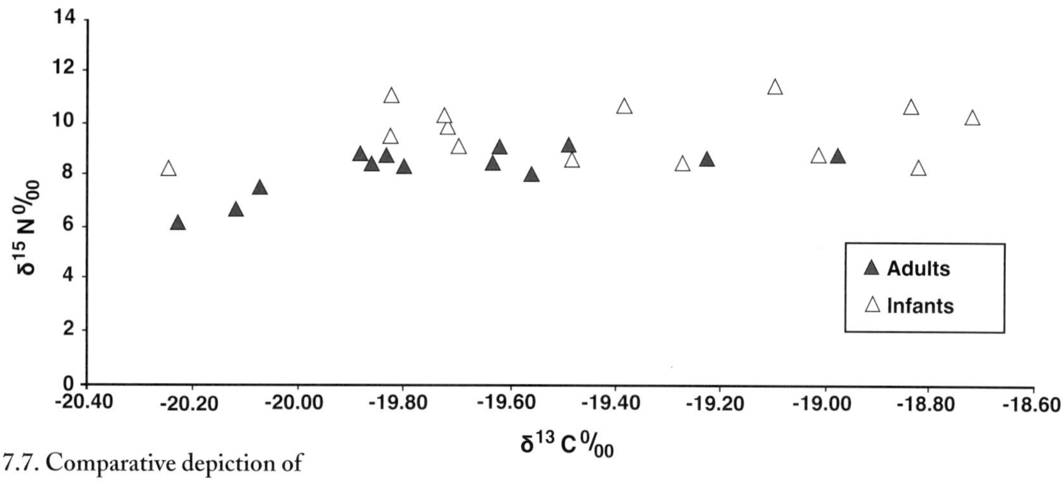

Figure 7.7. Comparative depiction of isotopic values of adults and infants from Ayios Dimitrios

minimum values of nitrogen in the two groups is about 2.4‰ and 2.0‰, respectively. This means that the infants are one trophic level above the adults, leading to the conclusion that the infants were still breastfeeding. The 2-year-old child, however, had $\delta^{15}N$ values in the range of the adults, implying that this individual was weaned around the age of his/her death. The time that supplementary food was introduced to the infants' diet cannot be further discerned, however, because this is the only sample in this age-range with values similar to adults. The only observation that can be positively made is that at the age of 1, infants were still breastfeeding, and there is an indication (albeit from only one individual) that weaning was completed by the age of 2. Additionally, a 6-month-old infant has similar values to the adults, possibly indicating that this particular individual never breastfed. Perhaps this was a contributing factor in the death of this infant.

Summarizing all the available data, the Ayios Dimitrios population of 51 individuals falls within the expected range in terms of sex and age-at-death distribution, with high infant and perinatal mortality, and at the lower end for stature for both males and females. The sample exhibits multiple skeletal and dental stress indicators, including high prevalence of anemic conditions, especially cribra orbitalia, as well as osteoarthritis and dental infection rates. This pathological profile suggests a consistent pattern of population stress and suboptimal living conditions. The aforementioned observations relate to the health status, nutrition, and lifestyle of the population and indicate that this group was stressed by a long-term nutritional deficiency, most probably associated with a protein-poor diet. This finding is supported by the isotopic analysis, which indicated a C_3 diet primarily based on plant protein. Such a diet could lead to the observed nutritional deficiencies as it is poor in iron, essential vitamins, and amino acids, and it is highly cariogenic, which is consistent with the high prevalence of caries and antemortem tooth loss.

In terms of mortuary practice, although the Ayios Dimitrios graves stress individuality, they are also organized into groups, which may correspond to burial mounds or burial circles, so that the collective character, continuity, and lineage is emphasized. All sex and age categories are represented, implying family organization and inclusion. Subadult underrepresentation is consistently observed during the preceding LBA in all Mycenaean chamber tomb and tholos cemeteries, and it is characteristic of this time period, implying that young children and infants received differential funerary treatment during the LBA. It is significant to note that this burial custom changes with the advent of the Iron Age when, progressively, all age groups become more equally represented and there is no differential access to cemeteries.[26] Except for the mortuary practices mentioned above, lack of differentiation is also evident in the prevalence of pathological conditions in males and females, as well as in diet and access to animal protein.

CONCLUSION

The isotopic analysis of the population of Ayios Dimitrios indicates that the main sources of dietary protein for this assemblage were terrestrial plant and animal C_3 resources, despite the site's close proximity to the sea. The occasional use of marine resources and C_4 plants by this population cannot be totally eliminated as a possibility, as this would be invisible to isotope analysis.

The Ayios Dimitrios sample is of significant archaeological importance, as it provides information for human health status and dietary reconstruction based on carbon and nitrogen stable isotopes in this little-known time period. In conclusion, stable isotope analysis results, high infant mortality, low stature estimates, anemic and pathological lesions, as well as high dental infection rates, all point to a very stressed and malnourished population based on a protein-poor C_3 plant diet. The present study provides reference and motivation for future work on Iron Age populations, a time when living conditions seemed to deteriorate and social diversification appeared to decrease compared to the Bronze Age. Very few human osteological series have been published for this time period, but a number of studies are currently in progress, and, hopefully, the database on this period will grow significantly in the following years. At this point in time, the database is sparse and the population size is very small and may not be representative. It is therefore currently difficult to draw more generalized conclusions.

ACKNOWLEDGMENTS

We would like to thank the three anonymous reviewers for their valuable and constructive comments, which significantly improved this manuscript. We are grateful to Maria-Foteini Papakonstantinou and Maria Sipsi for permitting access to the osteological material from Ayios Dimitrios and providing all the means to facilitate the study of this material. We would also like to thank Konstantinos Mpeltsios, University of Ioannina, Department of Materials Science and Engineering, Ioannina, Greece, for giving the opportunity and the means to work on this project as a Master's thesis, and Mike Richards, Max Planck Institute for Evolutionary Anthropology, Department of Human Evolution, Leipzig, Germany, for accepting to work on the skeletal material for the Master's thesis in the lab of the Institute. Finally, we would like to thank Olaf Nehlich, Max Planck Institute for Evolutionary Anthropology, Department of Human Evolution, Leipzig, Germany, for the help he provided during the work at the lab.

26. Morris 1992; Pappi and Triantaphyllou 2011.

REFERENCES

Buikstra, J. E., and D. H. Ubelaker, eds. 1994. *Standards for Data Collection from Human Skeletal Remains. Proceedings of a Seminar at the Field Museum of Natural History, Organized by Jonathan Haas* (Arkansas Archeological Survey Research Series 44), Fayetteville.

Coldstream, J. N. 1977. *Geometric Greece,* London.

Desborough, V. R., d'A. 1972. *The Greek Dark Ages,* London.

Dupras, T. L., H. P. Schwarcz, and S. I. Fairgrieve. 2001. "Infant Feeding and Weaning Practices in Roman Egypt," *American Journal of Physical Anthropology* 115, pp. 204–212.

Fuller, B. T., T. I. Molleson, D. A. Harris, L. T. Gilmour, and R. E. M. Hedges. 2006. "Isotopic Evidence for Breastfeeding and Possible Adult Dietary Differences from Late/Sub-Roman Britain," *American Journal of Physical Anthropology* 129, pp. 45–54.

Hedges, R. E. M., and L. M. Reynard. 2007. "Nitrogen Isotopes and the Trophic Level of Humans in Archaeology," *JAS* 34, pp. 1240–1251.

Hornblower, S., and A. Spawforth. 2003. *Oxford Classical Dictionary,* 3rd ed. rev., New York.

Keenleyside, A., H. Schwarcz, and K. Panayotova. 2006. "Stable Isotopic Evidence of Diet in a Greek Colonial Population from the Black Sea," *JAS* 33, pp. 1205–1215.

Kotsonas, A. 2009. "Central Greece and Crete in the Early Iron Age," in *Αρχαιολογικό Έργο Θεσσαλίας και Στερεάς Ελλάδας 2. Πρακτικά επιστημονικής συνάντησης, Βόλος 16.3–19.3.2006,* ed. A. Mazarakis-Ainian, Volos, pp. 1051–1065.

Morris, I. 1992. *Death-Ritual and Social Structure in Classical Antiquity,* Cambridge.

Müldner, G., and M. P. Richards. 2005. "Fast or Feast: Reconstructing Diet in Later Medieval England by Stable Isotope Analysis," *JAS* 32, pp. 39–48.

Ortner, D. J., and W. G. J. Putschar. 1985. *Identification of Pathological Conditions in Human Skeletal Remains,* Washington, D.C.

Papakonstantinou, M.-F., and M. Sipsi. 2009. "Το γεωμετρικό νεκροταφείο στη θέση Άγιος Δημήτριος Καινούργιου Νομού Φθιώτιδος," in *Αρχαιολογικό Έργο Θεσσαλίας και Στερεάς Ελλάδας 2. Πρακτικά επιστημονικής συνάντησης, Βόλος 16.3–19.3.2006,* ed. A. Mazarakis-Ainian, Volos, pp. 1029–1042.

Papathanasiou, A., E. Panagiotopoulou, K. Beltsios, M.-F. Papakonstantinou, and M. Sipsi. 2013. "Inferences from the Human Skeletal Material of the Early Iron Age Cemetery at Agios Dimitrios, Fthiotis, Central Greece," *JAS* 40, pp. 2924–2933.

Papathanasiou, A., L. Schepartz, and M. P. Richards. 2006. "Stable Isotope Analyses in Neolithic and Bronze Age Greece: An Overview," *16th European Meeting of the Paleopathology Association,* Santorini.

Pappi, E., and S. Triantaphyllou. 2011. "Mortuary Practices and the Human Remains: A Preliminary Study of the Geometric Graves in Argos, Argolid," in *The "Dark Ages" Revisited. Acts of an International Symposium in Memory of William D. E. Coulson, University of Thesssaly, Volos, 14–17 June 2007,* ed. A. Mazarakis-Ainian, Volos, pp. 717–732.

Prowse, T. L. 2001. "Isotopic and Dental Evidence for Diet from the Necropolis of Isola Sacra (1st–3rd Centuries A.D.), Italy" (diss. McMaster Univ.).

Resnick, D. 1995. *Diagnosis of Bone and Joint Disorders,* 3rd ed., Philadelphia.

Richards, M. P., B. T. Fuller, and T. I. Molleson. 2006. "Stable Isotope Palaeodietary Study of Humans and Fauna from the Multi-Period (Iron Age, Viking, and Late Medieval) Site of Newark Bay, Orkney," *JAS* 33, pp. 122–131.

Richards, M. P., and R. E. M. Hedges. 1999a. "How Chemical Analysis of Human Bones Can Tell Us the Diets of People Who Lived in the Past (and Other Contributions)," in *Minoans and Mycenaeans: Flavours of Their Time. National Archaeological Museum, 12 July–27 November 1999,* ed. Y. Tzedakis and H. Martlew, Athens p. 214.

———. 1999b. "Stable Isotope Evidence for Similarities in the Types of Marine Foods Used by Late Mesolithic Humans at Sites along the Atlantic Coast of Europe," *JAS* 26, pp. 717–722.

Schepartz, L. A., S. Miller-Antonio, and J. M. A. Murphy. 2009. "Differential Health among the Mycenaeans of Messenia: Status, Sex, and Dental Health at Pylos," in *New Directions in the Skeletal Biology of Greece,* (*Hesperia* Suppl. 43, OWLS 1), ed. L. A. Schepartz, S. C. Fox, and C. Bourbou, Princeton, pp. 155–174.

Snodgrass, A. M. 1971. *The Dark Age of Greece: An Archaeological Survey of the Eleventh to the Eighth Centuries B.C.,* Edinburgh.

Triantaphyllou, S. 2001. *A Bioarchaeological Approach to Prehistoric Cemetery Populations from Central and Western Greek Macedonia* (*BAR-IS* 976), Oxford.

Triantaphyllou, S., M. P. Richards, C. Zerner, and S. Voutsaki. 2008. "Isotopic Dietary Reconstruction of Humans from Middle Bronze Age Lerna, Argolid, Greece," *JAS* 35, pp. 3028–3034.

Vika, E., and M. P. Richards. 2003. "Παλαιοδιατροφή: Τα σταθερά ισότοπα και η αρχαιολογία," *Τέχνες και αρχαιολογία* 89, pp. 82–86.

CHAPTER 8

Diet and the Polis: An Isotopic Study of Diet in Athens and Laurion during the Classical, Hellenistic, and Imperial Roman Periods

by Anna Lagia

The increasingly common use of techniques of stable isotope analysis in bioarchaeology has provided the means to explore dietary variability within regional and temporal frameworks. In combination with analyses of dietary practices derived from the same context as the samples, it is now possible to evaluate the data interregionally, chronologically, and comparatively, while providing a solid basis to use bioarchaeology to begin informing historical narratives current in the Mediterranean sphere. In the past 15 years the analysis of diet in diverse historic-period sites in the eastern Mediterranean has demonstrated an important regional differentiation within a rather narrow framework of a specific, primarily terrestrial, diet.[1] With the exception of the coastal site of Leptiminus in the Roman period, where the importance of fish farming in the economy is also traceable in human chemical signatures,[2] appraisal of the contribution of marine products in the overall diet, from an isotopic perspective, has been less clear.[3]

Investigation of diet isotopically in historic and prehistoric sites of the Greek mainland supports the picture of a temporal increase in the amount of protein in the diet through a trend of more positive $\delta^{15}N$ signatures.[4] The Early Iron Age, however, at least in Thessaly, appears impoverished compared to sites of the Late Bronze Age and the Classical period, matching accounts of contracted economic growth in this period.[5] In Thebes, a decline in nitrogen values in the Hellenistic compared to the Classical period was attributed to historical processes that weakened the mainland and led to a "reversion to a 'rural' type diet, similar to that of the Bronze Age."[6] Regarding the Roman period, no isotopic evidence concerning diet has been thus far available from the Greek mainland. By contrast, evidence on the Italian peninsula and Roman provinces abounds and informs our

1. Schwarcz, Dupras, and Fairgrieve 1999; Dupras, Schwarcz, and Fairgrieve 2001; Garvie-Lok 2001; Triantaphyllou 2001, pp. 133–139; Henneberg and Henneberg 2003; Prowse et al. 2004, 2005, 2008; Keenleyside, Schwarcz, and Panayotova 2006; Bourbou and Richards 2007; Craig et al. 2009; Rutgers et al. 2009; Keenleyside et al. 2009; Vika, Aravantinos, and Richards 2009; Bourbou et al. 2011, 2013; Vika 2011; Fuller et al. 2012; Killgrove and Tykot 2013; Gregoricka and Sheridan 2013.

2. Keenleyside et al. 2009.

3. Vika and Theodoropoulou 2012; cf. Bourbou et al. 2011.

4. Triantaphyllou 2001, p. 135; Vika 2011, p. 1161.

5. Papathanasiou et al. 2013, p. 2928.

6. Vika 2011, p. 1162.

picture of communities with a significant protein input, including large aquatic resources but also important differentiations in isotopic signatures because of local climatic and cultural-historical conditions.[7]

The present study investigates diet in Athens in the three periods—the Classical, Hellenistic, and Imperial Roman—in which the *polis* (city-state) comprised the basic organizing structure of communities in the Greek peninsula and beyond. Analysis focuses on two cemeteries at the periphery of the *asty* (urban center)—the Kerameikos and the Plateia Kotzia cemeteries—that, at least to some extent, seem to have been used for the burial of different socioeconomic strata. The study investigates diet in terms of stable carbon and nitrogen isotopes between cemeteries, sexes, and adult age groups. Moreover, it attempts to evaluate change in diet temporally within this broad temporal span in which fundamental changes took place in the sovereignty, administration, and organization of the *polis* that served as a model for other *poleis,* not least of which were those that were democratically governed.[8] In the appraisal of diet in the Classical period, a small sample from the cemetery of Laurion associated with people who worked in the silver mines near Athens is contrasted to the urban samples.

HISTORICAL BACKGROUND

Athens comprised one of the largest *poleis* in classical antiquity, in terms of both hinterland as well as estimated overall population. It covered an area approximately 2,400–2,500 km^2, corresponding to the entire Attic landscape,[9] and it is estimated that its population reached a peak before the Peloponnesian War (430 B.C.) of 250,000 to 350,000 residents, including approximately 60,000 adult male citizens and 100,000 slaves.[10] The end of the Peloponnesian war found the population decimated to approximately half of its size, leading in all likelihood to "the most conspicuous case of severe but strictly localized [demographic] attrition" in Greco-Roman history.[11] The 4th century B.C. comprises a period of economic recovery, although it never again reaches the scale of the preceding century. Social and political rearrangements are believed to have taken place in this period, allowing social groups to emerge that had hitherto been politically marginalized.[12] Athens, however, was no longer at the height of its supremacy—the term "empire," an attribute to describe its hegemony, is only applicable in the 5th century B.C.[13]

The Macedonian threat found Athens weakened from the rivalries with other city-states. Oligarchic and democratic regimes alternated; however, always under the close supervision or control of Macedonian forces. A significant migratory movement was initiated with the foundation of new cities in the East and Egypt, while changes in the political system sharpened social conflicts.[14] The battle of Pydna in 168 B.C. marked the end of the Macedonian domination in the peninsula and the increasing presence of Rome.[15] Athens eschewed wars that involved other Greek *poleis* and aligned itself with Rome, relishing its privileged status. The alignment of the *polis* with Mithridates against Rome in 86 B.C., however, ended in siege of the

7. E.g. Richards et al. 1998; Schwarcz, Dupras, and Fairgrieve 1999; Dupras, Schwarcz, and Fairgrieve 2001; Prowse et al. 2004, 2005, 2008; Fuller et al. 2006; Müldner and Richards 2007; Keenleyside et al. 2009; Craig et al. 2009; Redfern, Hamlin, and Beavan Athfield 2010; Redfern, Millard, and Hamlin 2012; Killgrove and Tykot 2013.
8. Hansen 2006, p. 124.
9. Garnsey 1988, p. 92.
10. Garnsey 1988, p. 89; Scheidel 2007, p. 45.
11. Scheidel 2007, p. 58.
12. Burke 1992, p. 201.
13. Morris 2009.
14. Habicht 1998, pp. 65–70; Migeotte 2009, p. 22.
15. Habicht 1998, p. 281.

city and its ultimate submission to the rising empire. In 27 B.C. Achaia officially became a province of the Roman Empire. The centuries that followed are characterized as grim by economic historians, notwithstanding periods of favoritism and recovery inferred by the construction of splendid public buildings, especially in the 2nd century A.D.[16] Behind their facades intense social discontent seems to be cloaked with mounting socioeconomic disparities.[17] Results of archaeological survey suggest increased nucleation and contraction of the countryside.[18] The raid of the Heruli in 267 A.D., conventionally taken to mark the end of Roman domination in the city, was detrimental for Athens, which was never again fully rebuilt, although it continued its course to the 6th century A.D., albeit with a much reduced area and restricted revival.[19]

In discussing the standard of living in Athens, few details are known of regional variability among population groups or changes through time. The prevailing view, until recently, that the world of the Greek city-states was relatively poor, has been effectively replaced by the perception that the economy of the Greco-Roman world was one of prosperity for a large part of the population.[20] Supporting this view is evidence on housing, mortality, morbidity, trade, food, sanitation, public health conditions, and welfare institutions.[21] In the same way, past assessments of a poor, primarily cereal-based diet in the Greek and Roman world have been supplanted by accounts of an extraordinarily diverse diet, ranging from various kinds of meat, fowl, and marine resources to vegetables, fruits, nuts, pulses, cereals, spices, and condiments, derived through farming, stock rearing, and hunting, or through importation of fresh, salted, or dried foods.[22] In addition the diet was supplemented with honey, eggs, dairy products, wine, and olive oil. This new understanding of ancient regimes, nevertheless, needs to be appraised temporally and in terms of its regional distribution and variation. For Morris[23] the ancient economy was not static, as has been frequently thought, but instead followed a much more complex pattern of cyclical growth and contraction.

THE CEMETERIES

The Kerameikos cemetery has been central to funerary analyses in Greece because of its rich textual and archaeological record. Systematic excavations at the site were initiated in 1870 by the Greek Archaeological Society and continued since 1913 by the German Archaeological Institute. These excavations have identified part of one of the most important cemeteries in Athens, used from the Submycenaean to late Roman periods.[24] The importance of the cemetery, as attested in written sources,[25] was related to its location outside the Dipylon and Sacred Gates, both well-known for

16. Day 1942, pp. 165–169.
17. Day 1942, 272; Cartledge 1997, p. 10.
18. Alcock 1993, p. 78.
19. Zachariadou 2008; Eleutheratou 2008.
20. Morris, Scheidel, and Saller 2007.
21. Morris 2004, 2005; Kron 2005, 2011; Silver 2007; Ober 2010; Scheidel 2010.
22. Dalby 1996, pp. 57–92; Kron 2012; for comparative evidence from the zooarchaeological record in Athens, see MacKinnon 2014.
23. Morris 2004, 2005.
24. Knigge 1991, pp. 14–34.
25. *Thuc.* 2.34.5; Xenophon *Hell.* 2.4.33; Paus. 1.29.3, 36.3.

their association with major city processions. Extensive excavations at the Kerameikos throughout the 20th century have revealed magnificent burial monuments along the two main roadways that are identified with burials of eminent politicians, families, and war dead.[26] A correlation between the magnificence of these burials and their location along the main roadways is evident. Relative proximity to the Athenian city walls was important in the positioning of high-status burials. This has been confirmed by recent excavations in the outlying areas of the Kerameikos cemetery in which the absence of luxurious monuments or monumental funerary stelae is partially explained by the distance between the burials and the nucleus of the cemetery near the city walls.[27]

The preservation of skeletal material from diverse grave contexts, believed to represent upper social strata, and the tight chronology of a number of the graves, renders the Kerameikos cemetery an important site for bioarchaeological analysis. The curation of skeletal material from prominent graves appears to encourage a bias in favor of "high-status" burials and facilitates comparison between upper and lower social strata in Classical Athens.

The cemetery of Plateia Kotzia has a different excavation history, since it was excavated by the 3rd Ephorate of Prehistoric and Classical Antiquities over a three-year period between 1985 and 1988.[28] This cemetery, dating from the Protogeometric to Roman periods, is located approximately 40 m from the ancient city wall of Athens, most likely outside the Acharnian Gate in an area between two main roads that led to the northern suburbs of the city. After abandonment of the area as a cemetery in the late Roman period, ceramic workshops were constructed over the burial ground. Grave types and the furnishing of burials suggest that the cemetery was used by ordinary people from demes of the urban center.[29] Graves tend to cluster temporally in different areas of the cemetery, although burials of different time periods are also found throughout the site.

The cemetery at Laurion is situated in the deme of Thorikos, approximately 40 km southeast of Athens, and is associated with the metallurgical center of ancient Maroneia, one of the most important centers of ancient Laureotike.[30] It was excavated between 1980 and 1982 through rescue operations conducted by the 2nd Ephorate of Prehistoric and Classical Antiquities. Most of the graves at Laurion date to around the middle of the 5th century B.C. (470–440 B.C.). The cemetery is believed by the investigator, based on its proximity to the mines, the poverty of the grave goods, the simplicity and "careless" layout of the graves themselves, and other traits of the mortuary program, to have been used for the burial of slaves who worked in the numerous silver and lead mines of the locality. Documentary sources attest that slaves had the right to live with their families in the vicinity of their work, to maintain their religious rights, and to be buried with their families.[31]

26. Knigge 1991, pp. 11–13.
27. Parlama and Stampolidis 2000, p. 268.
28. Karagiorga-Stathakopoulou 1988, p. 102; Zachariadou and Kyriakou 1988.
29. Karagiorga-Stathakopoulou 1988, pp. 101–102.
30. Salliora-Oikonomakou 1985, pp. 131–132.
31. Salliora-Oikonomakou 1985, p. 130; Kurtz and Boardman 1994, p. 187.

THE CHRONOLOGICAL FRAMEWORK

The following chronology is used in this paper: Classical: 475–336 B.C.; Early Hellenistic: end of 4th–3rd century B.C.; Late Hellenistic: 2nd–1st century B.C.; Imperial Roman: 1st–3rd century A.D.

For the most part, this analysis proceeds by following the traditional chronological scheme that compares the Classical to the Hellenistic and Imperial Roman periods. Temporal changes are further examined according to a scheme that offers periodization based on socioeconomic attributes rather than politico-military events. Economic historians have argued that in many ways the 3rd century B.C., that is, the Early Hellenistic period, shows more similarities with the preceding hundred years than with the following two hundred[32] (the Late Hellenistic period). Likewise, analysts of surface survey data suggest that "the Classical and Early Hellenistic periods 'belong together,' as do the later Hellenistic and Early Roman epochs," and that both pairs differ dramatically from the Late Roman era.[33] Therefore, temporal analysis of the diet of the population will also proceed by comparing an "Early phase"—which in this chapter includes the Classical and Early Hellenistic periods—and a "Late phase"—which here includes the Late Hellenistic and Imperial Roman periods. A major advantage of this approach is the possibility of exploring changes that may be occurring within the Hellenistic period.[34]

SAMPLES AND STATISTICAL METHODS

A total of 80 samples (for the most part long bones [73%], ribs, and cranial fragments) from the three aforementioned cemeteries (Kerameikos, Plateia Kotzia, and Laurion) were analyzed at Bradford University, Department of Archaeological Sciences.[35] Due to diagenetic alterations, only 52 (64.2%) of these samples met the criteria for C:N ratio preservation and percent collagen yield.[36] Collagen preserved better at the Kerameikos (73.3%) than at Laurion (66.6%) and Plateia Kotzia (57.8%). Nonparametric statistical tests were used to conduct stable isotopic ratio comparisons of two (Mann-Whitney *U* test) or more (Kruskal-Wallis H test) groups on a single variable. The Spearman's correlation coefficient ρ was used to examine if there is a statistically significant correlation between carbon and nitrogen isotopic values.

RESULTS

As Figure 8.1 and Table 8.1 reveal, the adult $\delta^{13}C$ data range from -20.3‰ to -17.5‰ (average -19.1‰ ± 0.5), while the adult $\delta^{15}N$ values range from 4.0 to 12.7‰ (average 9.9 ± 1.2‰). Most of the stable isotope values cluster around the overall mean values, which suggest a rather restricted range of foods from terrestrial resources. A small number of individuals, however, appear to lie outside the denser concentration of values. These comprise three groups: (1) four individuals on the far right side of Figure 8.1, which exhibit a more positive correlation between $\delta^{13}C$ and $\delta^{15}N$ than the rest, suggestive of a more significant input of marine foods in their diet; (2) two

32. Reger 2007, p. 460.
33. Alcock 1993, p. 218.
34. Although difficulties in precise dating for this period may hamper such an attempt. Alcock 1993, p. 49; Morris 1997, p. 130; Reger 2007, p. 462.
35. For the methodology, see Chap. 2.
36. De Niro 1985; van Klinken 1999.

TABLE 8.1. STABLE ISOTOPE VALUES FOR HUMANS AT ATHENS AND LAURION

Sample ID	*Grave*	*Period*	*Sex*	*Age**	*% C*	$\delta^{13}C$ *(‰ VPDB)*	*% N*	$\delta^{15}N$ *(‰ AIR)*	*C:N*	*% Coll.*
ALB 26	Kerameikos 1 (PO3)	Late Archaic	F	YA	22.7	-19.4	7.5	9.4	3.5	0.2
ALB 13	Kerameikos 2 (WRb 29)	Classical	M	MA	42.5	-19.5	14.6	12.2	3.4	1.8
ALB 16	Kerameikos 3 (Eck 70)	Classical	F	MA	42.4	-19.0	15.2	11.5	3.3	1.3
AL 4A	Kerameikos 4 (Me 16)	Hellenistic	F	YA	44.1	-19.5	15.2	9.3	3.4	0.9
AL 3A	Kerameikos 5 (Me 8)	Hellenistic	F?	YA	45.5	-19.4	15.6	9.1	3.4	3.6
ALB 14	Kerameikos 6 (Eck 6)	Hellenistic	M	OA	37.5	-19.7	12.9	9.9	3.4	0.5
ALB 11	Kerameikos 7 (HL 32)	Hellenistic	M	OA?	34.8	-17.6	12.2	10.9	3.3	0.5
ALB 7	Kerameikos 8 (Dion 11)	Hellenistic	M	OA	43.6	-19.3	15.4	10.8	3.3	2.3
ALB 8	Kerameikos 9 (Dion 15)	Hellenistic	F	YA	41.4	-19.4	13.0	10.4	3.7	0.2
ALB 9	Kerameikos 10 (Dion 16)	Hellenistic	M	Ad	42.6	-19.4	15.0	9.3	3.3	1.2
ALB 20	Kerameikos 11 (SA 7)	Hellenistic	M	OA	43.6	-19.4	15.4	8.8	3.3	1.6
ALB 21	Kerameikos 12 (SA 12)	Hellenistic	M	MA?	40.4	-19.2	14.3	10.0	3.4	0.9
ALB 22	Kerameikos 13 (SA 16)	Hellenistic	F	MA	43.3	-19.1	15.0	9.6	3.4	1.8
ALB 25	Kerameikos 14 (SA 31)	Hellenistic	M	YA	44.5	-18.8	15.7	8.5	3.3	2.9
ALB 10	Kerameikos 15 (QO 29)	Hellenistic	M	MA	44.7	-19.1	15.9	10.0	3.3	3.9
ALB 19	Kerameikos 16 (HL 23)	Hellenistic	F	YA	42.0	-19.1	14.9	10.3	3.3	2.2
AL 7BA	Kerameikos 17 (QO 13)	Hellenistic	F	OA	44.8	-17.6	15.8	10.3	3.3	3.5
ALB 12	Kerameikos 18 (QO 22)	Hellenistic	F?	OA	45.1	-19.0	15.7	9.4	3.3	2.6
AL 5A	Kerameikos 19 (QO 24)	Imperial Roman	M	YA	40.7	-18.9	13.9	9.0	3.4	1.7
AL 27B	Kerameikos 20 (HL 2)	Imperial Roman	M	MA	44.7	-18.7	16.0	10.7	3.3	5.5
AL 6A	Kerameikos 21 (HL 13)	Imperial Roman	F?	YA	44.8	-19.1	15.5	9.3	3.4	2.6
ALB 18	Kerameikos 22 (QO 26)	Imperial Roman	F?	OA	43.8	-19.2	15.1	8.8	3.4	5.2
AL 42	Plateia Kotzia 99	Late Archaic	M	YA?	42.9	-19.3	15.0	10.5	3.3	1.5
AL 51	Plateia Kotzia 304	Classical	M?	Ad	33.8	-18.9	11.5	11.2	3.4	0.3
AL 47	Plateia Kotzia 166	Classical	F?	YA?	33.9	-19.0	11.7	10.5	3.4	0.3
AL 44	Plateia Kotzia 133	Classical	M?	MA?	28.5	-19.2	9.9	10.9	3.4	0.3
AL 53	Plateia Kotzia 366	Classical	F	MA	42.3	-19.4	14.9	10.0	3.3	1.8
AL 16A	Plateia Kotzia 146	Classical	M	MA	38.4	-19.6	13.0	9.9	3.4	0.4
AL 41	Plateia Kotzia 91B	Classical	F?	Ad	41.0	-19.5	14.7	10.2	3.3	4.8
AL 14	Plateia Kotzia 84	Classical	F	MA	42.5	-19.0	16.0	10.8	3.1	1.6
AL 49	Plateia Kotzia 278	Classical	F?	MA	35.4	-19.3	12.4	10.2	3.3	0.6
AL 57	Plateia Kotzia 409	Hellenistic	M	OA	40.1	-18.7	14.0	11.3	3.4	1.6
AL 59	Plateia Kotzia 433	Hellenistic	M	OA	43.9	-17.9	15.5	11.3	3.3	8.3
AL 61	Plateia Kotzia 396	Hellenistic	F	YA	41.1	-18.9	14.4	9.4	3.3	1.9
AL 60	Plateia Kotzia 450	Hellenistic	F	MA	43.0	-19.3	14.9	10.3	3.4	1.1
AL 36	Plateia Kotzia 136A	Hellenistic	F?	YA?	45.0	-19.2	15.9	8.9	3.3	1.8
AL 35	Plateia Kotzia 39B	Imperial Roman	M	Ad	42.8	-19.0	15.0	9.9	3.3	0.6
AL 11A	Plateia Kotzia 63A	Imperial Roman	M	Ad	42.8	-19.8	14.6	4.0	3.4	0.8
AL 15A	Plateia Kotzia 117	Imperial Roman	M	OA	41.0	-18.8	14.3	9.6	3.4	1.0
AL 52	Plateia Kotzia 338	Imperial Roman	M	Ad	35.6	-18.9	12.6	9.7	3.3	1.5

TABLE 8.1—*Continued*

Sample ID	*Grave*	*Period*	*Sex*	*Age**	*% C*	*δ¹³C (‰ VPDB)*	*% N*	*δ¹⁵N (‰ AIR)*	*C:N*	*% Coll.*
AL 18	Plateia Kotzia 362	Imperial Roman	M	MA	45.1	-19.3	16.4	9.8	3.2	1.3
AL 56	Plateia Kotzia 402	Imperial Roman	M	MA?	42.7	-18.8	15.0	10.5	3.3	1.7
AL 10A	Plateia Kotzia 24	Imperial Roman	F?	YA?	41.6	-19.0	14.3	8.8	3.4	0.9
AL 37	Plateia Kotzia 59C	Imperial Roman	F	Ad	41.1	-19.3	14.7	10.5	3.3	3.4
AL 13A	Plateia Kotzia 68	Imperial Roman	F	YA	45.0	-19.2	15.8	8.8	3.3	3.4
AL 31	Plateia Kotzia 20B	Imperial Roman	M?	Ad	40.7	-18.8	14.5	10.2	3.3	2.4
AL 32	Plateia Kotzia 21	Undated	M	OA	42.8	-19.0	15.1	9.5	3.3	2.2
AL 45	Plateia Kotzia 55	Hellenistic	I	infant	41.9	-19.5	14.6	3.4	3.4	2.4
ALB 1	Laurion 19	Classical?	F?	YA	41.9	-20.3	14.2	12.7	3.4	1.7
ALB 2	Laurion 38	Classical?	M	YA	35.3	-19.6	11.3	8.8	3.6	0.3
ALB 4	Laurion 41	Classical?	F	YA	23.7	-20.2	6.4	10.9	4.3	0.3
ALB 5	Laurion 50	Classical?	M	YA	40.4	-17.5	13.8	9.3	3.4	1.5

*Ad = adult; YA = young adult; MA = middle adult; OA = older adult; I = infant; M = male; F = female.

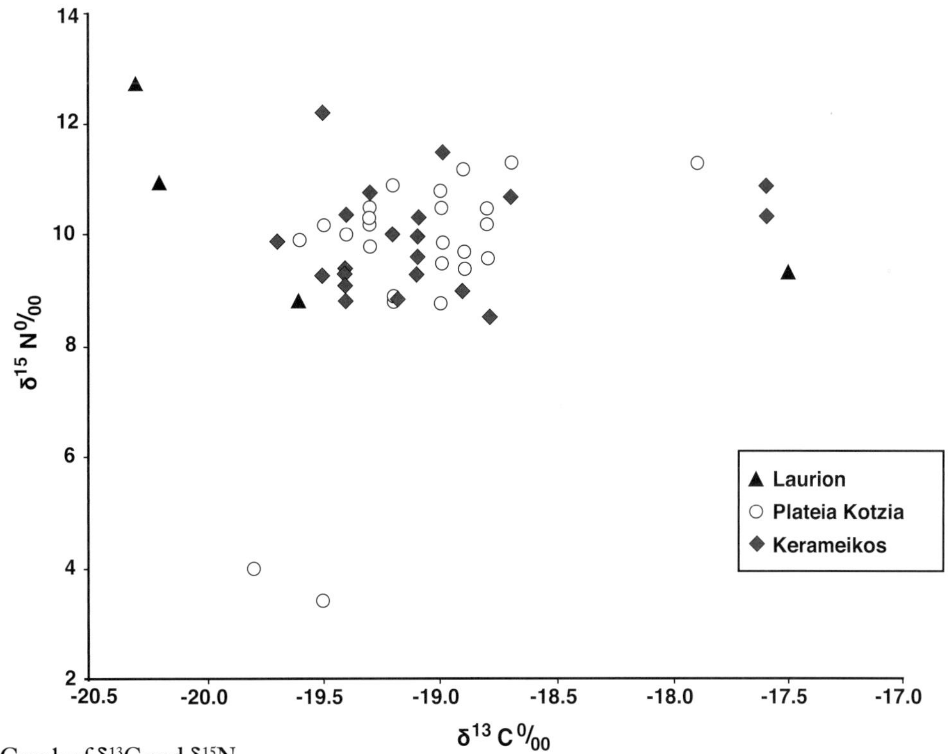

Figure 8.1. Graph of δ^{13}C and δ^{15}N values for collagen from humans buried at the Kerameikos, Plateia Kotzia, and Laurion cemeteries (all periods combined)

on the far left side of Figure 8.1 with elevated δ^{15}N that does not correlate with the rather negative δ^{13}C, suggestive of a higher level of terrestrial protein; and (3) two individuals on the bottom of Figure 8.1 with extremely low δ^{15}N, suggestive of an essentially herbivorous diet.

Overall, the isotope data indicate the consumption of a primarily terrestrial C_3 diet, with a significant contribution from animal products.

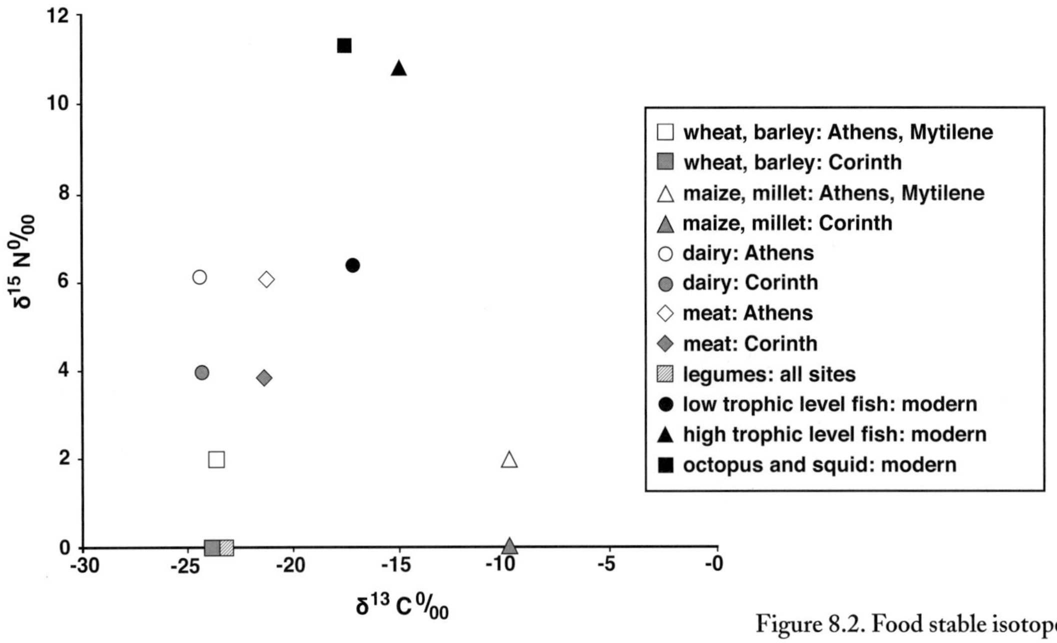

Figure 8.2. Food stable isotope values. After Garvie-Lok 2001, p. 331, fig. 7.7 and tables 7.8–7.10

Carbon isotopic values suggest that terrestrial protein mostly from staples such as barley and wheat formed the bulk of the Athenian diet, while $\delta^{15}N$ values show that meat and/or dairy products were an important part of the diet. The extent to which either meat or dairy products contributed to the diet cannot be determined based on current evidence given that dairy products have been shown to be isotopically similar to meat. Moreover, their $\delta^{15}N$ signatures often differ significantly between neighboring sites as close as Athens and Corinth[37] (Fig. 8.2) as well as within species from the same site.[38]

A positive correlation between the $\delta^{13}C$ and $\delta^{15}N$ values ($n = 47$, Spearman's $\rho = 0.245$, $p = 0.097$) suggests some contribution of marine sources in the Athenian diet not readily visible by the rather negative $\delta^{13}C$ values. Interestingly, it has been shown that marine sources of low trophic level are relatively close isotopically to terrestrial resources (Fig. 8.2), and when small amounts of low trophic level fish are incorporated into a basically C_3 terrestrial diet, they can go largely undetected.[39] This masking effect of small fish in a primarily terrestrial diet is further complicated by the substantial variability in fish signatures—superseding regional and temporal differentiation—attested recently for Greece.[40]

When diet is inferred from the present analysis of human samples from Athens and Laurion (Fig. 8.3) through conversion of the isotopic data for collagen—assuming collagen enrichment, with respect to diet, of 5‰ in $\delta^{13}C$ and of 3‰ in $\delta^{15}N$[41]—and is compared to fauna from Medieval

37. Garvie-Lok 2001, tables 7.1, 7.2.

38. Hedges and Reynard 2007, p. 1242, draw attention to the considerable variation in faunal $\delta^{15}N$ values within and between sites and species hampering comparisons between animal and human isotopes. Noteworthy, faunal $\delta^{15}N$ values from Athens are exceptionally high and variable (Garvie-Lok 2001, pp. 314–316).

39. Garvie-Lok 2001, pp. 326–332, 355–356.

40. Vika and Theodoropoulou 2012, pp. 1624–1625.

41. Schwarcz and Schoeninger 1991, although more recent views on nitrogen enrichment suggest up to 5‰ or 6‰, Hedges and Reynard 2007, p. 1241; O'Connell et al. 2012, p. 431.

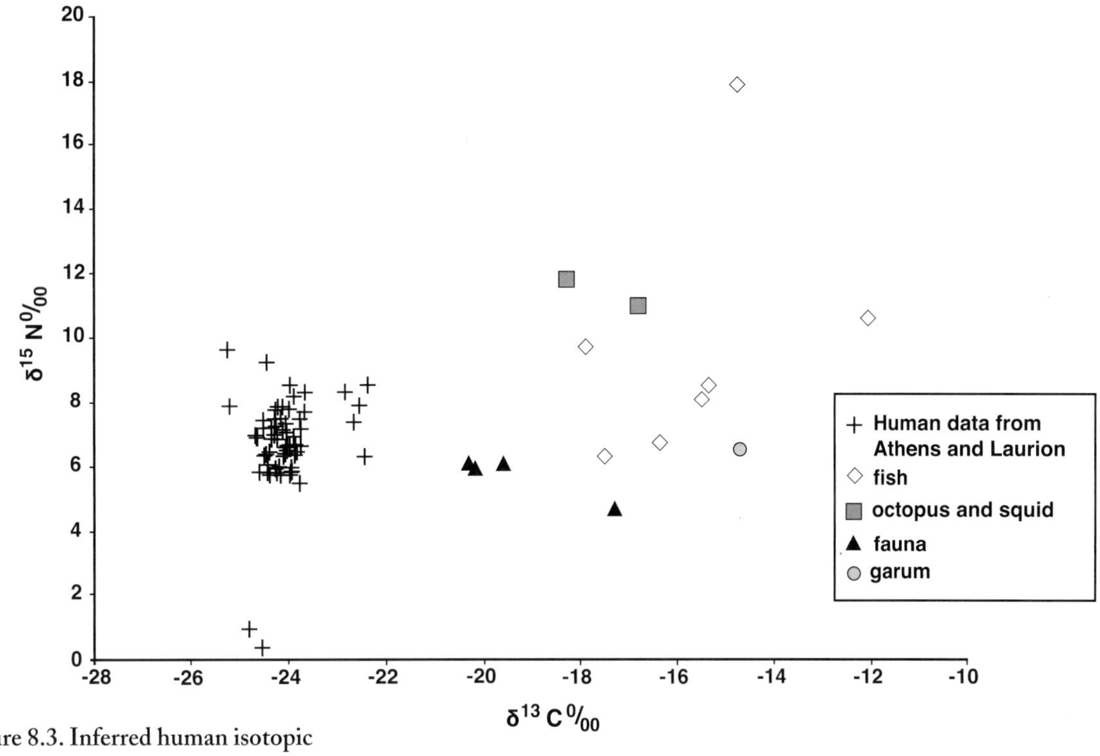

Figure 8.3. Inferred human isotopic compositions of diets from Athens and Laurion (+) with possible protein sources. After Garvie-Lok 2001, tables 7.5 and 7.9 and Prowse et al. 2004, table 4

Athens, modern marine foods from the Aegean,[42] and *garum* from low trophic level fish,[43] the $\delta^{15}N$ values from Athens match those of terrestrial herbivores and of low trophic level fish (e.g., sardines, anchovies, gilt-headed sea-bream) and *garum,* but not of high trophic level fish.[44] With the exception of the two very low nitrogen values of the outliers in Group 3 (p. 125), legumes do not appear clearly in the diet despite abundant textual and archaeobotanical evidence supporting the opposite.[45] It is possible, however, that their presence in the diet is indirectly detectable through an overall lowering of nitrogen values and increased isotopic variability at a site.[46]

Patterning among Cemeteries

When samples from the two Athenian cemeteries (the Kerameikos and Plateia Kotzia) are compared for individual $\delta^{13}C$ and $\delta^{15}N$ values during all periods combined, no significant differences are observed between the two cemeteries (Table 8.2, Fig. 8.1). Likewise, when the Classical Laurion sample is tested against Classical samples from the Kerameikos and Plateia Kotzia cemeteries, no significant differences are observed for either isotope.

42. Garvie-Lok 2001, tables 7.9, 7.10.

43. The analysis by Prowse et al. (2004) of 10 *garum* samples—two dating to the end of the Classical and early Hellenistic periods and the rest to the 2nd century A.D.—from Olbia, Sardinia, and Grado in the area of the northern Adriatic Sea produced lower $\delta^{15}N$ values than expected for marine herbivores (mean $\delta^{13}C$ = -14.7‰, SD = 0.6; mean $\delta^{15}N$ = 6.5‰, SD = 1.7) suggesting they were prepared from low trophic level fish.

44. The contribution of high trophic level fish in the diet has been argued to explain the high nitrogen values at Isola Sacra (Prowse et al. 2004).

45. Legumes as the poor man's meat: Garnsey 1999, p. 15; their use in the economy: Davies 2007, p. 344; for archaeobotanical evidence, see Sarpaki 1992; Bookidis et al. 1999; Hansen 2000; Megaloudi 2006; Margaritis 2014.

46. As, e.g., suggested by Craig et al. 2009, p. 579, to explain the depleted nitrogen values of one group at Velia.

TABLE 8.2. COMPARISON OF STABLE ISOTOPE VALUES FOR ADULTS AT ATHENS AND LAURION

Cemetery	*N*	*Mean*	*SD*	*Range*	*Min*	*Max*	*Comparison*[3]	*p*
$\delta^{15}N$ (‰ AIR)								
Kerameikos	22	9.9	0.9	3.7	8.5	12.2	Ker/PK[1]	0.24
Plateia Kotzia	25	9.9	1.4	7.3	4.0	11.3	–	–
PK without outlier	24	10.1	0.7	2.5	8.8	11.3	–	–
Laurion	4	10.4	1.8	3.9	8.8	12.7	Ker/PK/Lau[2]	0.51
Total	51	9.9	1.2	8.7	4.0	12.7	–	–
$\delta^{13}C$ (‰ VPDB)								
Kerameikos	22	-19.1	0.5	2.1	-19.7	-17.6	Ker/PK[1]	0.33
Plateia Kotzia	25	-19.1	0.4	1.9	-19.8	-17.9	–	–
PK without outlier	24	-19.1	0.3	1.7	-19.6	-17.9	–	–
Laurion	4	-19.4	1.3	2.8	-20.3	-17.5	Ker/PK/Lau[2]	0.18
Total	51	-19.1	0.5	2.8	-20.3	-17.5	–	–

[1] Mann-Whitney $U = 210.5$ for $\delta^{15}N$; $U = 219.5$ for $\delta^{13}C$.

[2] Kruskal-Wallis $H = 1.362$ for $\delta^{15}N$; $H = 3.384$ for $\delta^{13}C$.

[3] Comparisons do not include the outlier from Plateia Kotzia.

TABLE 8.3. STABLE ISOTOPE VALUES FOR ADULTS AT ATHENS BY PERIOD

	Total Sample				*Kerameikos*				*Plateia Kotzia*			
Period	*N*	*Mean*	*SD*	*Range*	*N*	*Mean*	*SD*	*Range*	*N*	*Mean*	*SD*	*Range*
$\delta^{15}N$ (‰ AIR)												
Classical	12	10.6	0.8	2.8	3	11.0	1.5	2.8	9	10.5	0.4	1.3
Hellenistic	20	9.9	0.8	2.8	15	9.8	0.7	2.4	5	10.2	1.1	2.4
Imperial Roman	13	9.7	0.7	1.9	4	9.5	0.9	1.9	9	9.8	0.63	1.7
Total	45	10.0	0.8	3.7	22	9.9	0.9	3.7	23	10.1	0.7	2.5
$\delta^{13}C$ (‰ VPDB)												
Classical	12	-19.3	0.2	0.7	3	-19.3	0.3	0.5	9	-19.2	0.2	0.7
Hellenistic	20	-19.0	0.6	2.1	15	-19.0	0.6	2.1	5	-18.8	0.6	1.4
Imperial Roman	13	-19.0	0.2	1.9	4	-19.0	0.2	0.5	9	-19.0	0.2	0.5
Total	45	-19.1	0.4	2.1	22	-19.1	0.5	2.1	23	-19.1	0.4	1.7

The Laurion sample, however, appears particularly interesting because: (1) it shows unexpectedly high nitrogen signatures for a "slave population;" and (2) the range of values is rather broad both for nitrogen and carbon, unlike those from the Kerameikos and Plateia Kotzia cemeteries for the same period (Table 8.3). This observation is illustrated in Figure 8.1, where only one of four values from Laurion clusters with the majority of values from the Kerameikos and Plateia Kotzia, while the rest lie outside the main clustering of the two Athenian cemeteries.

With respect to overall $\delta^{15}N$ signatures (Table 8.2), the cemetery at Plateia Kotzia shows a broader range of values compared to that at the Kerameikos. When the Plateia Kotzia values are examined individually, it becomes apparent this result is due to two outliers with very low nitrogen signatures: an infant and a young adult male.

Outliers

The very low nitrogen isotopic signature ($\delta^{15}N$ = 3.4‰; $\delta^{13}C$ = -19.5‰) of the 1-year-old infant is striking, since at this age one might expect $\delta^{15}N$ to be raised due to nursing. It is likely, therefore, that the infant subsisted on weaning food of low quality protein before death.[47] Analyses of the diets of infants and young children from diverse sites and periods in Greece and Italy have shown rather elevated $\delta^{15}N$ values in these age groups compared to those of the adults.[48] Interestingly, in Imperial Roman Isola Sacra, although infants had considerably higher nitrogen values ($\delta^{15}N$ mean 12.5 ± 1.9‰),[49] children aged 5–15 years had the lowest nitrogen values ($\delta^{15}N$ mean 10.4 ± 0.8‰) and large carbon apatite and collagen spacing, "confirming that they were consuming a predominantly C_3 plant based diet, i.e., a simple and frugal diet in accordance with their [low] status in Roman society."[50]

The second very low $\delta^{15}N$ value (4.0‰) from the Plateia Kotzia cemetery, that from a young adult male, is indicative of an essentially herbivorous diet. Such low values have only occasionally been identified in Neolithic, Early Iron Age, and medieval samples from Greece[51] and Anatolia,[52] and most likely reflect a greater reliance on cereals and legumes in the diet. It is of interest that at Imperial Roman Isola Sacra and Velia, the great within and between site isotopic variability has allowed the differentiation of subsamples in the populations in terms of protein consumption. At Isola Sacra, the less privileged group comprised the minority of the analyzed population and was spatially separated from the main group situated at the port.[53] At Velia, on the other hand, high protein variability was observed within a single cemetery located at the port, in which the less privileged formed the largest group.[54] Based on this evidence, it needs to be considered whether the occurrence of one adult in Athens with an essentially vegetarian diet, in addition to the malnourished infant, is evidence of the presence of an unprivileged subgroup within the population that is not clearly visible isotopically within the small sample analyzed.

Temporal Comparison: Classical, Hellenistic, and Imperial Roman Periods

Figure 8.4 and Table 8.3 illustrate that a moderate amount of protein was consumed in the ancient Athenian and Laurion diets during all periods represented by the samples. Interestingly, for the Athenian cemeteries individually, and on average for all three, the Classical period exhibits the highest maximum, minimum, and mean values for $\delta^{15}N$ (Fig. 8.5).

Temporal comparison among the three chronological periods shows statistically significant differences for the two Athenian cemeteries combined and in each individually: $\delta^{15}N$ values decrease from the Classical

47. Katzenberg, Herring, and Saunders 1996.

48. Garvie-Lok 2001, p. 374; Bourbou and Richards 2007, p. 69; Triantaphyllou et al. 2008, p. 3032.

49. Prowse et al. 2008, p. 300.

50. Prowse et al. 2005, p. 10.

51. Papathanasiou 2001, table 22; Triantaphyllou 2001, table 7.10; Bourbou et al. 2011, fig. 3; Papathanasiou et al. 2013, table 3; see also Chaps. 2–11.

52. Lösch, Grupe, and Peters 2006.

53. It is referred to as the inland ANAS cemetery comprising 14 individuals vs. 105 from the port; see Prowse et al. 2004, p. 262.

54. One hundred unprivileged vs. 17 individuals for the privileged group; see Craig et al. 2009, p. 578.

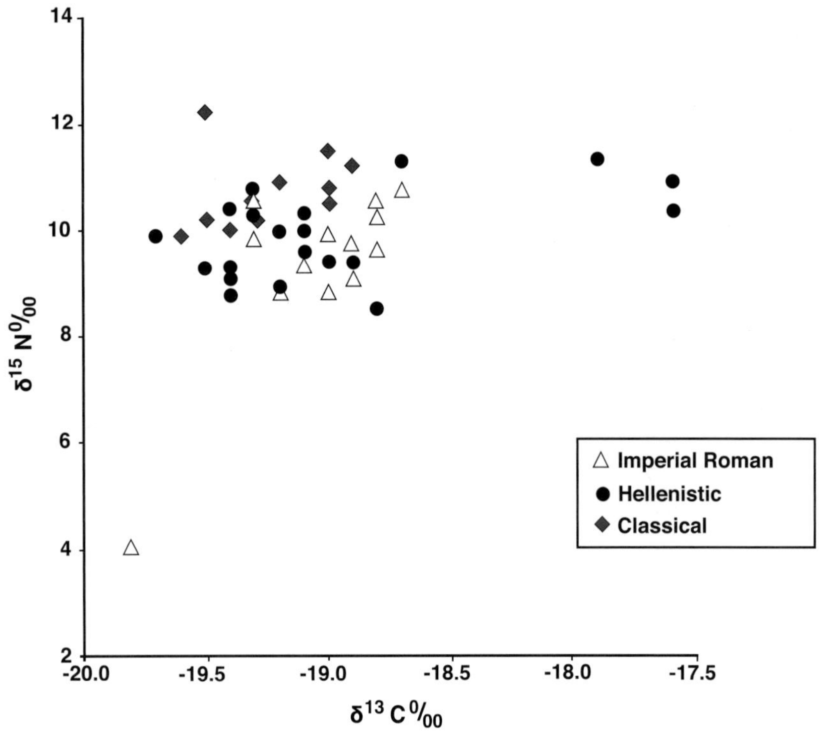

Figure 8.4. Temporal comparison of $\delta^{13}C$ and $\delta^{15}N$ collagen values in Athens

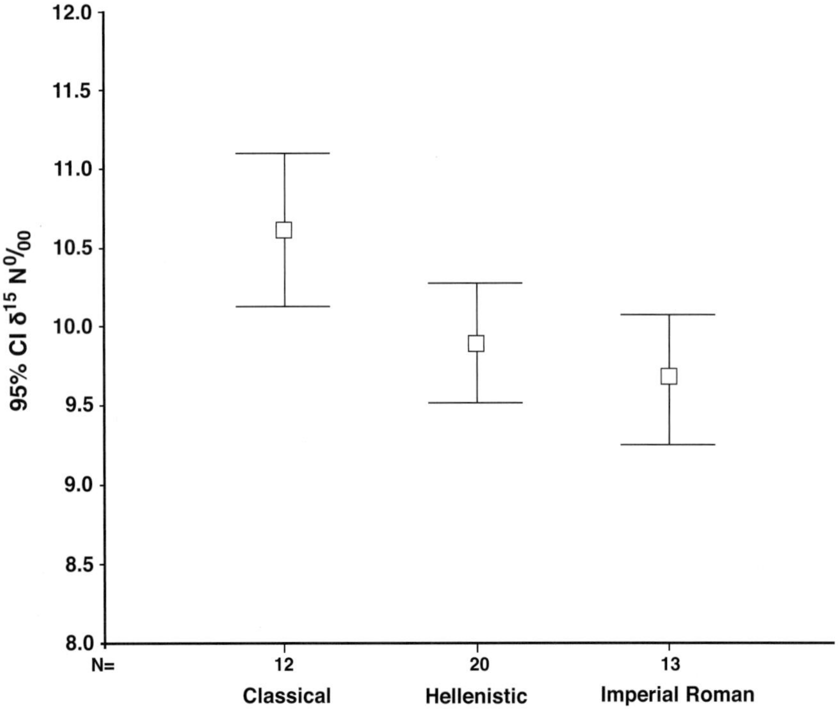

Figure 8.5. Temporal comparison of mean $\delta^{15}N$ values in Athens

through the Imperial Roman periods, while $\delta^{13}C$ values become somewhat more positive. Although not statistically significant for the Kerameikos cemetery, most likely because of the small subsamples per period, the trend is clear in both cemeteries (Fig. 8.6).

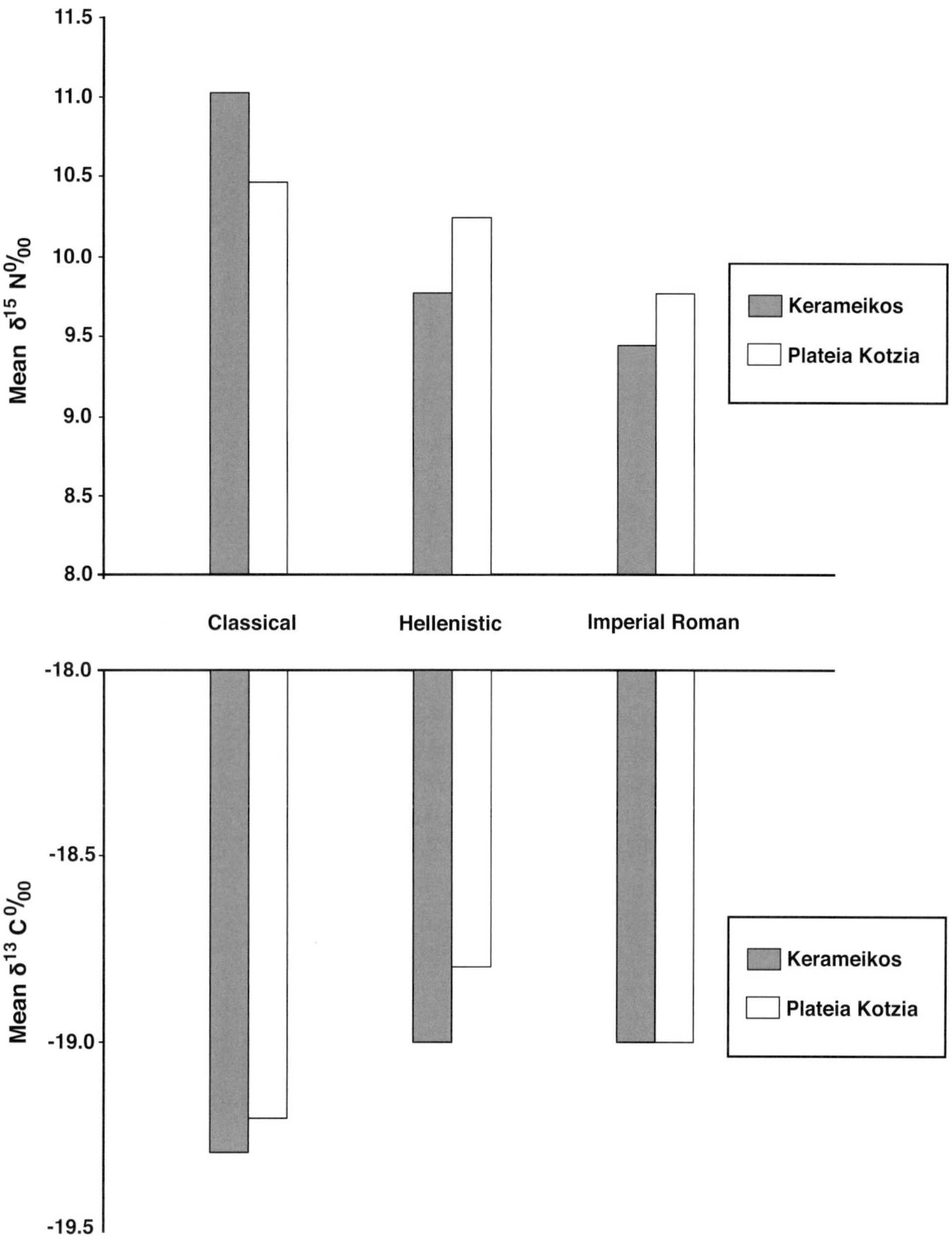

Figure 8.6. Comparison of mean $\delta^{15}N$ and $\delta^{13}C$ values for each chronological period per Athenian cemetery

When periods are compared with each other for the cemeteries combined, significant differences are noted for $\delta^{15}N$ between the Classical and the Hellenistic (Mann-Whitney $U = 64.5$, $p = 0.03$) and the Classical and the Imperial Roman periods (Mann-Whitney $U = 27.5$, $p = 0.004$), but not between the Hellenistic and the Imperial Roman periods (Mann-Whitney $U = 108.5$, $p = 0.27$). In terms of $\delta^{13}C$, statistically significant differences are noted only between the Classical and the Imperial Roman periods (Mann-Whitney $U = 43$, $p = 0.03$). Moreover, in the Hellenistic period the range of $\delta^{13}C$ values is much broader than that of the other periods in each cemetery individually as well as in the cemeteries combined (Table 8.3).

TABLE 8.4. STABLE ISOTOPE VALUES FROM EARLY AND LATE SAMPLES AT ATHENS

	Total Sample				*Kerameikos*				*Plateia Kotzia*			
Phase	*N*	*Mean*	*SD*	*Range*	*N*	*Mean*	*SD*	*Range*	*N*	*Mean*	*SD*	*Range*
$\delta^{15}N$ (‰ AIR)												
Early	27	10.2	0.9	3.7	15	10.0	1.0	3.7	12	10.5	0.6	1.9
Late	17	9.8	0.7	1.9	7	9.7	0.7	1.9	10	9.8	0.6	1.7
$\delta^{13}C$ (‰ VPDB)												
Early	27	-19.1	0.5	2.1	15	-19.2	0.5	2.1	12	-19.1	0.5	1.7
Late	17	-18.9	0.4	1.7	7	-18.8	0.6	1.6	10	-19.0	0.2	0.5

Temporal Comparison: Early/Late Periodization Scheme

Temporal changes are further examined according to a temporal scheme that identifies an "Early phase"—which includes the Classical and Early Hellenistic periods—and a "Late phase"—which includes the Late Hellenistic and Imperial Roman periods (see rationale above). This analysis is tentative due to the uneven spread of samples within each subphase and aims to complement the aforementioned analysis. When these Early and Late phases are compared with respect to $\delta^{13}C$ and $\delta^{15}N$ (Table 8.4), it is apparent that they differ significantly from each other for the two cemeteries combined ($\delta^{13}C$ Mann-Whitney $U = 160.5$, $p = 0.06$; $\delta^{15}N$ Mann-Whitney $U = 160$, $p = 0.05$). When examined individually, the trend is apparent in both cemeteries but is statistically significant with respect to $\delta^{15}N$ at the Plateia Kotzia cemetery (Mann-Whitney $U = 25.5$, $p = 0.01$) and to $\delta^{13}C$ at the Kerameikos (Mann-Whitney $U = 19.5$, $p = 0.02$).

It is surprising that the temporal difference observed previously for $\delta^{15}N$ at the Kerameikos cemetery is not statistically valid in this scheme. This, however, could be an artifact of the composition of the Early and Late samples at the Kerameikos: the majority (11/14) of the "Early" Kerameikos sample dates to the Early Hellenistic period, while the respective Plateia Kotzia sample comprises mostly (8/12) samples that date to the Early Classical period. For the Kerameikos, therefore, a comparison between the Early and Late phases is essentially one between the Early Hellenistic period and the Late Hellenistic/ Imperial Roman periods combined. The lack of a significant difference between the two may suggest that at the Kerameikos, a significant shift in diets took place prior to the Early Hellenistic, most likely during the Late Classical period, whereas at the Plateia Kotzia cemetery, a more gradual dietary change may have taken place that was more clearly visible in Imperial Roman times.

Sex and Age-at-Death Comparisons

No statistically significant differences for $\delta^{13}C$ and $\delta^{15}N$ are noted between the sexes for the cemeteries and chronological periods combined, nor when cemeteries are compared to each other or are examined individually (Table 8.5).

TABLE 8.5. STABLE ISOTOPE VALUES AT ATHENS BY SEX

	Total Sample				*Kerameikos*				*Plateia Kotzia*			
Sex	*N*	*Mean*	*SD*	*Range*	*N*	*Mean*	*SD*	*Range*	*N*	*Mean*	*SD*	*Range*
$\delta^{15}N$ (‰ AIR)												
Male	25	9.9	1.5	8.2	11	10.0	1.1	3.7	14	9.9	1.8	7.3
Male without outlier	24	10.2	0.9	3.7	–	–	–	–	13	10.3	0.7	1.8
Female	22	9.8	0.7	2.7	11	9.8	0.8	2.7	11	9.9	0.7	2.0
$\delta^{13}C$ (‰ VPDB)												
Male	25	-19.0	0.5	2.2	11	-19.1	0.6	2.1	14	-19.0	0.5	1.9
Male without outlier	24	-19.0	0.5	2.1	–	–	–	–	13	-18.9	0.4	1.7
Female	22	-19.1	0.4	1.9	11	-19.1	0.5	1.9	11	-19.2	0.2	0.6

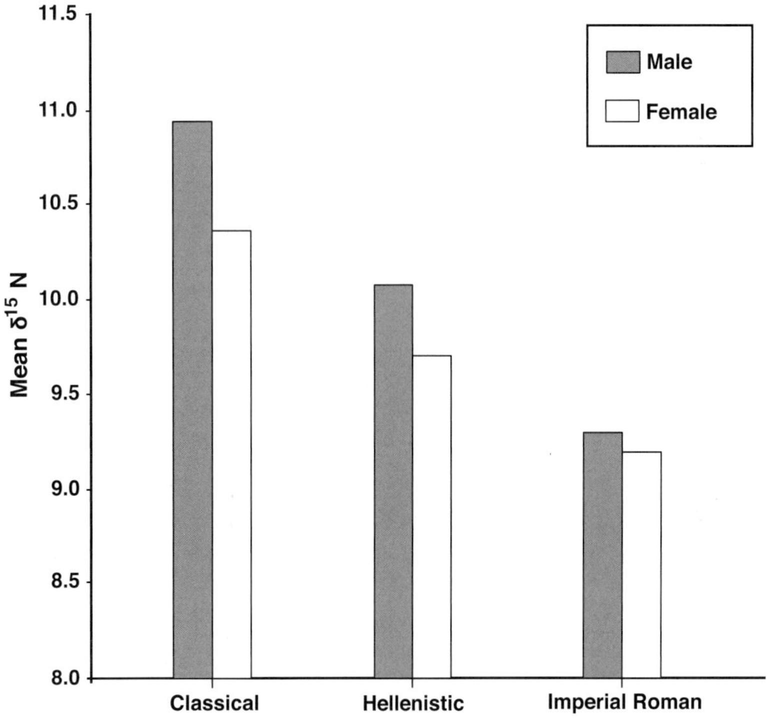

Figure 8.7. Histogram of mean $\delta^{15}N$ values for each sex per chronological period in Athens

Temporal comparison of nitrogen values between the sexes (Table 8.6, Fig. 8.7) demonstrates that the decrease in protein levels from the Classical to the Imperial Roman periods noted above is statistically significant for both men (Mann-Whitney $U = 5$, $p = 0.02$) and women (Mann-Whitney $U = 4.5$, $p = 0.03$). In all periods, males show consistently more positive mean nitrogen values than females.

Young adult age groups[55] differ significantly with respect to $\delta^{15}N$ from both the middle adults ($\delta^{15}N$ Mann-Whitney $U = 32.5$, $p = 0.003$) and the old adults ($\delta^{15}N$ Mann-Whitney $U = 44.5$, $p = 0.07$) for the chronological periods pooled (Table 8.7). In the middle and old adult age groups, mean nitrogen values are more positive than in the young adults, while carbon is more positive in the old adults (Fig. 8.8). The range of values for both isotopes is higher in the older age groups compared to young adults.

55. The adult age groups follow the Buikstra and Ubelaker (1994) scheme: Young Adults = 20–34 years; Middle Adults = 35–49 years; Old Adults = 50+.

TABLE 8.6. STABLE ISOTOPE VALUES AT ATHENS BY SEX AND CHRONOLOGICAL PERIOD

	Males				*Females*			
Period	*N*	*Mean*	*SD*	*Range*	*N*	*Mean*	*SD*	*Range*
$\delta^{15}N$ (‰ AIR)								
Classical	5	10.9	0.9	2.3	7	10.4	0.7	2.1
Hellenistic	10	10.1	1.0	2.8	10	9.7	0.6	1.5
Imperial Roman	9	9.3	2.0	6.7	5	9.2	0.7	1.7
$\delta^{13}C$ (‰ VPDB)								
Classical	5	-19.3	0.3	0.7	7	-19.2	0.2	0.5
Hellenistic	10	-18.9	0.7	2.1	10	-19.1	0.5	1.9
Imperial Roman	9	-19.0	0.3	1.1	5	-19.2	0.1	0.3

TABLE 8.7. STABLE ISOTOPE VALUES AT ATHENS BY AGE GROUP

	Total Sample			
Age Group	*N*	*Mean*	*SD*	*Range*
$\delta^{15}N$ (‰ AIR)				
Young Adult	14	9.4	0.7	2.0
Middle Adult	14	10.5	0.7	2.6
Older Adult	11	10.1	0.9	2.5
$\delta^{13}C$ (‰ VPDB)				
Young Adult	14	-19.2	0.2	0.7
Middle Adult	14	-19.2	0.3	0.9
Older Adult	11	-18.7	0.7	2.1

Noteworthy in the older age categories are the mean nitrogen values, which are higher for men than for women. This pattern is similar to that noted at Isola Sacra, where men have consistently elevated mean nitrogen in all age categories.[56] It is interesting that of the four individuals showing more positive carbon values combined with nitrogen (Group 1) in Figure 8.1, three are elderly (50+) of both sexes from "high" status burials of the Hellenistic period.

Patterning Based on Grave Type

No statistically significant differences are noted among grave types between cemeteries or temporal periods, although the temporal trend of higher isotopic values in the Classical period stands for each grave type as well. That is, in the Classical period, all graves have more elevated isotopic values than the respective types in Hellenistic and Imperial Roman times.

Of particular interest, nevertheless, are the statistically significant differences noted between males and females for earth pit graves (Fig. 8.9). Among these simple pits in the ground, males have more positive nitrogen values than females ($\delta^{15}N$ Mann-Whitney $U = 19.5$, $p = 0.04$).

56. Prowse et al. 2005, fig. 4; also compare to the Bronze Age; see Chap. 3.

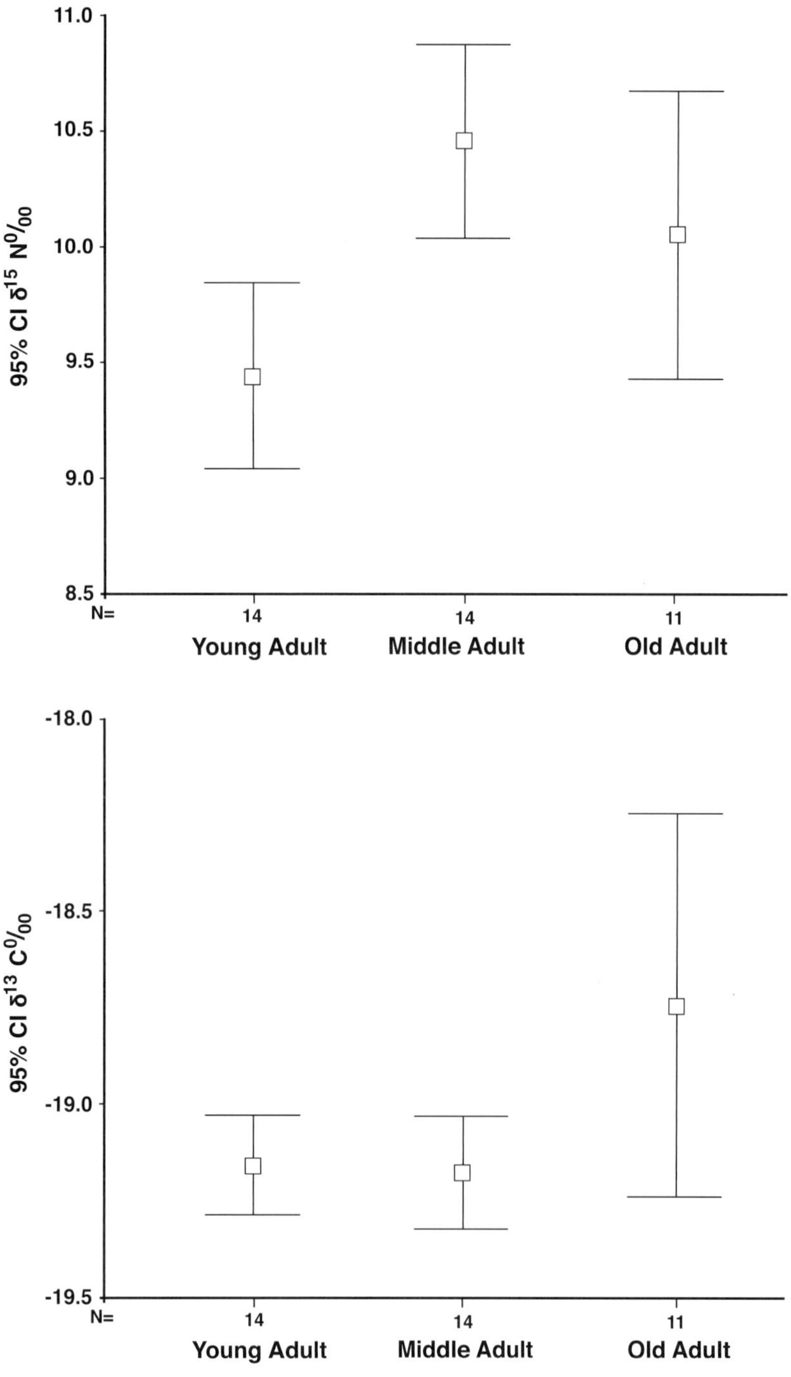

Figure 8.8. Comparison of $\delta^{15}N$ and $\delta^{3}C$ values by age group in Athens cemeteries

Comparison to Other Sites

Comparison of the three chronological periods in Athens to other sites of historic times reveals important variations in terms of region and temporal period (Table 8.8).

In comparison to Classical Makriyalos in central Macedonia,[57] Athens in all periods has significantly more positive mean $\delta^{15}N$ but similar mean $\delta^{13}C$ values (for the Classical period $\delta^{15}N$ Mann-Whitney $U = 0.000$, $p<0.001$; $\delta^{13}C$ Mann-Whitney $U = 35$, $p = 0.03$; for the Hellenistic period $\delta^{15}N$ Mann-Whitney $U = 13$, $p<0.000$; $\delta^{13}C$ Mann-Whitney $U = 96$,

57. Triantaphyllou 2001, table 7.10.

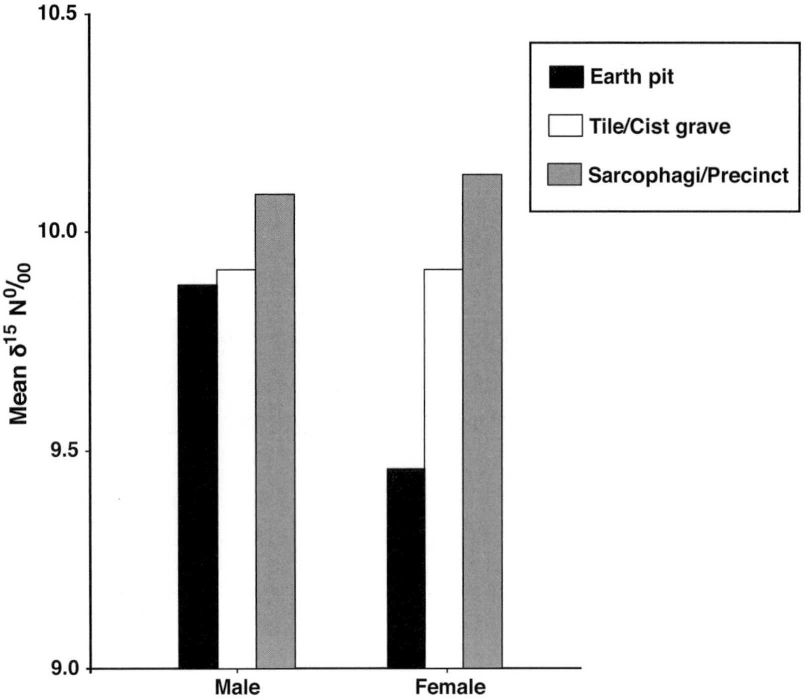

Figure 8.9. Comparison of δ^{15}N and δ^{13}C values by grave type in Athens cemeteries

p = 0.35; for the Roman period δ^{15}N Mann-Whitney $U = 26$, $p = 0.003$; δ^{13}C Mann-Whitney $U = 80.5$, $p = 0.86$). Whether these differences in protein levels express an overall different economy, and/or local faunal signatures, remains to be tested. It is important, however, that in this site too, in comparison to earlier periods of the prehistoric and protohistoric eras, the Classical period displays a significant increase in terms of protein inclusion in the diet. Makriyalos pertains to the few sites thus far analyzed in Greece where the inclusion of millet, a C_4 plant is discussed as realistically affecting human diet through nutrition or animal fodder.[58]

In comparison to Classical and Hellenistic Thebes,[59] the lack of statistically significant differences in both mean δ^{15}N and δ^{13}C values from the respective periods in Athens is notable (for the Classical period δ^{15}N Mann-Whitney $U = 175.5$, $p = 0.9$; δ^{13}C Mann-Whitney $U = 152$, $p = 0.43$; for the Hellenistic period δ^{15}N Mann-Whitney $U = 358$, $p = 0.23$; and δ^{13}C Mann-Whitney $U = 365.5$, $p = 0.28$). It seems likely that the concurrent decrease of nitrogen in the two cities is the result of parallel processes shaping their history. The relatively high nitrogen input at Thebes in the Classical period contrasting with the low carbon values was partially attributed to the contribution of freshwater fish in the diet[60]—a suggestion that has also been proposed to explain the more positive average nitrogen values at the early Christian cemetery of St. Callixtus situated near the Tiber.[61]

A comparison of Classical and Hellenistic Athens combined (n = 32) with the 5th to 2nd century B.C. coastal Apollonia, a Greek colony on the Black Sea, reveals a similar mean nitrogen input (Mann-Whitney $U = 812$, $p = 0.64$) but differences in terms of carbon isotope values (Mann-Whitney $U = 319$, $p<0.000$).[62] Likewise, the Classical period in Athens differs from Imperial Roman Isola Sacra in terms of δ^{13}C (Mann-Whitney $U = 131$, $p<0.000$) but not in terms of δ^{15}N (Mann-Whitney $U = 518.5$, $p = 0.41$).

58. See Triantaphyllou 2001, fig. 7.16, on temporal comparison, and p. 135 on millet.

59. Vika 2011, table 1.

60. Vika, Aravantinos, and Richards 2009, p. 1081.

61. Rutgers et al. 2009, p. 1130.

62. Keenleyside, Schwarcz, and Panayotova 2006, p. 1211.

TABLE 8.8. MEAN STABLE ISOTOPE VALUES FOR ADULTS AT HISTORIC-PERIOD SITES IN GREECE AND ITALY

Site	*Date*	*N*	*$\delta^{13}C \pm 1\sigma$ (‰ VPDB)*	*$\delta^{15}N \pm 1\sigma$ (‰ AIR)*
Athens	Classical	12	-19.3 ± 0.2	10.6 ± 0.8
Athens	Hellenistic	20	-19.0 ± 0.6	9.9 ± 0.8
Athens	Imperial Roman	14	-19.1 ± 0.3	9.3 ± 1.7
Makriyalos	Classical	12	-19.0 ± 0.3	8.2 ± 0.8
Apollonia	5th–2nd century B.C.	54	-18.5 ± 0.5	10.1 ± 0.8
Thebes	Classical	30	-19.2 ± 0.4	10.7 ± 1.0
Thebes	Hellenistic	44	-19.2 ± 0.6	9.6 ± 0.7
Rome (Casal Bertone)	Imperial Roman	23	-18.2 ± 0.6	10.0 ± 1.5
Rome (Castellaccio Europarco)	Imperial Roman	7	-18.5 ± 0.6	9.5 ± 1.3
Isola Sacra	Imperial Roman	102	-18.8 ± 1.2	10.8 ± 1.2
ANAS	Imperial Roman	14	-19.4 ± 0.4	9.5 ± 1.8
Velia-Group I	Imperial Roman	100	-19.5 ± 0.2	8.2 ± 0.7
Velia-Group II	Imperial Roman	17	-19.3 ± 0.3	11.2 ± 1.3
Rome (catacombs)	mid 3rd–early 5th century A.D.	22	- 19.8 ± 0.4	10.6 ± 0.8
Nemea	5th–6th century A.D.	22	-19.0 ± 0.8	8.6 ± 0.6
Eleutherna	6th–7th century A.D.	27	-18.9 ± 0.6	8.2 ± 1.4
Messene	6th–7th century A.D.	21	-19.2 ± 0.3	8.7 ± 0.6

Sources: Makriyalos: Triantaphyllou 2001, table 7.10; Apollonia: Keenleyside, Schwarcz, and Panayotova 2006, table 2; Thebes: Vika 2011, table 1; Rome (Casal Bertone and Castellaccio Europarco): Killgrove and Tykot 2013, table 5; Isola Sacra and ANAS: Prowse et al. 2004, table 2; Velia: Craig et al. 2009, table 1; Rome (catacombs): Rutgers et al. 2009, p. 1129; Nemea: Garvie-Lok 2001, table 8.8; Eleutherna and Messene: Bourbou et al. 2011, table 1.

At both sites the more positive carbon values were attributed to the contribution of higher trophic level marine foods in the diet. In contrast to the Classical period, average isotopic values in the Hellenistic and Imperial Roman periods in Athens are significantly lower compared to the Imperial Roman Isola Sacra values (for the Hellenistic period δ^{15}N Mann-Whitney U = 501.5, $p<0.000$; δ^{13}C Mann-Whitney U = 550.5, $p = 0.001$; for the Roman period δ^{15}N Mann-Whitney U = 240, $p<0.000$; δ^{13}C Mann-Whitney U = 356, $p = 0.002$).[63] Instead, in terms of δ^{15}N they seem more similar to inland ANAS a contemporaneous and neighboring cemetery to Isola Sacra (for the Hellenistic period δ^{15}N Mann-Whitney U = 131.5, $p = 0.77$; for the Roman δ^{15}N Mann-Whitney U = 87.5, $p = 0.64$).

An even more nuanced impression of substantial variations in the Greco-Roman world is offered by a comparison of the Athenian diets to Imperial Roman Velia in Southern Italy:[64] Classical Athens is more similar in terms of nitrogen and carbon isotopes to the higher status Group II at Velia (δ^{15}N Mann-Whitney U = 81, $p = 0.35$; δ^{13}C Mann-Whitney U = 84.5, $p = 0.44$), which comprised a minority that consumed more meat and fish. Hellenistic and Imperial Roman Athens, however, have lower mean nitrogen values and more positive carbon values compared to the higher status group (for the Hellenistic period δ^{15}N Mann-Whitney U = 69, $p = 0.002$; δ^{13}C Mann-Whitney U = 112, $p = 0.08$; for the Roman period δ^{15}N Mann-Whitney

63. Prowse et al. 2004, table 2.
64. Craig et al. 2009, table 2.

$U = 29.5$, $p<0.000$; $\delta^{13}C$ Mann-Whitney $U = 58.5$, $p = 0.02$) but significantly more positive isotopic values than the lower status Group I, which is believed to have consumed more cereals and legumes (for the Hellenistic period $\delta^{15}N$ Mann-Whitney $U = 429.5$, $p<0.000$; $\delta^{13}C$ Mann-Whitney $U = 105.5$, $p<0.000$; for the Roman period $\delta^{15}N$ Mann-Whitney $U = 180$, $p<0.000$; $\delta^{13}C$ Mann-Whitney $U = 206$, $p<0.000$).

Comparison of Athens with two cemeteries of the Roman period situated closer to Rome than Isola Sacra reveals a rather different picture compared to that from the port but also from Velia. Roman Athens but also the Classical and Hellenistic periods differ from adult values from Rome in terms of average $\delta^{13}C$ values only (for the Classical period $\delta^{15}N$ Mann-Whitney $U = 144$, $p = 0.316$; $\delta^{13}C$ Mann-Whitney $U = 28.5$, $p<0.000$; for the Hellenistic period $\delta^{15}N$ Mann-Whitney $U = 277.5$, $p = 0.656$; $\delta^{13}C$ Mann-Whitney $U = 121.5$, $p<0.000$; for the Roman period $\delta^{15}N$ Mann-Whitney $U = 158$, $p = 0.327$; $\delta^{13}C$ Mann-Whitney $U = 53$, $p<0.000$). The different consistency and location of the two cemeteries (the Casal Bertone and the Castellaccio Europarko comprise a mausoleum and two different necropoleis situated 1.5 km and 12 km from the city wall, respectively)[65] makes it difficult to appraise similarities in terms of mean nitrogen with Athens. Interpreting respective disparities from the port of Rome and Velia, Killgrove and Tykot draw attention to their different location which, in the case of Isola Sacra, situated near the Tyrrhenian Sea, would provide the population with aquatic resources, while, in the case of Velia situated 400 km away from Rome, may explain the significant differences in overall faunal and human signatures.[66]

While factors taking place locally seem to be of paramount importance in shaping local isotopic signatures, it is noteworthy that Athens in all periods investigated in this study shows elevated mean $\delta^{15}N$ values in comparison to diverse sites of the early Christian period.[67]

CONCLUSIONS

The observable increase of the protein level from prehistoric to historic times has been shown in different studies of the Greek peninsula.[68] The present investigation reinforces this view, demonstrating a higher protein input in the three historic periods in Athens. Within the three periods examined, the protein level was shown to be higher in the Classical period in comparison to the Hellenistic and Imperial Roman periods. In the latter, discrepancies between the cemeteries were mitigated. The study of osseous lesions that characterize periods of stress during growth has demonstrated a corresponding decline of health gradually from Classical to Imperial Roman times.[69] For women, a more uniform condition became apparent through the ratio of anemic lesions changing minimally through time. The analyses reinforce the impression, therefore, that the statistically significant differences in protein level between the three chronological periods under investigation are a result of diet rather than exogenous factors that may affect $\delta^{15}N$ values such as drought and famine, as argued for the 4th century B.C.; climatic warming, as suggested for the Roman period;[70] or endogenous factors related to growth, disease, or pregnancy.[71]

65. Killgrove and Tykot 2013, p. 29.
66. Killgrove and Tykot 2013, pp. 32–33.
67. Garvie-Lok 2001; Bourbou et al. 2011.
68. Triantaphyllou 2001, p. 135; Vika 2011, p. 1161.
69. Lagia 2014.
70. Camp 1982; Kuniholm 1990; Bianchi and McCave 1999, p. 516.
71. White and Armelagos 1997; Katzenberg and Lovell 1999; Fuller et al. 2005; Waters-Rist and Katzenberg 2010.

The decline in nitrogen values between the Classical and Hellenistic periods has also been observed at Thebes and has been attributed to disruption of normal patterns of food production as a result of the Macedonian presence, the development of the Hellenistic centers in the east, and the ensuing emigration.[72] In terms of the economy, it is claimed that the depleted nitrogen was the result of exploiting a stressed agropastoral environment that had reached its capacity in the Classical period, ultimately steering the population to a diet akin to that of the Bronze Age. If this model applies to some extent to Athens as well, we can clearly add a multitude of factors such as the loss of revenue from tribute *poleis* that was no longer available to the city, the repeated destruction of the Attic landscape by Macedonian troops and Greek coalitions, the occupation of Piraeus, and constitutional changes that lead to the development of an oligarchy among the citizen body.[73]

The diachronic comparison of dietary change that identifies greater similarities in the socioeconomic sphere between the Classical and Early Hellenistic periods versus Late Hellenistic and Imperial Roman times[74] suggests that the decrease of the protein level in the "elite" cemetery at the Kerameikos may have taken place earlier than in the "commoners" cemetery of Plateia Kotzia. In the latter, the decline in protein appears to have been more gradual. In both cemeteries, nevertheless, the lower nitrogen values are marked in the Late Hellenistic and Imperial Roman periods. Salient changes are observable as well in the cultural sphere, where there are also changes in mortuary practices.[75]

Multiple graves increase significantly in numbers from this period onward, betraying the occurrence of substantial changes in the perception of collective identity and the way this is expressed in the funerary landscape. Athens, now only part of an enormous empire and situated peripheral to new economic and cultural centers of development, seems steeped in trends of social conflict originating in the Hellenistic period, while needing to redefine its identity in the framework of a global empire.[76] Elements of emulation or assimilation into wider-ranging ways of life seem to find expression at many levels. Surveys draw attention to the presence of mausolea and large villas dotting the Attic landscape,[77] hitherto unknown in terms of monumentality and individualism. The cemetery of the Kerameikos demonstrates a similar monumentality along the Dromos that centuries earlier had been used for state burials.[78] Analysis of diet based on grave types suggests that socioeconomic differences may have been articulated by grave type and sex, where simple pit graves of women in particular, have lower protein signatures compared to those buried in built graves and sarcophagi.

The analysis of bone samples from Laurion, a cemetery potentially belonging to workers of the silver mines in the Classical period, demonstrated that their diet was not inferior to that of residents from the urban center. The range, however, of protein sources seems to have been broader, like respective dietary signatures from the Hellenistic period from the urban cemeteries. Relevant dietary ranges in other contexts have been attributed to greater diversity in the dietary spectrum due to the presence of different ethnic groups in the sample or greater social stratification.[79] In Greece, demographic mobility is believed to have accelerated between 350 and

72. Vika 2011.
73. Habicht 1998, p. 65.
74. Alcock 1993, 218; Reger 2007, p. 460.
75. Lagia 2007, 2015.
76. Alcock 1993, pp. 16–17.
77. Alcock 1993, p. 67.
78. Stroszeck 2008.
79. Garvie-Lok 2001, pp. 456–457; Rutgers et al. 2009, p. 1133; Gregoricka and Sheridan 2013, p. 70; Killgrove and Tykot 2013, p. 36.

250 B.C., when a number of institutions and agreements between *poleis* "facilitated movements of persons or groups."[80] It has been suggested that Athens had a greater cosmopolitanism after the reacquisition of Delos in 167 B.C.,[81] and an increase in the movements of people from rural areas to urban centers was seen then, driven by growing economic pressures and social polarities.[82]

The significant reduction of the protein level in the Imperial Roman period suggests a deterioration of the standard of living during the empire. This assessment conflicts with the proposed model of a generalized improvement in the standard of living in the Roman period, and instead corroborates studies from other Roman provinces where the standard of living seems to have declined in the Roman period.[83] Comparison with other parts of the empire offers the possibility to refine our understanding of socioeconomic disparities that are otherwise difficult to access through historical sources. Classical Athens seems thus far to be more similar in terms of protein inclusion with the higher strata of Roman sites in Isola Sacra and Velia.[84] By contrast, the Hellenistic and Imperial Roman periods are more in line with weaker strata in these sites. Respective evidence nevertheless from the city of Rome bears witness of the multiplicity of factors that may have shaped dietary patterns between and within regions.[85]

Classical Thebes, with a significant contribution of freshwater fish, differs in terms of diet from Roman Isola Sacra and Leptiminus where large marine resources are apparent in the diet.[86] In several sites and chronological periods the contribution of marine foods to the diet seems weakly represented. It has been noted, however, that this could be due to the difficulty of tracing marine foods, especially small fish such as sardine or anchovy, within a basically terrestrial diet, if their presence does not exceed 20%–30% of the diet.[87] Likewise, the contribution of pulses in the diet of ordinary people, attested in written sources, is only indirectly inferred from a generalized decrease in nitrogen levels.[88]

In recent years, attention has been drawn to the difficulties of quantifying the importance of terrestrial or marine protein in diets, because of our poor understanding of the degree of fractionation between diet and collagen, which may lead to an overestimation of the contribution of high trophic level foods such as meat, milk, and fish in the diet.[89] Likewise, the effects on $\delta^{15}N$ of factors relating to human physiology, including disease and growth patterns, and of dynamics related to climatic stress and husbandry strategies are not yet well understood.[90] The homogenous picture of an elevated protein in the Classical period in sites as diverse as Makriyalos, Thebes, and Athens vis-à-vis earlier periods, in addition to a corresponding decrease in the Hellenistic period in Athens and Thebes, suggest that the causes of these changes were transformations in the living standards rather than random factors related to individual physiology or economic strategy. If manuring is suggested to explain the increase of nitrogen in the Classical period, it should also concern the numerous sites from which cereals were systematically imported in Athens, plus the concurrent change of $\delta^{15}N$ values in two *poleis* of the Hellenistic period.

Although the question of dietary quality and its variable distribution among populations in classical antiquity cannot be addressed yet with

80. Davies 2006, pp. 78–79; Reger 2007, pp. 461–475.

81. Day 1942, p. 76; Habicht 1998, p. 378.

82. Alcock 1993, pp. 215–217.

83. E.g., Redfern, Hamlin, and Beavan Athfield 2010; Redfern and DeWitte 2011.

84. Prowse et al. 2004; Craig et al. 2009.

85. Killgrove and Tykot 2013, p. 33.

86. Prowse et al. 2004; Keenleyside et al. 2009; Vika, Aravantinos, Richards 2009.

87. Garvie-Lok 2001, pp. 326–332; Hedges 2004, p. 37.

88. E.g., Lösch, Grupe, and Peters 2006, p. 190; Vika 2011, p. 1159.

89. O'Connell et al. 2012, p. 432.

90. Fraser et al. 2011; and references in note 71, above.

precision, it is clear that the investigation of larger samples from diverse sites, the study of the entire food web, and cognizance of the many factors that affect isotopic values will provide a basis for conducting comparative analyses and ultimately a more nuanced understanding of ancient standards of living.

ACKNOWLEDGMENTS

I would like to thank Mike Richards for conducting the stable isotope analysis during his appointment at Bradford University and Ian Morris and the Department of Classics of Stanford University for supporting and funding the initial stages of this work. For permissions to take samples from the cemeteries of Plateia Kotzia, the Kerameikos, and Laurion, I would like to thank Liana Parlama and Olga Zahariadou of the 3rd Ephorate of Prehistoric and Classical Antiquities; Jutta Stroszeck, Rudolf Stichel, and Karin Braun of the German Archaeological Institute at Athens; Christos Kittas and Theodoros Pitsios from the Museum of Anthropology of the University of Athens; and the late Maria Salliora-Oikonomakou of the 2nd Ephorate of Prehistoric and Classical Antiquities. Finally, I would like to thank the reviewers and Donald Haggis for valuable comments on the text.

REFERENCES

Alcock, S. E. 1993. *Graecia capta: The Landscapes of Roman Greece,* Cambridge.

Bianchi, G. C., and I. N. McCave. 1999. "Holocene Periodicity in North Atlantic Climate and Deep-Ocean Flow South of Iceland," *Nature* 397, pp. 515–517.

Bookidis, N., J. Hansen, L. Snyder, and P. Goldberg. 1999. "Dining in the Sanctuary of Demeter and Kore at Corinth," *Hesperia* 68, pp. 1–54.

Bourbou, C., B. T. Fuller, S. J. Garvie-Lok, and M. P. Richards. 2011. "Reconstructing the Diets of Greek Byzantine Populations (6th–15th Centuries A.D.) Using Carbon and Nitrogen Stable Isotope Ratios," *American Journal of Physical Anthropology* 146, pp. 569–581.

———. 2013. "Nursing Mothers and Feeding Bottles: Reconstructing Breastfeeding and Weaning Patterns in Greek Byzantine Populations (6th–15th Centuries A.D.) Using Carbon and Nitrogen Stable Isotope Ratios," *JAS* 40, pp. 3903–3913.

Bourbou, C., and M. P. Richards. 2007. "The Middle Byzantine Menu: Palaeodietary Information from Isotopic Analysis of Humans and Fauna from Kastella, Crete," *International Journal of Osteoarchaeology* 17, pp. 63–72.

Buikstra, J. E., and D. H. Ubelaker, eds. 1994. *Standards for Data Collection from Human Skeletal Remains. Proceedings of a Seminar at the Field Museum of Natural History, Organized by Jonathan Haas* (Arkansas Archeological Survey Research Series 44), Fayetteville.

Burke, E. M. 1992. "The Economy of Athens in the Classical Era: Some Adjustments to the Primitivist Model," *TAPA* 122, pp. 199–226.

Camp, J. McK., II. 1982. "Drought and Famine in the 4th Century B.C." in *Studies in Athenian Architecture, Sculpture, and Topography: Presented to Homer A. Thompson* (*Hesperia* Suppl. 20), Princeton, pp. 9–17.

Cartledge, P. 1997. "Introduction," in *Hellenistic Constructs: Essays in Culture, History, and Historiography* (Hellenistic Culture and Society 26), ed. P. Cartledge, P. Garnsey, and E. Gruen, Berkeley, pp. 1–19.

Craig, O. E., M. Biazzo, T. C. O'Connell, P. Garnsey, C. Martinez-Labarga, R. Lelli, L. Salvadei, G. Tartaglia, A. Nava, L. Renò, A. Fiammenghi, O. Rickards, and L. Bondioli. 2009. "Stable Isotopic Evidence for Diet at the Imperial Roman Coastal Site of Velia (1st and 2nd Centuries A.D.) in Southern Italy," *American Journal of Physical Anthropology* 139, pp. 572–583.

Dalby, A. 1996. *Siren Feasts: A History of Food and Gastronomy in Greece,* London.

Davies, J. K. 2006. "Hellenistic Economies," in *The Cambridge Companion to The Hellenistic World,* ed. G. R. Bugh, Cambridge, pp. 73–93.

———. 2007. "Classical Greece: Production," in Scheidel, Morris, and Saller 2007, pp. 333–361.

Day, J. 1942. *An Economic History of Athens under Roman Domination,* New York.

DeNiro, M. J. 1985. "Postmortem Preservation and Alteration of *in vivo* Bone Collagen Isotope Ratios in Relation to Palaeodietary Reconstruction," *Nature* 317, pp. 806–809.

Dupras, T. L., H. P. Schwarcz, and S. I. Fairgrieve. 2001. "Infant Feeding and Weaning Practices in Roman Egypt," *American Journal of Physical Anthropology* 115, pp. 204–212.

Eleutheratou, S. 2008. "Στοιχεία πολεοδομικής και οικιστικής οργάνωσης από το νότιο τμήμα της αρχαίας Αθήνας," in *Η Αθήνα κατά τη ρωμαϊκή εποχή: Πρόσφατες ανακαλύψεις, νέες έρευνες. Akten des Symposions im Benakimuseum Oktober 2006* (Benaki Museum Suppl. 4), ed. S. Vlizos, Athens, pp. 185–205.

Fraser, R. A., A. Bogaard, T. Heaton, M. Charles, G. Jones, B. T. Christensen, P. Halstead, I. Merbach, P. R. Poulton, D. Sparkes, and A. K. Styring. 2011. "Manuring and Stable Nitrogen Isotope Ratios in Cereals and Pulses: Towards a New Archaeobotanical Approach to the

Inference of Land Use and Dietary Practices," *JAS* 38, pp. 2790–2804.

Fuller, B. T., B. De Cupere, E. Marinova, W. Van Neer, M. Waelkens, and M. P. Richards. 2012. "Isotopic Reconstruction of Human Diet and Animal Husbandry Practices during the Classical–Hellenistic, Imperial, and Byzantine Periods at Sagalassos, Turkey," *American Journal of Physical Anthropology* 149, pp. 157–171.

Fuller, B. T., J. L. Fuller, N. E. Sage, D. A. Harris, T. C. O'Connell, and R. E. M. Hedges. 2005. "Nitrogen Balance and δ^{15}N: Why You're Not What You Eat during Nutritional Stress," *Rapid Communications in Mass Spectrometry* 19, pp. 2497–2506.

Fuller, B. T., T. L. Molleson, D. A., Harris, L. T. Gilmour, and R. E. Hedges. 2006. "Isotopic Evidence for Breastfeeding and Possible Adult Dietary Differences from Late/Sub-Roman Britain," *American Journal of Physical Anthropology* 129, pp. 45–54.

Garnsey, P. 1988. *Famine and Food Supply in the Graeco-Roman World: Responses to Risk and Crisis*, Cambridge.

———. 1999. *Food and Society in Classical Antiquity,* Cambridge.

Garvie-Lok, S. J. 2001. "Loaves and Fishes: A Stable Isotope Reconstruction of Diet in Medieval Greece" (diss. Univ. of Calgary).

Gregoricka, L. A., and S. G. Sheridan. 2013. "Ascetic or Affluent? Byzantine Diet at the Monastic Community of St. Stephen's, Jerusalem from Stable Carbon and Nitrogen Isotopes," *JAnthArch* 32, pp. 63–73.

Habicht, C. 1998. *Ελληνιστική Αθήνα,* Athens.

Hansen, J. M. 2000. "Palaeoethnobotany and Palaeodiet in the Aegean Region: Notes on Legume Toxicity and Related Pathologies," in *Palaeodiet in the Aegean. Papers from a Colloquium Held at the 1993 Meeting of the Archaeological Institute of America in Washington D.C.* (Wiener Laboratory Monograph 1), ed. S. J. Vaughan and W. D. E. Coulson Oxford, pp. 13–27.

Hansen, M. H. 2006. *Polis: An Introduction to the Ancient Greek City-State,* Oxford.

Hedges, R. E. M. 2004. "Isotopes and Red Herrings: Comments on Milner et al. and Lidén et al.," *Antiquity* 78, pp. 34–37.

Hedges, R. E. M., and L. M. Reynard. 2007. "Nitrogen Isotopes and the Trophic Level of Humans in Archaeology," *JAS* 34, pp. 1240–1251.

Henneberg, R. J., and M. Henneberg. 2003. "The Diet of the Metapontine Population as Reconstructed from the Physical Remains," in *Living Off the Chora: Diet and Nutrition at Metaponto,* ed. J. C. Carter, Austin, pp. 29–36.

Karagiorga-Stathakopoulou, T. 1988. "Δημόσια έργα και ανασκαφές στην Αθήνα τα τελευταία πέντε χρόνια," *ΗΟΡΟΣ* 6, pp. 87–108.

Katzenberg, M. A., D. A. Herring, and S. R. Saunders. 1996. "Weaning and Infant Mortality: Evaluating the Skeletal Evidence," *Yearbook of Physical Anthropology* 39, pp. 177–199.

Katzenberg, M. A., and N. C. Lovell. 1999. "Stable Isotope Variation in Pathological Bone," *International Journal of Osteoarchaeology* 9, pp. 316–324.

Keenleyside, A., H. Schwarcz, and K. Panayotova. 2006. "Stable Isotopic Evidence of Diet in a Greek Colonial Population from the Black Sea," *JAS* 33, pp. 1205–1215.

Keenleyside, A., H. Schwarcz, L. Stirling, and N. Ben Lazreg. 2009. "Stable Isotopic Evidence for Diet in a Roman and Late Roman population from Leptiminus, Tunisia," *JAS* 36, pp. 51–63.

Killgrove, K., and R. H. Tykot. 2013. "Food for Rome: A Stable Isotope Investigation of Diet in the Imperial period (1st–3rd Centuries A.D.)," *JAnthArch* 32, pp. 28–38.

Knigge, U. 1991. *The Athenian Kerameikos: History-Monuments-Excavations*, Athens.

Kron, G. 2005. "Anthropometry, Physical Anthropology, and the Reconstruction of Ancient Health, Nutrition, and Living Standards," *Historia: Zeitschrift für Alte Geschichte* 54, pp. 68–83.

———. 2011. "The Distribution of Wealth at Athens in Comparative Perspective," *ZPE* 179, pp. 129–138.

———. 2012. "Standards of Living, Wealth" in *The Encyclopedia of Ancient History*, ed. R. S. Bagnall, K. Brodersen, C. B. Champion, A. Erskine, and S. R. Huebner, Oxford.

Kuniholm, P. I. 1990. "Archaeological Evidence and Non-Evidence for Climatic Change," *Philosophical Transactions of the Royal Society of London. Series A: Mathematical and Physical Sciences* 330 1615, pp. 645–655.

Kurtz, D. C., and J. Boardman. 1994. *Greek Burial Customs,* Athens.

Lagia, A. 2007. "Notions of Childhood in the Classical *Polis:* Evidence from the Bioarchaeological Record," in *Constructions of Childhood in Greek and Roman Antiquity,* ed. J. Rutter and A. Cohen (*Hesperia* Suppl. 41), Princeton, pp. 293–306.

———. 2014. "Health Inequalities in the Classical City: A Biocultural Approach to Socioeconomic Differentials in the Polis of Athens during the Classical, Hellenistic and Imperial Roman Periods," in *Corps, travail et statut social*, ed. A.-C. Gillis, Villeneuve d'Ascq, pp. 95–115.

———. 2015. "The Potential and Limitations of Bioarchaeological Investigations in Classical Contexts in Greece: An Example from the *Polis* of Athens," in *Classical Archaeology in Context: Theory and Practice in Excavation in the Greek World*, ed. D. C. Haggis and C. M. Antonaccio, Berlin, pp. 149–173.

Lösch, S., G. Grupe, and J. Peters. 2006. "Stable Isotopes and Dietary Adaptations in Humans and Animals at Pre-Pottery Neolithic Nevalı Çori, Southeast Anatolia," *American Journal of Physical Anthropology* 131, pp. 181–193.

MacKinnon, M. 2014. "Animals, Economics, and Culture in the Athenian Agora: Comparative Zooarchaeological Investigations," *Hesperia* 83, pp. 189–255.

Margaritis, E. 2014. "The Kapeleio at Hellenistic Krania: Food Consumption, Disposal, and the Use

of Space," *Hesperia* 83, pp. 103–121.
Megaloudi, F. 2006. *Plants and Diet in Greece from Neolithic to Classic Periods: The Archaeobotanical Remains* (*BAR-IS* 1516), Oxford.
Migeotte, L. 2009. *The Economy of the Greek Cities*, Berkley.
Morris, I. 1997. "Periodization and the Heroes: Inventing a Dark Age," in *Inventing Ancient Culture*, ed. M. Golden and I. Toohey, New York, pp. 96–131.
———. 2004. "Economic Growth in Ancient Greece," *Journal of Institutional and Theoretical Economics* 160, pp. 709–742.
———. 2005. "Archaeology, Standards of Living, and Greek Economic History," in *The Ancient Economy. Evidence and Models*, ed. S. Manning and I. Morris, Stanford, pp. 91–126.
———. 2009. "The Greater Athenian State," in *The Dynamics of Ancient Empires*, ed. I. Morris and W. Scheidel, Oxford, pp. 99–177.
Morris, I., W. Scheidel, and R. Saller. 2007. "Introduction," in Scheidel, Morris, and Saller 2007, pp. 1–12.
Müldner, G., and M. P. Richards. 2007. "Stable Isotope Evidence for 1,500 Years of Human Diet at the City of York, UK," *American Journal of Physical Anthropology* 133 pp. 682–697.
Ober, J. 2010. "Wealthy Hellas," *TAPA* 140, pp. 241–286.
O'Connell, T. C., C. J. Kneale, N. Tasevska, and G. G. C. Kuhnle. 2012. "The Diet-Body Offset in Human Nitrogen Isotopic Values: A Controlled Dietary Study," *American Journal of Physical Anthropology* 149, pp. 426–434.
Papathanasiou, A. 2001. *A Bioarchaeological Analysis of Neolithic Alepotrypa Cave, Greece* (*BAR-IS* 961), Oxford.
Papathanasiou, A., E. Panagiotopoulou, K. Beltsios, M-F. Papakonstantinou, and M. Sipsi. 2013. "Inferences from the Human Skeletal Material of the Early Iron Age Cemetery at Agios Dimitrios, Fthiotis, Central Greece," *JAS* 40, pp. 2924–2933.
Parlama, L., and C. Stampolidis. 2000. *Η πόλη κάτω από την πόλη*, Athens.
Prowse, T. L., S. R. Saunders, H. P. Schwarcz, P. Garnsey, R. Macchiarelli, and L. Bondioli. 2008. "Isotopic and Dental Evidence for Infant and Young Child Feeding Practices in an Imperial Roman Skeletal Sample," *American Journal of Physical Anthropology* 137, pp. 294–308.
Prowse, T. L., H. P. Schwarcz, S. R. Saunders, R. Macchiarelli, and L. Bondioli. 2004. "Isotopic Paleodiet Studies of Skeletons from the Imperial Roman-Age Cemetery of Isola Sacra, Rome, Italy," *JAS* 31, pp. 259–272.
———. 2005. "Isotopic Evidence for Age-Related Variation in Diet from Isola Sacra, Italy," *American Journal of Physical Anthropology* 128, pp. 2–13.
Redfern, R. C., and S. N. DeWitte. 2011. "A New Approach to the Study of Romanization in Britain: A Regional Perspective of Cultural Change in Late Iron Age and Roman Dorset Using the Siler and Gompertz-Makeham Models of Mortality," *American Journal of Physical Anthropology* 144, pp. 269–285.
Redfern, R. C., C. Hamlin, and N. Beavan Athfield. 2010. "Temporal Changes in Diet: A Stable Isotope Analysis of Late Iron Age and Roman Dorset, Britain," *JAS* 37, pp. 1149–1160.
Redfern, R. C., A. R. Millard, and C. Hamlin. 2012. "A Regional Investigation of Subadult Dietary Patterns and Health in Late Iron Age and Roman Dorset, England," *JAS* 39, pp. 1249–1259.
Reger, G. 2007. "Hellenistic Greece and Western Asia Minor," in Scheidel, Morris, and Saller 2007, pp. 460–483.
Richards, M. P., R. E. M. Hedges, T. I. Molleson, and J. C. Vogel. 1998. "Stable Isotope Analysis Reveals Variations in Human Diet at the Poundbury Camp Cemetery Site," *JAS* 25, pp. 1247–1252.
Rutgers, L. V., M. van Strydonck, M. Boudin, and C. van der Linde. 2009. "Stable Isotope Data from the Early Christian Catacombs of Ancient Rome: New Insights into the Dietary Habits of Rome's

Early Christians," *JAS* 36, pp. 1127–1134.
Salliora-Oikonomakou, M. 1985. "Αρχαίο νεκροταφείο στην περιοχή Λαυρίου," *ArchDelt* 40, A′, pp. 90–132.
Sarpaki, A. 1992. "The Palaeoethnobotanical Approach: The Mediterranean Triad, or Is It a Quartet?" in *Agriculture in Ancient Greece. Proceedings of the Seventh International Symposium at the Swedish Institute at Athens, 16–17 May, 1990* (*ActaAth* 4°, 42), ed. B. Wells, Stockholm, pp. 61–76.
Scheidel, W. 2007. "Demography," in Scheidel, Morris, and Saller 2007, pp. 38–86.
———. 2010. "Human Development and Quality of Life in the Long Run: The Case of Greece," Princeton/Stanford Working Papers in Classics, http://www.princeton.edu/~pswpc/pdfs/scheidel/091006.pdf (accessed October 22, 2012).
Scheidel, W., I. Morris, and R. P. Saller, eds. 2007. *The Cambridge Economic History of the Greco-Roman World*, Cambridge.
Schwarcz, H. P., T. L. Dupras, and S. I. Fairgrieve. 1999. "^{15}N Enrichment in the Sahara: In Search of a Global Relationship," *JAS* 26, pp. 629–636.
Schwarcz, H. P., and M. J. Schoeninger. 1991. "Stable Isotope Analyeses in Human Nutritional Ecology," *Yearbook of Physical Anthropology* 34, pp. 283–321.
Silver, M. 2007. "Roman Economic Growth and Living Standards: Perceptions versus Evidence," *Ancient Society* 37, pp. 191–252.
Stroszeck, J. 2008. "Römische Gräber und Grabbauten vor dem Dipylon," in *Η Αθήνα κατά τη ρωμαϊκή εποχή: Πρόσφατες ανακαλύψεις, νέες έρευνες. Akten des Symposions im Benakimuseum Oktober 2006* (Benaki Museum Suppl. 4), ed. S. Vlizos, Athens, pp. 291–309.
Triantaphyllou, S. 2001. *A Bioarchaeological Approach to Prehistoric Cemetery Populations from Central and Western Greek Macedonia* (*BAR-IS* 976), Oxford.
Triantaphyllou, S., M. P. Richards, C. Zerner, and S. Voutsaki. 2008. "Isotopic Dietary Reconstruction of Humans from Middle Bronze Age Lerna, Argolid, Greece," *JAS* 35, pp. 3028–3034.
van Klinken, G. J. 1999. "Bone Collagen Quality Indicators for Paleodietary and Radiocarbon Measurements," *JAS* 26, pp. 687–695.
Vika, E. 2011. "Diachronic Dietary Reconstructions in Ancient Thebes, Greece: Results from Stable Isotope Analyses," *JAS* 38, pp. 1157–1163.
Vika, E., V. Aravantinos, and M. P. Richards. 2009. "Aristophanes and Stable Isotopes: A Taste for Freshwater Fish in Classical Thebes, (Greece)?" *Antiquity* 83, pp. 1076–1083.
Vika, E., and T. Theodoropoulou. 2012. "Re-Investigating Fish Consumption in Greek Antiquity: Results from $\delta^{13}C$ and $\delta^{15}N$ Analysis from Fish Bone Collagen," *JAS* 39, pp. 1618–1627.
Waters-Rist, A. L., and M. A. Katzenberg. 2010. "The Effect of Growth on Stable Nitrogen Isotope Ratios in Subadult Bone Collagen," *International Journal of Osteoarchaeology* 20, pp. 172–191.
White, C. D., and G. J. Armelagos. 1997. "Osteopenia and Stable Isotope Ratios in Bone Collagen of Nubian Female Mummies," *American Journal of Physical Anthropology* 103, pp. 185–199.
Zachariadou, O. 2008. "Η ανατολική περιοχή της Αθήνας κατά τη ρωμαϊκή περίοδο," in *Η Αθήνα κατα τη ρωμαϊκή εποχή: Πρόσφατες ανακαλύψεις, νέες έρευνες. Akten des Symposions im Benakimuseum Oktober 2006* (Benaki Museum Suppl. 4), ed. S. Vlizos, Athens, pp. 153–166.
Zachariadou, O., and D. Kyriakou. 1988. "Πλατεία Κοτζιά," *ArchDelt* 43, B′1, pp. 22–29.

CHAPTER 9

Stable Isotope Evidence for Infant Feeding Practices in the Greek Colony of Apollonia Pontica

by Cynthia S. Kwok and Anne Keenleyside

Weaning, the process whereby an infant gradually becomes independent of breast milk and begins to consume complementary food,[1] is a fundamental step in an individual's life. The importance of breast milk to a growing infant cannot be diminished, since it provides food and nourishment for the first few months of life, as well immunological factors such as T- and B-lymphocytes, immunoglobulins, and anti-staphylococcus factors necessary for proper immune system development.[2] Furthermore, as the majority of infants in the past were likely breastfed due to the lack of safe alternatives, the significance of weaning cannot be overstated.

A number of studies have utilized stable isotope analysis on bones and teeth to determine the timing and duration of weaning in past populations.[3] Particular attention has been paid to stable nitrogen isotope analysis, as tissues of breastfeeding infants have been demonstrated to be enriched in ^{15}N in comparison to their mothers.[4] The premise behind the use of stable nitrogen isotopes in weaning studies is that nitrogen exhibits a "trophic level effect" in which organisms higher on the food chain will show a step-wise elevation in their $\delta^{15}N$ values by approximately 2‰ to 4‰ compared to their diet.[5] Since infants feeding exclusively on breast milk are considered to be ingesting their mothers' tissues, theoretically they are one trophic level higher and should accordingly have $\delta^{15}N$ values that are elevated by

1. The term "weaning" has been used differently within the literature, and it is not always clear whether it refers to the onset of weaning, if it is a gradual process, or if it means the complete cessation of breastfeeding (see Dettwyler and Fishman 1992, pp. 172–175, for a discussion of the terminology of weaning). In this paper, weaning is considered a gradual process, as outlined by Herring, Saunders, and Katzenberg 1998, p. 425. The term "weaned" represents when the infant has completely ceased breastfeeding, and "weaning age" signifies the age at which subadults are completely weaned.

2. Hanson and Winberg 1972, p. 846; Slade and Schwarcz 1987, p. 352; see also Metsecky, Blair, and Ogra 1991.

3. Fogel, Tuross, and Owsley 1989, pp. 111–116; Katzenberg and Pfeiffer 1995, pp. 225–231; Katzenberg, Herring, and Saunders 1996, pp. 187–192; Schurr 1997, pp. 922–925; Herring, Saunders, and Katzenberg 1998, pp. 432–437; Wright and Schwarcz 1998; 1999, pp. 1163–1169; Dupras, Schwarcz, and Fairgrieve 2001; Richards, Mays, and Fuller 2002, pp. 207–209; Fuller, Molleson, et al. 2006; Bourbou and Richards 2007; Dupras and Tocheri 2007, pp. 68–72; Kwok 2007, pp. 97–125; Prowse et al. 2008, pp. 300–305; Triantaphyllou et al. 2008, pp. 3030–3033; Bourbou and Garvie-Lok 2009, pp. 77–83; Keenleyside et al. 2009, p. 60.

4. Fogel, Tuross, and Owsley 1989, p. 114.

5. DeNiro and Epstein 1981, p. 343; Minagawa and Wada 1984, p. 1139; Schoeninger and DeNiro 1984, p. 635.

approximately 2‰ to 4‰ in comparison to the values of their mothers.[6] Once weaning begins and breast milk is no longer the main source of dietary protein, the $\delta^{15}N$ values of the child will decrease. The end of the weaning period is signaled when the infant $\delta^{15}N$ values approach the mean adult female values, which indicates the consumption of "adult foods."[7]

Previous biochemical weaning studies in the classical world have been largely confined to the Roman period, while ancient Greek weaning practices remain unexplored. This paper investigates infant feeding practices in the Greek colony of Apollonia Pontica, located on the Black Sea coast of Bulgaria, using stable nitrogen and carbon isotope analyses of bone collagen from subadult skeletal remains. The objectives of this study are to determine: (1) the general age at which weaning was initiated and terminated, (2) the types of foods onto which these subadults were weaned, (3) the composition of the childhood diet, (4) the degree to which the subadult diet differed from that of the adults, and (5) the health of these children in relation to their diet.

ANCIENT LITERARY EVIDENCE FOR INFANT FEEDING PRACTICES

Few sources exist in the Greco-Roman literature that detail infant feeding practices and the childhood dietary regimen. Most sources date to the Roman period, while Greek references to weaning practices are few in number. However, considering that Roman practices were heavily influenced by Greek culture, particularly medicine, it is assumed that both cultures shared similar practices.[8]

Medical texts and wet-nursing contracts from Egypt provide much of the evidence for weaning practices in Greco-Roman society. As detailed in the medical treatises of Hippocrates, Soranus, and Galen, weaning was a gradual process that, on average, began at 6 months and ended around 2 to 3 years of age.[9] In addition to chronological age, other indicators used to signal the onset and duration of weaning were the state of dental emergence, the environment, and the sex of the child. Weaning was to be terminated when the majority of the deciduous dentition had emerged, while winter or spring were considered to be the best seasons for successful weaning since the weather was consistent. Autumn was thought to be the worst season because of the fluctuating temperatures and increased prevalence of illnesses, believed to be brought on by a change in people's daily habits. Furthermore, girls were reportedly weaned 6 months later than boys, owing to their "weaker nature," although physicians such as Soranus

6. Fogel, Tuross, and Owsley 1989, p. 114; Schurr 1997, p. 920; Dupras, Schwarcz, and Fairgrieve 2001, p. 205; Fuller, Richards, and Mays 2003, p. 1673.

7. Fogel, Tuross, and Owsley 1989, pp. 113–114.

8. King 2001, p. 32.

9. Fildes 1988; Garnsey 1998, p. 268; Gal. *On Hygiene* 1.9, trans. R. M. Green, Springfield 1951; Sor. *Gyn.* 2.46.115, trans. O. Temkin, Baltimore 1956.

disagreed with this characterization, noting that girls in some cases were "stronger and fleshier" than boys.[10]

Variation in weaning practices is also documented, with some children being weaned earlier, and others later than the recommended age. Greco-Roman period Egyptian wet-nursing contracts detail weaning ages ranging from 6 months to 3 years.[11] In comparison, Soranus[12] documents mothers weaning their children onto cereals as early as 40 days after birth. In contrast, he also recommends that children who fall ill after weaning revert to breastfeeding.[13]

Typical weaning foods in antiquity were milk and cereal dishes that were prepared as porridges or gruel and flavored with milk, honey, or salt.[14] Wheat and barley breadcrumbs softened with wine, water, or milk were also considered to be suitable complementary foods, while Soranus[15] recommended that older children be given spelt and an egg that could be "sipped."

HISTORICAL AND ARCHAEOLOGICAL CONTEXT OF APOLLONIA PONTICA

Located on the Black Sea coast of Bulgaria on the site of the modern day town of Sozopol, Apollonia Pontica was founded in 610 B.C. by colonists from the city of Miletus in Asia Minor (Fig. 9.1). The site had a number of desirable characteristics, including an excellent harbor, plentiful fishing, and an easily defensible position.[16] Its strategic location contributed to its establishment as a major trading center, and the colony flourished during the 5th to 4th centuries B.C., with a population estimated to have numbered 3,000 inhabitants.[17]

In 1938, the chance discovery of a grave in a necropolis located 2.5 km south of Sozopol (Fig. 9.2) prompted scholars to direct their attention to this area, and from 1946 to 1949, large-scale excavations uncovered over 800 graves.[18] Since 1992, renewed excavations of what is now known as the Kalfata-Budjaka necropolis by Kristina Panayotova of the Institute of Archaeology in Sofia have recovered 1,500 additional graves consisting of simple, unlined pits, stone cists, tile graves, wooden coffins, pithoi, and amphoras. Most of the burials are typical Greek inhumations containing single individuals laid out on their backs in an extended position. The remains date from the second half of the 5th century to the middle of the 3rd century B.C. based on associated grave goods.[19] Palaeopathological and stable isotope analyses have been conducted on some of these remains,[20] and the samples utilized in this study are derived from the most recent excavations.

10. Fildes 1986, p. 7; Hippoc. *On Dentition* 9.16, trans. W. H. S. Jones, London 1923; Sor. *Gyn.* 2.46.115, trans. O. Temkin, Baltimore 1956.
11. Garnsey 1998, p. 268.
12. Sor. *Gyn.* 2.46.115, trans. O. Temkin, Baltimore 1956.
13. Sor. *Gyn.* 2.48.117, trans. O. Temkin, Baltimore 1956; see also Chaps. 7, 10.
14. Brothwell and Brothwell 1998, pp. 187–188; Gal. *On the Powers of Food* 3, trans. M. Grant, London, 2000.
15. Sor. *Gyn.* 2.46.115, trans. O. Temkin, Baltimore, 1956.
16. Nedev and Panayotova 2003, p. 8.
17. Danov 1948, as cited in Nedev and Panayotova 2003.
18. Venedikov 1963, p. 372.
19. Nedev and Panayotova 2003.
20. Keenleyside and Panayotova 2005; 2006, pp. 373–382; Keenleyside, Schwarcz, and Panayotova 2006.

Figure 9.1. Location of Sozopol (ancient Apollonia Pontica), Bulgaria

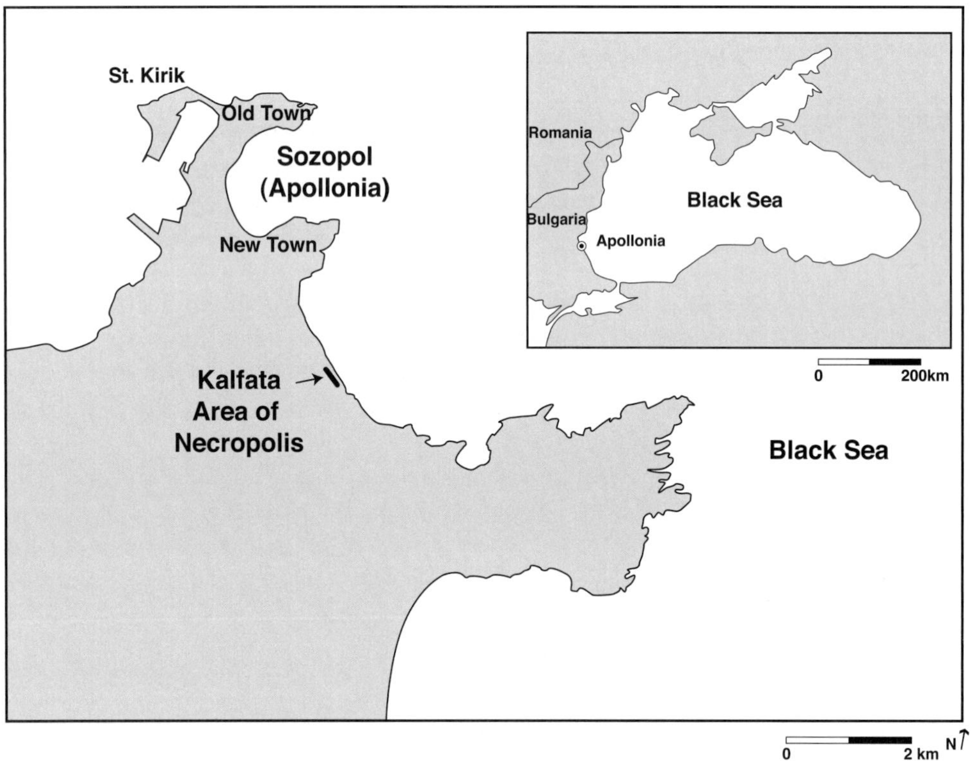

Figure 9.2. Location of the Kalfata area of the necropolis at Apollonia Pontica

MATERIALS AND METHODS

Subadult age was estimated using standard morphological criteria, specifically: (1) the degree of dental development,[21] (2) the state of dental eruption,[22] (3) the diaphyseal length,[23] and (4) the stage of epiphyseal fusion.[24] Any pathological lesions observed in the remains were recorded in order to assess the health impact of weaning on the individuals examined in this study.

Bone samples were obtained from 64 subadults ranging in age from 3 months to 15 years. Collagen was extracted using the Longin method, as modified by DeNiro and Epstein, and Chisholm and colleagues.[25] The samples were analyzed using a Thermo Finnigan Delta Plus XP mass spectrometer, and the precision of analysis was ± 0.1‰ for $\delta^{13}C$ and ± 0.2‰ for $\delta^{15}N$. The $\delta^{13}C$ and $\delta^{15}N$ values were measured relative to VPDB and AIR standards, respectively. The degree of preservation and integrity of collagen were examined using collagen yields, atomic carbon to nitrogen ratios (atomic C:N), and carbon (%C) and nitrogen (%N) content. The collagen yields ranged from 0.4% to 15.2%, and all atomic C:N ratios fell within the acceptable range of 2.9 to 3.6, as defined by DeNiro,[26] indicating adequate sample preservation. In addition, the %C and %N of the samples were within the range of satisfactory values (%C: 15% to 37%; %N: 5% to 17%),[27] with the exception of one sample (burial no. 5072–27), which had a carbon content that fell below this range[28] (Table 9.1). The sample was consequently excluded from further analysis.

RESULTS: STABLE ISOTOPE DATA BY AGE

The $\delta^{15}N$ values for the Apollonian subadults ranged from 8.4‰ to 12.5‰ (mean = 10.2‰ ± 1.1‰) (Table 9.1). When the values were plotted against the estimated age-at-death for each individual in the sample (Fig. 9.3), infants less than 1 year of age exhibited the highest $\delta^{15}N$ signatures, averaging approximately 1‰ higher than the adult female mean (Mann-Whitney, $U = 16$, $p<0.0001$), suggesting that they were feeding on breast milk near the time of their death. The $\delta^{15}N$ values decline around 6 months to 1 year of age, signaling that breast milk no longer made up the majority of their protein source and that weaning had begun. Subadults between the ages of 2 and 4 possessed $\delta^{15}N$ values that are not significantly different from the adult female mean, indicating that weaning was completed by this age.

The $\delta^{13}C$ values ranged from -17.3‰ to -19.5‰ (mean = -18.5‰ ± 0.5‰) (Table 9.1). When these values are compared to the estimated age-at-death (Fig. 9.4), infants less than 1 year of age exhibit the highest $\delta^{13}C$ values, possibly reflecting a small trophic level shift in breastfeeding

21. Moorrees, Fanning, and Hunt 1963a, pp. 1495–1496; 1963b; Smith 1991.

22. Ubelaker 1989, p. 64.

23. Stloukal and Hanáková 1978, p. 65; Sundick 1978; Ubelaker 1989, pp. 70–71.

24. As cited in Buikstra and Ubelaker 1994, p. 43; see also Flecker 1932.

25. Longin 1971; DeNiro and Epstein 1981, pp. 342–343; Chisholm et al. 1983, p. 357.

26. DeNiro 1985, p. 808.

27. Ambrose 1990, p. 441; van Klinken 1999, pp. 690–691.

28. Ambrose 1990, p. 441.

TABLE 9.1. STABLE ISOTOPE VALUES AT APOLLONIA PONTICA (N = 64)*

Burial No.	*Age (years)*	*Tissue*	*δ¹⁵N (‰ AIR)*	*δ¹³C (‰ VPDB)*	*Atomic C:N*	*Collagen Yield (%)*	*% N*	*% C*
Birth to 1 Year								
247	0.25	ribs	12.3	-18.7	3.4	12.0	15.1	43.9
266–03	0.25	ribs	11.8	-17.9	3.3	2.2	13.0	37.4
363	0.25	long bones	12.3	-18.3	3.4	2.1	12.9	37.6
366	0.25	long bones	10.3	-18.2	3.3	1.3	13.3	37.4
372	0.25	ribs	11.1	-17.7	3.4	7.2	14.3	41.4
381	0.25	ribs	11.0	-18.7	3.4	9.7	15.3	44.0
8036–12	0.25	ribs	11.9	-17.7	3.4	9.2	15.4	44.9
225	0.25	ribs, cranial, clavicle	12.0	-18.9	3.4	13.9	15.4	45.0
449	0.75	ribs	11.9	-17.0	3.4	11.2	17.6	51.5
404	0.8	ribs	11.7	-18.6	3.3	8.9	14.5	41.3
Mean	*0.4*		*11.6*	*-18.2*	*3.4*	*7.8*	*14.7*	*42.4*
SD (±)	*0.2*		*0.6*	*0.6*	*0.05*	*4.5*	*1.4*	*4.4*
1–3 Years								
199	1.0	ribs	11.8	-18.2	3.3	13.4	16.1	46.2
266–05	1.5	ribs	10.3	-18.0	3.4	14.8	15.6	45.6
297A	1.5	ribs	11.7	-18.6	3.3	6.8	13.8	39.5
5072–9	1.5	ribs	12.5	-18.3	3.5	9.5	15.6	46.5
204	2.0	ribs	9.9	-18.6	3.3	1.4	13.6	38.4
239	2.0	ribs	9.8	-18.5	3.4	12.2	15.4	44.7
361	2.0	cranial	10.6	-18.2	3.3	5.6	15.0	42.8
243	2.5	ribs	9.4	-18.7	3.4	2.9	13.4	39.2
403	2.5	ribs	11.2	-18.3	3.3	8.9	14.1	39.7
237	3.0	ribs	12.0	-18.5	3.3	3.3	13.5	38.5
348	3.0	ribs, long bones	10.4	-18.3	3.3	1.3	15.2	43.3
426	3.0	ribs	10.2	-17.5	3.4	12.4	15.4	45.4
Mean	*2.1*		*10.8*	*-18.3*	*3.4*	*7.7*	*14.7*	*42.5*
SD (±)	*0.7*		*1.0*	*0.3*	*0.1*	*4.8*	*1.0*	*3.2*
3–5 Years								
295	3.25	ribs	10.4	-18.9	3.3	6.8	15.2	43.3
196	3.5	ribs	9.7	-19.4	3.4	8.2	15.6	45.2
393	3.5	ribs	9.3	-18.5	3.3	11.4	11.3	32.9
314	4.0	ribs	11.2	-18.8	3.4	7.4	14.4	41.8
380	4.0	ribs	8.6	-18.0	3.2	1.9	13.3	37.0
8036–11	4.0	ribs	10.4	-18.1	3.4	5.0	13.8	39.8
297B	4.5	cranial	11.8	-17.3	3.3	6.3	16.2	45.6
241	5.0	ribs	9.0	-18.5	3.3	1.2	13.6	38.7
273	5.0	ribs	9.1	-19.0	3.3	1.9	13.2	37.9
212B	5.0	ribs, cranial, clavicle	10.6	-18.5	3.3	8.0	16.7	45.5
Mean	*4.2*		*10.0*	*-18.5*	*3.3*	*5.8*	*14.3*	*40.8*
SD (±)	*0.7*		*1.0*	*0.6*	*0.1*	*3.3*	*1.6*	*4.2*

TABLE 9.1—*Continued*

Burial No.	*Age (years)*	*Tissue*	*$\delta^{15}N$ (‰ AIR)*	*$\delta^{13}C$ (‰ VPDB)*	*Atomic C:N*	*Collagen Yield (%)*	*% N*	*% C*
5–10 Years								
448A	5.5	ribs	8.4	-19.5	3.4	15.2	16.4	48.1
288	5.5	ribs	8.7	-18.8	3.3	8.8	14.0	39.5
355	5.5	ribs	9.0	-19.0	3.4	7.9	14.8	42.4
212A	6.0	ribs	9.3	-18.7	3.4	4.3	14.7	42.8
268	6.0	ribs	12.5	-18.4	3.4	14.4	15.5	44.7
320	6.0	ribs	9.5	-19.1	3.3	1.2	13.0	37.2
358	6.0	ribs, long bones	9.4	-18.8	3.3	2.6	13.7	39.4
395	6.0	ribs	9.5	-19.0	3.3	0.4	13.9	39.6
211	6.5	ribs	9.2	-18.9	3.3	5.5	14.7	41.9
5083–34	6.5	ribs	9.5	-19.0	3.3	4.8	14.7	41.9
202	6.5	ribs	10.2	-18.9	3.3	5.4	15.4	44.0
431	6.5	ribs	10.0	-18.5	3.4	14.2	15.9	46.5
208	7.0	ribs	8.8	-18.8	3.4	4.1	14.4	41.7
313	7.0	ribs	9.7	-18.7	3.4	5.6	14.1	40.7
440	7.0	ribs	9.4	-18.5	3.3	11.6	16.2	46.2
233	7.5	ribs	10.0	-18.5	3.4	5.5	14.9	43.0
428	7.5	ribs	9.6	-19.2	3.4	8.9	15.0	43.5
#?A	8.0	ribs	10.0	-18.9	3.3	10.4	16.8	47.5
5072–27	9.0	ribs	7.8	-18.9	3.1	11.0	5.3	14.4
443	10.0	ribs	10.1	-17.4	3.3	2.4	15.6	44.2
5072–20	10.0	ribs	10.1	-18.8	3.4	11.2	15.8	45.5
Mean	*6.9*		*9.6*	*-18.8*	*3.3*	*7.4*	*14.5*	*41.7*
SD (±)	*1.3*		*0.9*	*0.4*	*0.1*	*4.4*	*2.3*	*6.9*
10–15 Years								
Kal 400+	10.75	ribs	9.1	-18.9	3.4	7.1	14.6	42.4
294	11.0	ribs	10.1	-19.0	3.4	2.1	12.7	36.8
423	11.0	ribs	9.6	-18.5	3.3	5.4	15.4	44.1
5078–6	11.0	ribs	9.1	-18.6	3.4	13.4	15.9	46.8
413	12.0	ribs	9.2	-18.9	3.4	4.3	14.2	41.2
#?B	12.0	cranial	9.9	-18.9	3.5	7.8	15.3	45.4
5072–22	12.5	ribs	11.3	-19.0	3.3	12.2	15.7	44.9
448B	13	ribs	9.2	-18.0	3.4	10.6	16.1	46.6
5072–23	14	ribs	10.4	-18.5	3.4	8.5	15.0	43.9
5A	14	ribs	9.8	-18.6	3.4	8.1	15.5	44.9
5083–9	15	ribs	9.0	-18.8	3.4	11.1	15.0	43.7
Mean	*12.4*		*9.7*	*-18.7*	*3.4*	*8.2*	*15.0*	*43.7*
SD (±)	*1.5*		*0.7*	*0.3*	*0.1*	*3.4*	*1.0*	*2.8*
Total Mean	***5.4***		***10.2***	***-18.5***	***3.4***	***7.4***	***14.6***	***42.1***
Total SD (±)	***3.9***		***1.1***	***0.5***	***0.1***	***4.2***	***1.7***	***5.0***

*Burial no. 5072–27 was excluded from further stable isotope analysis due to diagenetic alteration and was therefore not included in the mean values.

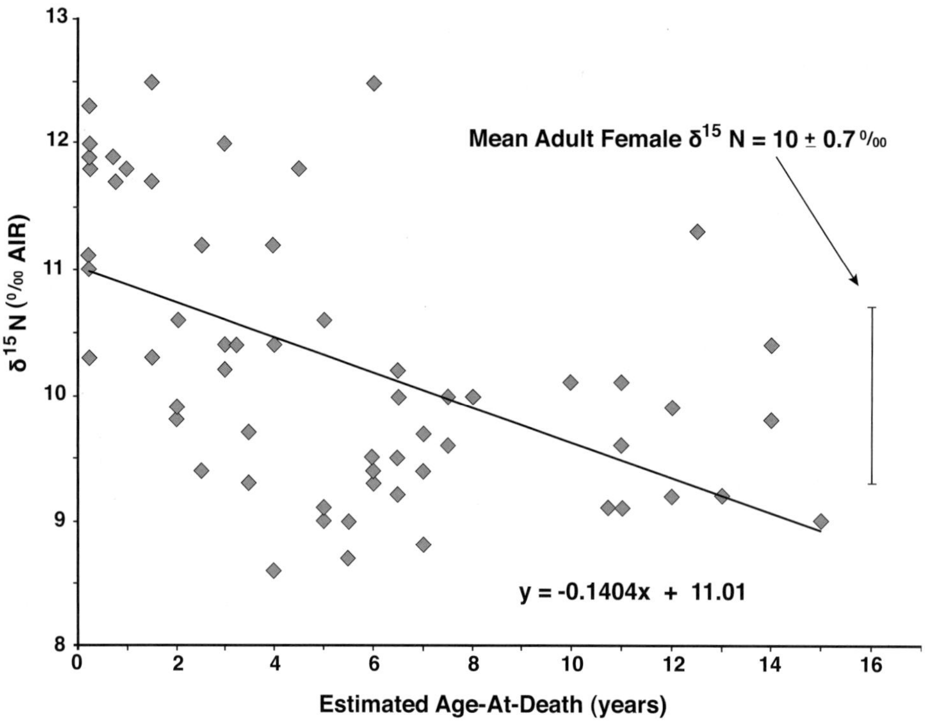
Mean Adult Female δ15 N = 10 ± 0.7 ‰
y = -0.1404x + 11.01
δ 15 N (‰ AIR)
Estimated Age-At-Death (years)
13
12
11
10
9
8
0
2
4
6
8
10
12
14
16

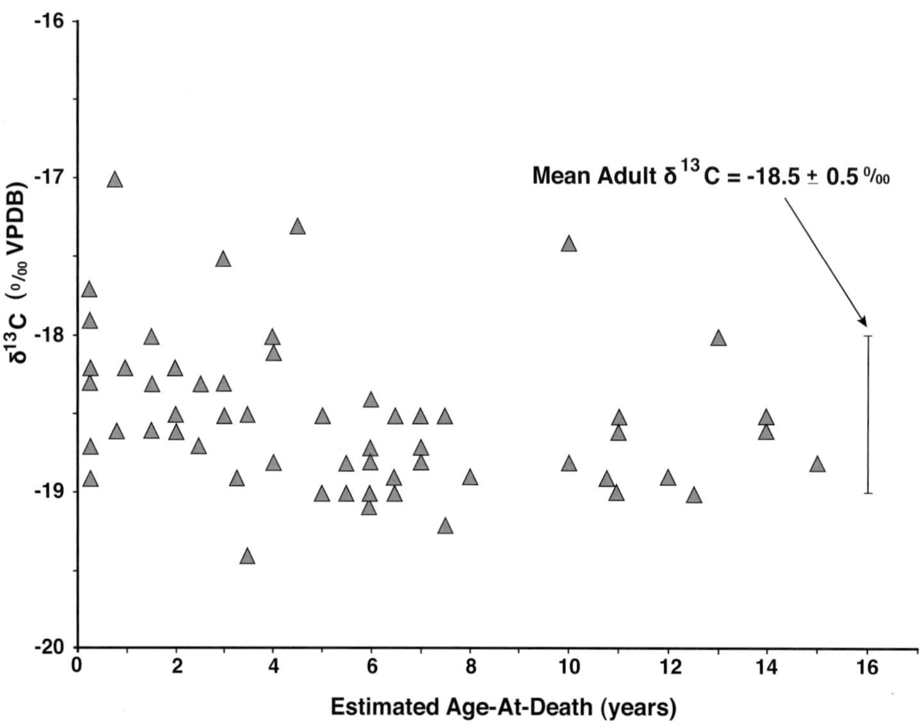
Mean Adult δ 13 C = -18.5 ± 0.5 ‰
δ13C (‰ VPDB)
Estimated Age-At-Death (years)
-16
-17
-18
-19
-20
0
2
4
6
8
10
12
14
16

Figure 9.3 *(opposite, top)*. **Subadult $\delta^{15}N$ values compared to estimated age-at-death (n = 63) at Apollonia Pontica. The mean adult female $\delta^{15}N$ value for Apollonia and the standard deviation (10 ± 0.7‰) is indicated by the vertical bar. A linear regression line is plotted to illustrate the weaning pattern and general decline in $\delta^{15}N$ values in older subadults.**

Figure 9.4 *(opposite, bottom)*. **Subadult $\delta^{13}C$ values compared to estimated age-at-death (n = 63) at Apollonia Pontica. The mean adult $\delta^{13}C$ values of males and females (n = 54) and standard deviation is indicated by the vertical bar.**

infants.[29] Between the ages of 1 and 3, the $\delta^{13}C$ values approach the adult mean, which also coincides with the onset of weaning and the consumption of complementary foods. Children between the ages of 5 and 10 display significantly more negative $\delta^{13}C$ values than the adult average (Mann-Whitney, U = 419, p = 0.008), which suggests the ingestion of a diet less enriched in ^{13}C. Past the age of 10, the $\delta^{13}C$ values are not significantly different from the adult mean, demonstrating that adolescents were consuming a diet similar to the adults.

DISCUSSION

Stable Nitrogen Isotopes and Weaning Practices

As suggested by the biochemical data, weaning at Apollonia began between 6 months and 1 year of age and was completed between the ages of 2 and 4, a finding that is consistent with the ancient literary evidence. The loose clustering of the $\delta^{15}N$ values suggests that not everyone practiced the same weaning regimen at Apollonia (Fig. 9.3). One 3-month-old subadult (burial no. 366) possessed a $\delta^{15}N$ signature that is close to the adult female mean, suggesting that this individual may have been weaned early. It is also possible that this infant was never breastfed. Based on skeletal remains alone, it is difficult to identify the reasons why certain infants were never breastfed or why they may have been breastfed for only a short period of time. Possibilities include the inability of some mothers to breastfeed or the death of the mother during childbirth, both of which were realities in the classical world. It was also common for mothers who had to work to support their family to wean their children earlier.[30] In addition, some mothers may not have wanted to breastfeed their child for fear of "growing old."[31]

Delayed weaning was also practiced at Apollonia, since some subadults (e.g., burial no. 268) continued to exhibit elevated $\delta^{15}N$ signatures past the weaning age. As noted earlier, late weaning may have been related to the health of the child.[32] Socioeconomic status of the family may also have played a role; infants may have continued to breastfeed if families could not procure alternative nutritious complementary foods.[33]

Archaeological Evidence for Weaning Practices

Infant feeding vessels have been recovered from child graves in Greece and elsewhere and can provide further information about weaning practices (Fig. 9.5).[34] During classical antiquity, two distinct sizes of infant feeders existed. A smaller vessel measuring 30 mm to 70 mm in height and 60 mm to 70 mm in diameter may have been used to feed neonates with supplementary fluids before or during the early phases of breastfeeding. In contrast, larger vessels measuring 95 mm to 125 mm in height and 100 mm to 150 mm in width may have functioned to administer supplementary fluids to older infants during weaning.[35]

It is possible to integrate the biochemical data with the archaeological evidence of feeding vessels, as they were buried with four of the subadults

29. Katzenberg 1993, pp. 54–55; Katzenberg, Saunders, and Fitzgerald 1993, p. 276; Richards, Mays, and Fuller 2002, pp. 207; Fuller, Fuller, et al. 2006, p. 288; Prowse et al. 2008, p. 302.

30. Garnsey 1998, pp. 269–269.

31. Sor. *Gyn.* 2.19.88, trans. O. Temkin, Baltimore 1956.

32. Sor. *Gyn.* 2.48.117, trans. O. Temkin, Baltimore 1956.

33. Garnsey 1998; see also Chap. 10.

34. Rosenthal 1936; Fildes 1986, p. 310.

35. Fildes 1981, p. 128; 1986, pp. 310–311.

Figure 9.5. Infant feeding vessel P 10282 from the Athenian Agora, located in the Agora Museum. Photo Craig Mauzy; courtesy Agora Excavations

analyzed in this study. Measurements of these artifacts indicate that all vessels are of the smaller type. The $\delta^{15}N$ values of two 3-month-old infants (burials nos. 266–03, 247) who were buried with these feeders suggest that they were breastfeeding prior to death. It can be reasoned, then, that these individuals were also fed supplementary fluids during the neonatal and post-neonatal period, likely consisting of honey and milk, as indicated in the ancient literature.[36]

A 2-year-old and a 4-year-old (burials nos. 204, 8036–11) were also buried with the smaller type of feeding vessel and have $\delta^{15}N$ values suggesting that they had already been weaned, were never breastfed, or were breastfed for only a short duration. It is possible that these individuals had to rely on these vessels for nutrition. Alternatively, they may have been used to feed supplementary fluids when these subadults were neonates and may have been buried with them for symbolic reasons.

Comparison to Other Greco-Roman Biochemical Studies

Few Greco-Roman biochemical weaning studies have been conducted, making it difficult to fully assess the larger temporal and spatial trends in infant feeding practices. Published studies are limited to 13 other investigations of samples spanning the Bronze Age to the Byzantine period[37] (Table 9.2). It is apparent that the onset and cessation of weaning is similar for all sites. This is not entirely surprising, given that Roman and Byzantine practices were heavily influenced by Greek customs. Furthermore, the infant feeding practices utilized by these populations, as inferred from the biochemical evidence, are all consistent with the ancient literature. It also appears that weaning may have been a standardized practice throughout the Greek and Roman world, although more studies are needed to examine this issue further.

36. Gal. *On the Powers of Food* 3, trans. M. Grant, London 2000; Sor. *Gyn.* 2.17.86, trans. O. Temkin, Baltimore 1956.

37. Garvie-Lok 2001, pp. 451–454; Fuller, Molleson, et al. 2006, p. 49; Bourbou and Richards 2007, pp. 69–70; Pennycook 2008, pp. 174–177; Prowse et al. 2008, pp. 304–305; Triantaphyllou et al. 2008, p. 3033; Bourbou and Garvie-Lok 2009, pp. 78–79; Keenleyside et al. 2009, p. 60.

TABLE 9.2. COMPARISON OF GRECO-ROMAN WEANING PRACTICES DERIVED FROM STABLE ISOTOPE ANALYSIS

Population	*Reference*	*Weaning Onset*	*Weaning Termination*
Lerna (Greece, ca. 2100–1700 B.C.)	Triantaphyllou et al. 2008, p. 3032	n/a	2.5–3 years
Apollonia (Bulgaria, 5th–3rd centuries B.C.)	Kwok 2007, pp. 118–120; current study	6 months to 1 year	2–4 years
Isola Sacra (Italy, 1st–3rd centuries A.D.)	Prowse et al. 2008, p. 306	<1 year	2–2.5 years
Dakhleh Oasis (Egypt, ca. A.D. 250)	Dupras, Schwarcz, and Fairgrieve 2001, p. 208	6 months	ca. 3 years
Leptiminus (North Africa, 4th–5th centuries A.D.)	Keenleyside et al. 2009, p. 58	n/a	ca. 3 years
Queenford Farm (Britain, 4th–6th centuries A.D.)	Fuller, Molleson, et al. 2006, p. 49	n/a	2–4 years
Stymphalos and Zakara (Greece, 5th–6th centuries A.D.)	Pennycook 2008, p. 172	n/a	2–3 years
Eleutherna (Crete, 6th–7th centuries A.D.)	Bourbou et al. 2013, p. 3909	n/a	ca. 4 years
Messene (Greece, 5th–7th centuries A.D.)	Bourbou et al. 2013, p. 3909	n/a	ca. 4 years
Sourtara (Greece, 6th–7th centuries A.D.)	Bourbou et al. 2013, p. 3909	n/a	ca. 4 years
Kastella (Crete, ca. 11th century A.D.)	Bourbou and Richards 2007, p. 69	n/a	ca. 2 years
Stylos (Crete, 11th–12th centuries A.D.)	Bourbou et al. 2013, p. 3909	n/a	ca. 4 years
Servia (Greece, 11th–15th centuries A.D.)	Garvie-Lok 2001, pp. 451–454; Bourbou and Garvie-Lok 2009, p. 78;	n/a	ca. 4 years
Petras (Crete, 12th–13th centuries A.D.)	Garvie-Lok 2001, pp. 451–454; Bourbou and Garvie-Lok 2009, p. 78;	n/a	ca. 4 years

Weaning Diet

To determine the type of diet onto which Apollonian infants were weaned, the $\delta^{13}C$ and $\delta^{15}N$ values of subadults between the ages of 2 and 5 years (n = 15, Fig. 9.6) were compared to isotope values of plants,[38] the few faunal remains recovered at Apollonia,[39] dairy products (including milk, yogurt, and cheese) extrapolated from the archaeological sheep and goat values, and

38. Stable isotope ratios of foods commonly consumed in the Mediterranean were reconstructed using published data and values from the Apollonia faunal remains. Average grain $\delta^{13}C$ values were derived from stable isotope data on modern species of wheat (*Triticum aestivum,* bread wheat), barley (*Hordeum* sp.), and millet (*Panicum miliaceum,* common millet); see Bender 1968, pp. 469–470; Smith and Epstein 1971, pp. 381–382; DeNiro and Epstein 1978, p. 497; Deines 1980, p. 334; Nakamura et al. 1982, p. 392. Plant $\delta^{13}C$ values were adjusted for the atmospheric shift (-1.5‰) associated with the Industrial Revolution (Friedli et al. 1986, p. 238; Marino and McElroy 1991, p. 127). The $\delta^{15}N$ values of wheat, barley, and millet were determined by subtracting the assumed dietary spacing of approximately 3‰–5‰ (Bocherens and Drucker 2003, p. 49) from the Apollonian faunal stable isotope ratios (Garvie-Lok 2001, pp. 309–340). Average stable isotope ratios of legumes (*Pisum sativum,* pea; *Archis hypogaea,* peanut) were reconstructed from published values; see Smith and Epstein 1971, pp. 381–382.

39. Faunal remains from Apollonia include sheep, goat, and unidentified ungulate. For the stable isotope values, see Keenleyside, Schwarcz, and Panayotova 2006, p. 1209.

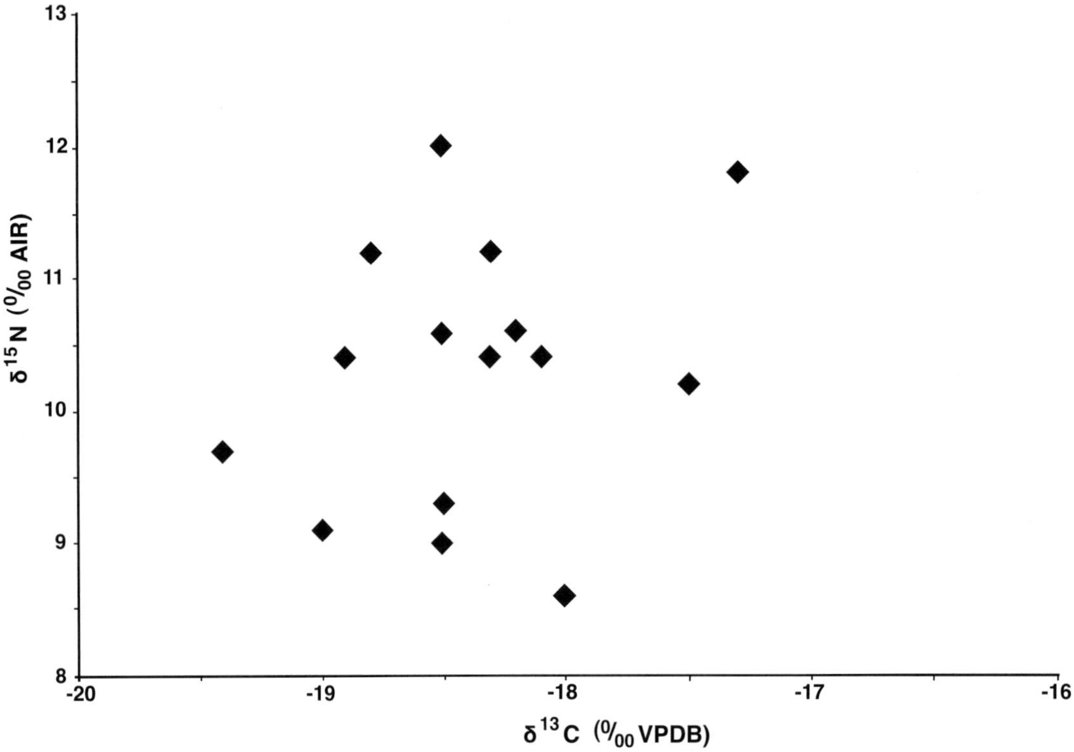
δ 15 N (0/00 AIR)
δ 13 C (0/00 VPDB)

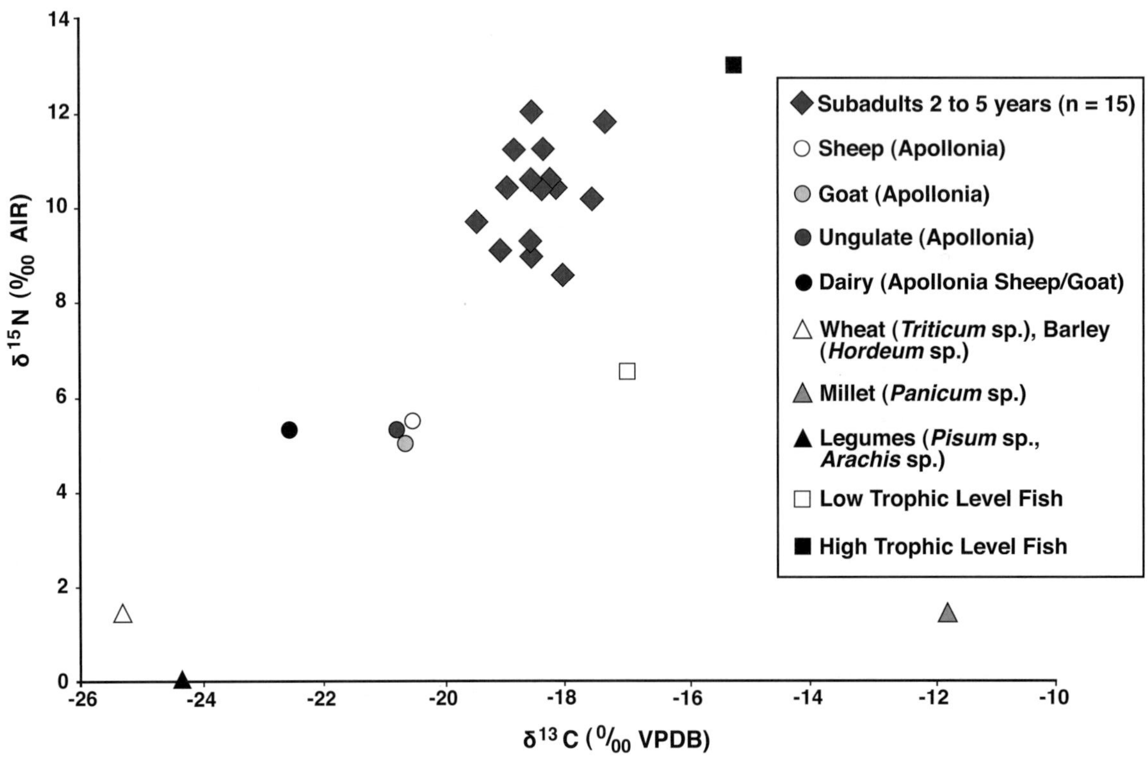
δ 15 N (0/00 AIR)
δ 13 C (0/00 VPDB)
Subadults 2 to 5 years (n = 15)
Sheep (Apollonia)
Goat (Apollonia)
Ungulate (Apollonia)
Dairy (Apollonia Sheep/Goat)
Wheat (Triticum sp.), Barley (Hordeum sp.)
Millet (Panicum sp.)
Legumes (Pisum sp., Arachis sp.)
Low Trophic Level Fish
High Trophic Level Fish

Figure 9.6 ***(opposite, top).*** **Variation in $\delta^{13}C$ and $\delta^{15}N$ values of subadults between 2 and 5 years of age (n = 15) from Apollonia Pontica**

Figure 9.7 ***(opposite, bottom).*** **Comparison of the isotope values of subadults (ages 2–5) with those of archaeological fauna from Apollonia (sheep, goat, unidentified ungulate). Stable isotope ratios of dairy, wheat, barley, and millet are roughly estimated from the Apollonian faunal values. The stable isotope values of legumes and high and low trophic level fish are derived from published data.**

other common Mediterranean foods reconstructed from other ecological and palaeodietary studies[40] (Fig. 9.7). From the $\delta^{13}C$ signatures, the weaning diet appears to have been composed primarily of terrestrial resources including meat, dairy, and C_3 plants such as cereals, fruits, and vegetables.

Meat was an expensive commodity in classical antiquity and was likely consumed only in small amounts during ceremonial sacrifices and other rituals.[41] Livestock, including sheep, goats, and oxen were raised in the Black Sea region,[42] and sheep astragali have been recovered from some of the graves at Apollonia. These animals, however, were employed primarily for their labor and milk, making terrestrial meat an unlikely component of the Apollonian weaning diet. Dairy products and C_3 plants were the more probable weaning foods, given the recommendations in the ancient literary texts that infants be weaned onto simple milk and cereal dishes.[43] Sheep or goat's milk was mixed with gruels and porridges or consumed on its own. Furthermore, porridges could also be consumed with an egg that could be "sipped."[44] Wine may have also been given to thirsty infants,[45] although there was some debate regarding the suitability of wine as a beverage for young children.[46] Furthermore, the tight clustering of $\delta^{13}C$ values suggests that the Apollonian subadults consumed a similar type of weaning diet. This is supported by the ancient literature, which describes a rather monotonous weaning diet of cereals and milk.

Childhood Diet

Stable isotope values of subadults over the age of 5 (n = 31) were utilized to determine the type of childhood diet, since weaning was generally completed by this age. The $\delta^{13}C$ and $\delta^{15}N$ signatures of these children (Fig. 9.8) were compared to the isotope values for modern plants[47] and typical Mediterranean foods[48] (Fig. 9.9). Generally, the Apollonian children consumed foods depleted in ^{13}C and ^{15}N compared to the adults, encompassing C_3 plants and dairy products, which is consistent with the ancient literary sources. Cereals, fruits, and vegetables were major staples of the Mediterranean diet and also made up a large part of the childhood diet. Grain was plentiful in the Black Sea colonies, which were major exporters of wheat in the trade with Athens and other Greek localities.[49]

40. Foods such as cheese, yogurt, and milk have similar isotopic values and are referred to generally as "dairy" products. Because the stable isotope values of whole milk are similar to the diet of dairying animals, it is possible to extrapolate the isotopic values of dairy products from that of archaeological fauna. A spacing of -3‰ was applied to bone collagen $\delta^{13}C$ values from archaeological sheep and goats from Apollonia to estimate dairy protein $\delta^{13}C$ values. In contrast, $\delta^{15}N$ values are considered to be similar to the bone collagen values of dairying animals; see Garvie-Lok 2001, pp. 321–326. Stable isotope values of common Mediterranean fish are derived from Jennings et al. 1997, p. 112; Garvie-Lok 2001, pp. 321–326; Prowse 2001, p. 132; Badalamenti et al. 2002, pp. 564–566.

41. Waterlow 1989, p. 6; Garnsey 1998, p. 267.

42. Ov. *Pont.* 1.8.41–62, trans. A. Elliot, Devon 2003.

43. Gal. *On the Powers of Food* 3, trans. M. Grant, London, 2000.

44. Sor. *Gyn.* 2.46.115, trans. O. Temkin, Baltimore 1956.

45. Sor. *Gyn.* 2.46.115, trans. O. Temkin, Baltimore 1956.

46. Arist. *Pol.* 1336a; Dalby 2003, p. 251.

47. See note 39. Bender 1968, pp. 469–470; Smith and Epstein 1971, pp. 381–382; DeNiro and Epstein 1978, p. 497; Deines 1980, p. 334; Nakamura et al. 1982, p. 392.

48. Jennings et al. 1997, p. 112; Garvie-Lok 2001, p. 328; Prowse 2001, p. 132; Badalamenti et al. 2002, pp. 564–566.

49. Polyb. 4.37.8–47.2, trans. M. Chambers, New York 1966; Waterlow 1989, p. 6

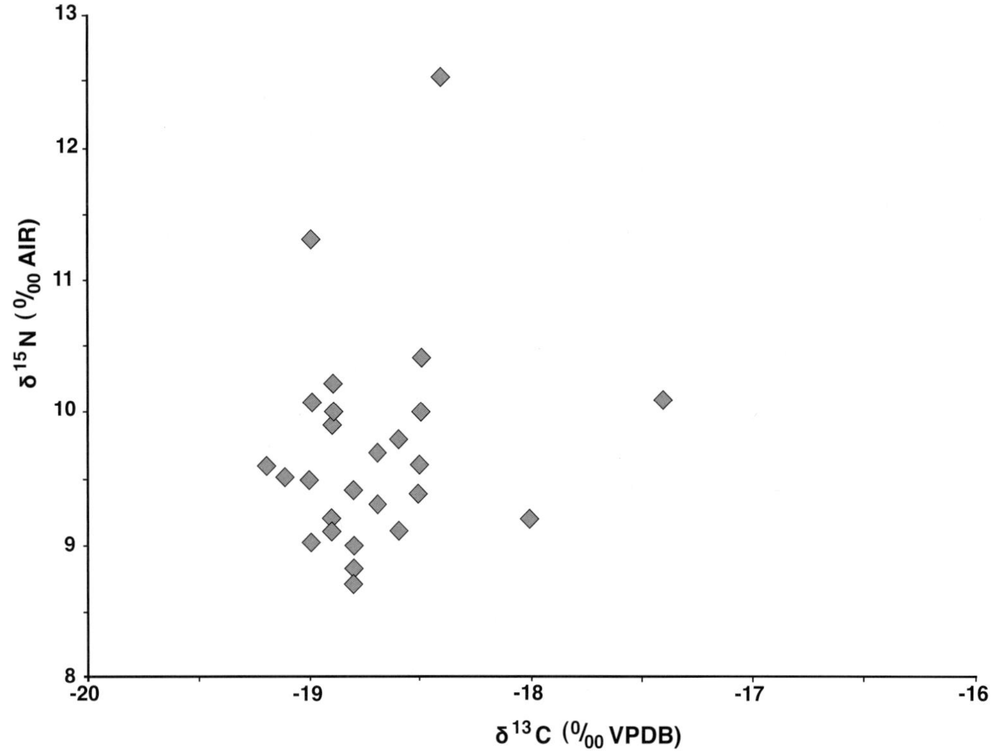
δ15N (0/00 AIR)
δ13C (0/00 VPDB)

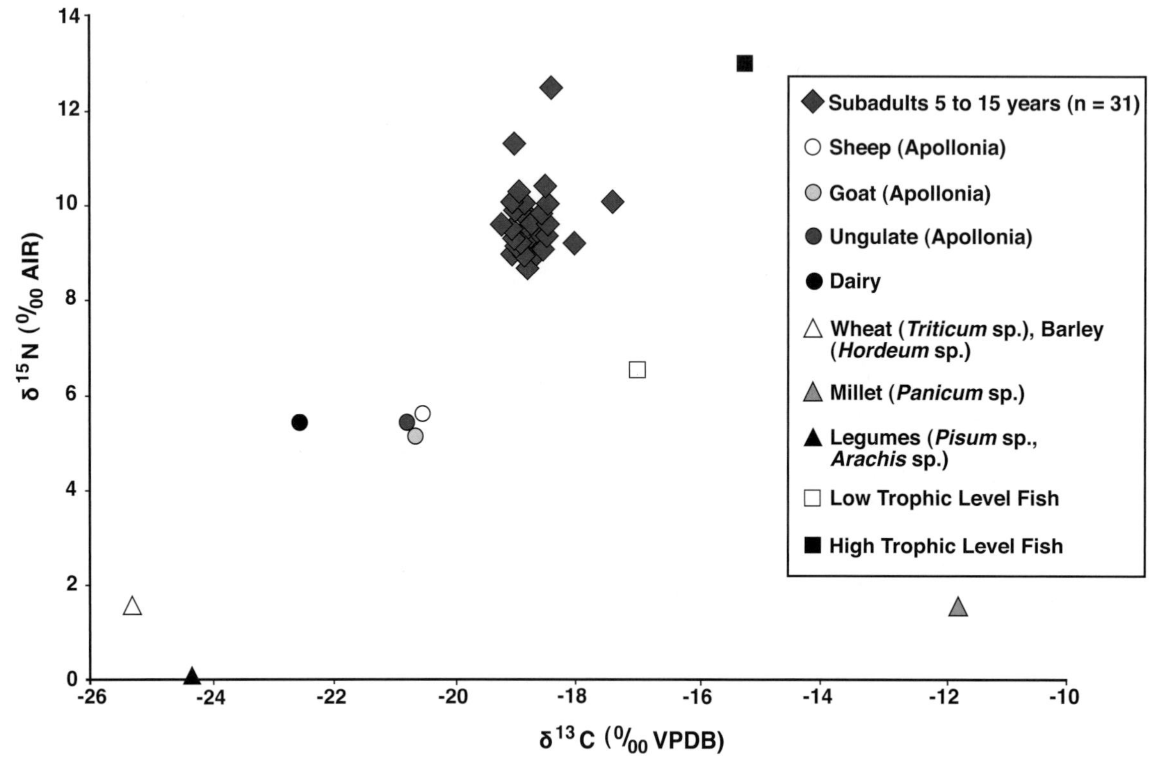
Subadults 5 to 15 years (n = 31)
Sheep (Apollonia)
Goat (Apollonia)
Ungulate (Apollonia)
Dairy
Wheat (*Triticum* sp.), Barley (*Hordeum* sp.)
Millet (*Panicum* sp.)
Legumes (*Pisum* sp., *Arachis* sp.)
Low Trophic Level Fish
High Trophic Level Fish
δ15N (0/00 AIR)
δ13C (0/00 VPDB)

Figure 9.8 *(opposite, top)*. **Variation in $\delta^{13}C$ and $\delta^{15}N$ values of subadults between 5 and 15 years of age (n = 31) at Apollonia Pontica**

Figure 9.9 *(opposite, bottom)*. **Comparison of isotope values of subadults between 5 and 15 years of age with those of archaeological fauna from Apollonia (sheep, goat, unidentified ungulate). Stable isotope ratios of dairy, wheat, barley, and millet are roughly estimated from the Apollonian faunal values. The stable isotope values of legumes and high and low trophic level fish are derived from published data.**

Wheat and barley were the principle cereals consumed by the ancient Greeks and were eaten in the form of bread, cakes, mash, gruel, and porridge.[50] Vegetables and fresh fruit were also consumed,[51] and shoots of figs were recommended as part of the childhood diet, although Galen advised that they be pre-masticated by a wet nurse before being offered to the child.[52] Dairy products in the form of milk and cheese were also eaten, while legumes were not an important component of the subadult diet at Apollonia.[53]

The stable isotope data do not support a diet based heavily on marine resources, as the subadult $\delta^{13}C$ and $\delta^{15}N$ values are mainly clustered around those typical of terrestrial resources. This is somewhat surprising given the geographical location of Apollonia, the variety of available fish in the Black Sea, and the archaeological evidence of fish grills.[54] In addition, the stable isotope analysis of adult remains from Apollonia revealed that their diet consisted of both marine and terrestrial resources.[55] Children may not have been consuming copious amounts of seafood due to individual preferences, or the fact that it was not seen as suitable food for young children.

Based on the isotope studies conducted to date, it is quite common for post-weaning subadults to display significantly lower $\delta^{15}N$ values compared to the adults, as this pattern has been observed in several different populations.[56] This has led some scholars to suggest that varying $\delta^{15}N$ values observed in children may reflect an imbalance in nitrogen due to growth.[57] This is unlikely to be the case for the Apollonian subadults for several reasons. First, when subadult diaphyseal lengths were plotted against age-at-death for the Apollonian children, a linear pattern was revealed, suggesting that the growth-effect on the subadult $\delta^{15}N$ values is negligible.[58] Secondly, Waters-Rist and Katzenberg[59] have recently demonstrated through stable isotope analysis of subadult long bones from a proto-historic Ontario cemetery that the growth-effect does not significantly affect subadult bone collagen $\delta^{15}N$ values. Thirdly, considering that most Greco-Roman written sources report a childhood diet based on cereals and dairy products, it is argued that the isotope values of the Apollonian children reflect a shift in dietary regimens.

Childhood Diet Compared to the Adult Diet

In Greek and Roman society, childhood was considered distinct from adulthood, and many ancient writers commented that children required a separate diet from adults.[60] Given this belief, the stable isotope data obtained from the Apollonian children were compared to those derived from a previous study of the adults[61] to determine whether any dietary differences

50. Fidanza 1979, pp. 81–82; Garnsey 1998, p. 15.
51. Fidanza 1979, p. 82; Braun 1991, pp. 43–54; Brothwell and Brothwell 1998, pp. 105–152; Grivetti 2001, p. 11.
52. Gal. *On the Powers of Food* 3, trans. M. Grant, London 2000.
53. Kwok 2007, p. 134.
54. Nedev and Panayotova 2003; Keenleyside, Schwarcz, and Panayotova 2006, p. 1207.
55. Keenleyside, Schwarcz, and Panayotova 2006, pp. 1210–1213.
56. Tuross and Fogel 1994, p. 287; Katzenberg, Herring, and Saunders 1996, p. 189; Schurr 1997, pp. 923–924; Richards, Mays, and Fuller 2002, pp. 207–208.
57. Schurr 1997, p. 924; Millard 2000, p. 53.
58. Richards, Mays, and Fuller 2002, pp. 207–208; Kwok and Keenleyside 2010, p. 143.
59. Waters-Rist and Katzenberg 2010, p. 187.
60. French 1988; Golden 1990, p. 8; Katz 2007, p. 123.
61. Keenleyside, Schwarcz, and Panayotova 2006, p. 1209.

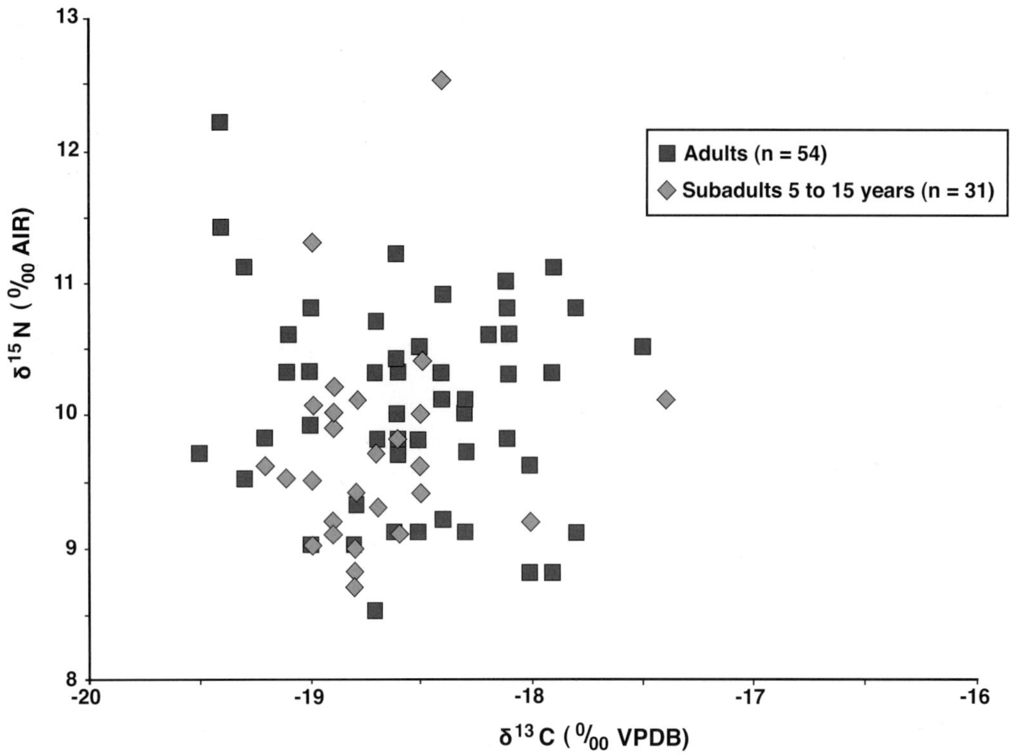
13
12
11
10
9
8
-20
-19
-18
-17
-16
δ15N (0/00 AIR)
δ13C (0/00 VPDB)
Adults (n = 54)
Subadults 5 to 15 years (n = 31)

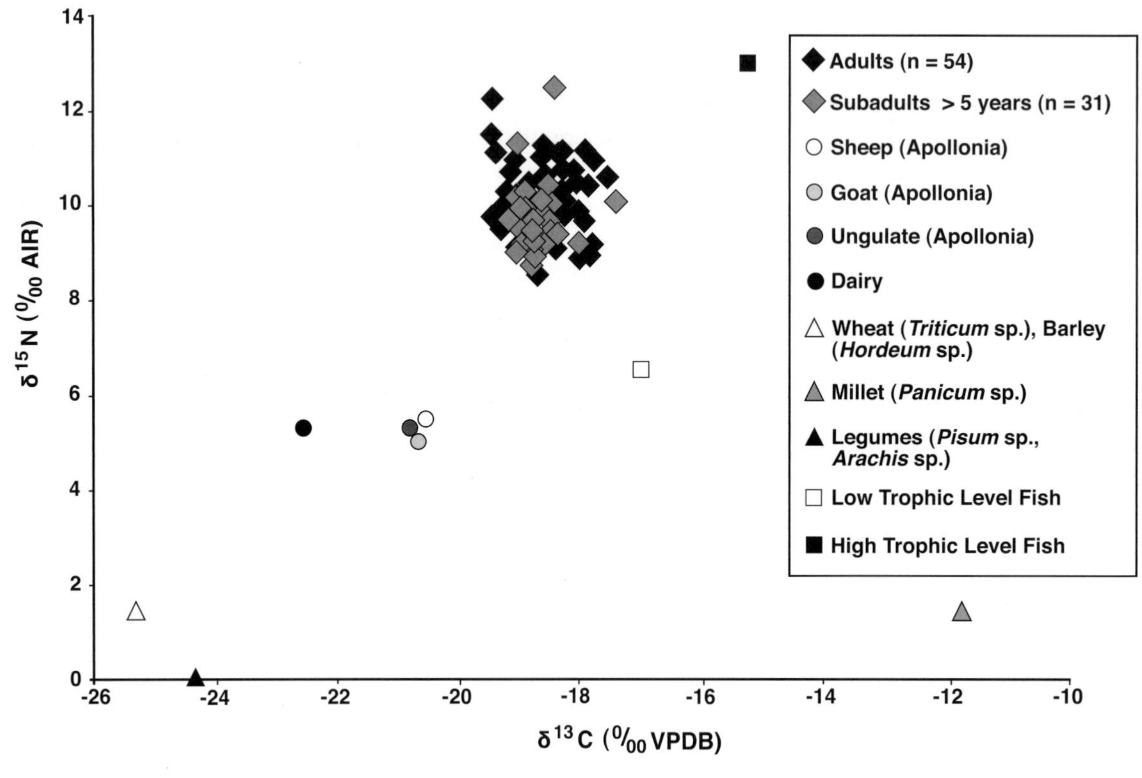
14
12
10
8
6
4
2
0
-26
-24
-22
-20
-18
-16
-14
-12
-10
δ15N (0/00 AIR)
δ13C (0/00 VPDB)
Adults (n = 54)
Subadults > 5 years (n = 31)
Sheep (Apollonia)
Goat (Apollonia)
Ungulate (Apollonia)
Dairy
Wheat (Triticum sp.), Barley (Hordeum sp.)
Millet (Panicum sp.)
Legumes (Pisum sp., Arachis sp.)
Low Trophic Level Fish
High Trophic Level Fish

Figure 9.10 *(opposite, top)*. **Comparison of adult and subadult (5 to 15 years of age) $\delta^{13}C$ and $\delta^{15}N$ values from Apollonia Pontica**

Figure 9.11 *(opposite, bottom)*. **Comparison of isotope values of subadults (5 to 15 years of age) and adults with those of archaeological fauna from Apollonia (sheep, goat, unidentified ungulate). Stable isotope ratios of dairy, wheat, barley, and millet are roughly estimated from the Apollonian faunal values. The stable isotope values of legumes and high and low trophic level fish are derived from published data.**

existed. While an overlap in isotope values can be seen between the juveniles and adults (Figs. 9.10, 9.11), children between the ages of 5 and 10 years possessed significantly lower $\delta^{13}C$ and $\delta^{15}N$ signatures (Mann-Whitney, $U = 758.5$, $p = 0.001$; $U = 419.0$, $p = 0.004$, respectively). These findings suggest that juveniles ingested fewer marine resources and more cereals, fruits, and vegetables than adults. This dietary difference is consistent with the ancient literature. Galen, for example, recommended that juveniles be fed first on cereals before being gradually introduced to other food items.[62]

Health Implications Associated with Weaning and Dietary Practices

Child-rearing practices in antiquity were often detrimental to the health of the infant, contributing to "early death, stunted development, or selective malnutrition."[63] Among the Romans, for instance, the health of infants could be jeopardized by practices such as withholding colostrum from newborns, using wet nurses, weaning infants too early or too late, and weaning them onto nutritionally inadequate foods.[64] Certain Roman physicians considered a mother's colostrum to be harmful to her child, and purges consisting of honey, goat's milk, and water were given instead to expel an infant's meconium.[65]

Once a child is weaned, the following period is critical to the survival of that individual. The transition to non–breast milk foods is associated with increased risk of weanling diarrhea due to exposure to pathogens in food and water and the loss of passive immunity.[66] In agricultural populations, weaning infants onto a high carbohydrate diet of foods such as porridge can result in protein-deficiency disease and other nutritional disorders.[67]

A number of skeletal stress indicators have been linked to weaning stress, including enamel hypoplasia and cribra orbitalia.[68] Cribra orbitalia, a porous lesion on the orbital roof, may result from a number of health problems, including eye infections, rickets, and acquired and genetic anemias.[69] Within Mediterranean populations, this condition has typically been attributed to thalassemia and/or iron-deficiency anemia,[70] and at Apollonia Pontica, the nature and distribution of these lesions suggest that it most likely resulted from the latter.[71]

In the sample analyzed in this study, infants younger than 6 months of age did not exhibit any signs of cribra orbitalia (Table 9.3, Fig. 9.12), likely reflecting the fact that the majority of these individuals were feeding on breast milk during this time, as early breast milk contains a higher amount

62. Garnsey 1998, p. 267.
63. Garnsey 1999, p. 106.
64. Garnsey 1999, p. 53, 107–108.
65. Gal. *On the Powers of Food* 3, trans. M. Grant, London, 2000; Sor. *Gyn.* 2.17.86, trans. O. Temkin, Baltimore 1956; Fildes 1986, p. 81; Garnsey 1998, pp. 263–264.
66. Katzenberg, Herring, and Saunders 1996, p. 180; Herring, Saunders, and Katzenberg 1998, p. 427.
67. Brothwell and Brothwell 1998, p. 187.
68. Corruccini, Handler, and Jacobi 1985, pp. 706–707; Blakey, Lelise, and Reidy 1994, p. 378; Fairgrieve and Molto 2000, p. 328; Dupras, Schwarcz, and Fairgrieve 2001, p. 209.
69. Wapler, Crubézy, and Schultz 2004, pp. 336–338.
70. Angel 1966, p. 762; 1971, pp. 77–84; Papathanasiou, Larsen, and Norr 2000; p. 224; Keenleyside and Panayotova 2006, p. 382; Stravopodi et al. 2009, pp. 257–258.
71. Stuart-Macadam 1991, p. 104; Hershkovitz et al. 1997, pp. 222–224; Ortner 2003, pp. 364–367; Keenleyside and Panayotova 2006, p. 382.

TABLE 9.3. FREQUENCY OF APOLLONIA PONTICA SUBADULTS WITH CRIBRA ORBITALIA

	Frequency		*Active Lesions*	
Age-at-Death Category (years)	*No. Affected/ No. Observed*	*%*	*No. Active/ No. Affected*	*%*
Birth to 0.5	0/4	0	0/0	0
0.5 to 1	0/2	0	0/0	0
1 to 2	1/2	50.0	0/1	0
2 to 3	1/4	25.0	0/1	0
3 to 4	3/4	75.0	0/3	0
4 to 5	2/4	50.0	1/2	50.0
5 to 10	9/13	69.2	4/9	44.4
10 to 15	7/11	63.6	6/7	85.7
Total	23/44	52.3	11/23	47.8

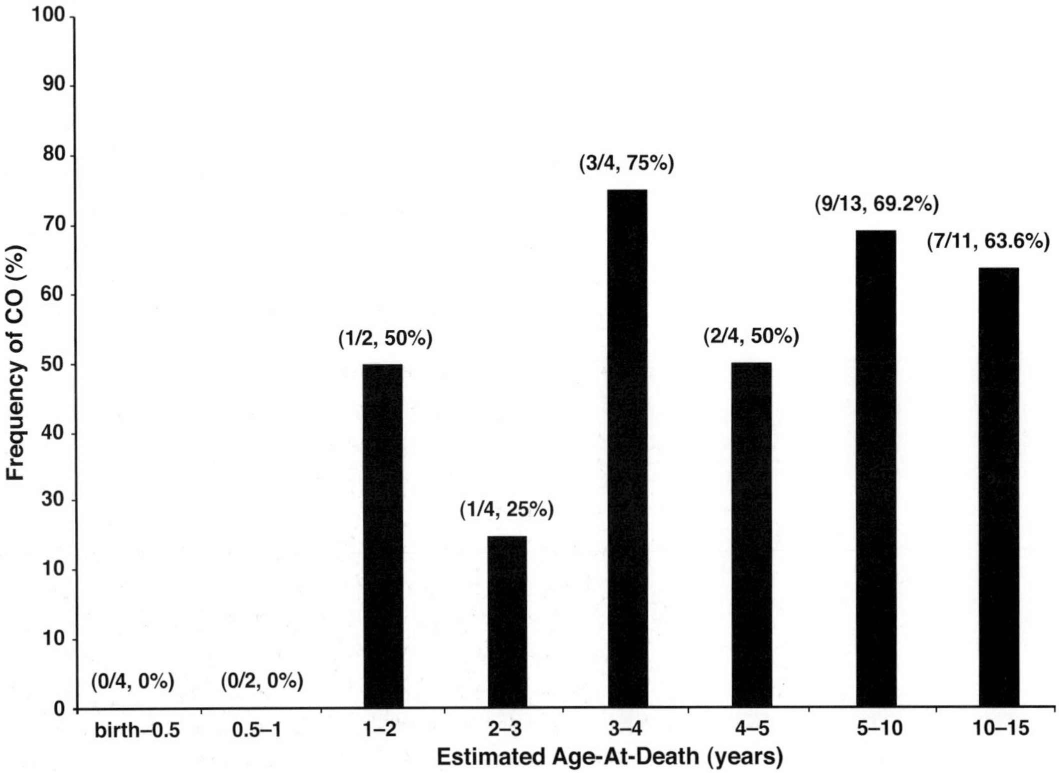

Figure 9.12. Frequency of cribra orbitalia in the subadult sample (n = 44, individuals with observable eye orbits) from Apollonia Pontica

of iron compared to mature breast milk.[72] Additionally, infants would also have been relying on their in-utero accumulated iron reserves at this age.[73]

Similarly, individuals less than 12 months of age did not show any evidence of cribra orbitalia (Table 9.3, Fig. 9.12). Surprisingly, a 1-year-old subadult (burial no. 199) affected with cribra orbitalia also possessed a $\delta^{15}N$ value that suggests that he or she was consuming breast milk prior to death. By 6 months of age, iron reserves in the body become depleted.

72. Feely et al. 1983, p. 445.
73. Dewey et al. 1998, p. 878.

At the same time, the demands for iron increase and breast milk is unable to meet these demands.[74] Combined with a weaning diet of cereals and dairy, the chance of developing nutritional disorders such as iron-deficiency anemia increases due to the presence of phytates in cereals, which inhibit the intestinal absorption of iron,[75] and the fact that infants do not absorb iron from animal milk as well as they do from human milk.[76] A complete reliance by infants on sheep's and goat's milk for nutrition can also result in folate deficiency and megaloblastic anemia, which can manifest as cribra orbitalia in young children.[77]

Interestingly, the highest frequency of cribra orbitalia (75%) is observed in subadults between the ages of 3 and 4 years (Table 9.3, Fig. 9.12). The occurrence of these lesions could represent the end of weaning and the transition onto post-weaning foods, as the $\delta^{15}N$ values of the affected subadults are similar to the adult female mean.

Subadults over 5 years of age also exhibited a high prevalence of cribra orbitalia (over 60%), reflecting their heavy reliance on cereals (Table 9.3, Fig. 9.12). The consumption of wine may also have been a contributing factor. Wine contains polyphenols, and excessive consumption of this beverage can interfere with the body's ability to absorb iron.[78] Parasitic and bacterial infections, common in antiquity,[79] may also have led to the development of anemia in the Apollonian children. Crowded living conditions, heavy traffic associated with a port town, and poor sanitation would have increased the risk of infectious disease. Parasites can be introduced through a variety of foods and can cause severe gastrointestinal bleeding, resulting in anemia.[80] *Escherichia coli* bacteria can also be introduced by drinking spoiled, unpasteurized milk, and the incorporation into the weaning diet of foods such as honey, a known source of *Clostridium botulinum,* can be a significant source of infection.[81] In addition, the use of feeding vessels to dispense milk can promote bacterial growth if not properly cleaned.[82]

CONCLUSIONS

This study is the first to examine infant feeding practices in a Greek colony on the Black Sea coast. The stable isotope evidence is consistent with the ancient literary sources and indicates that weaning was a gradual process at Apollonia. Weaning foods ingested by infants consisted mostly of dairy products and cereals, while the childhood diet was composed of cereals, fruits, vegetables, and dairy foods. Dietary differences were noted between the older children and the adults of Apollonia. Children ate mostly cereals, fruits, and vegetables, whereas the adults ingested more meat and marine resources. This observation is consistent with the belief held in Greek and Roman society that children were a separate social group from adults. Furthermore, the health of the Apollonian children was compromised by a diet based heavily on cereals and dairy products, which increased their susceptibility to developing anemia. Similarities in weaning practices between Apollonia and several Roman populations, as indicated by previous biochemical studies, suggest that weaning may have been a standardized practice in classical antiquity.

74. Dewey et al. 1998, p. 878; Krebs 2000, p. 358S; Griffin and Abrams 2001, p. 405.
75. Garn 1992, p. 55.
76. Vaughan, Weber, and Kemberling 1979, pp. 2304–2305.
77. Fairgrieve and Molto 2000, p. 329, but see Walker et al. 2009, pp. 115–116, for a discussion of the etiology of cribra orbitalia.
78. Garn 1992, p. 53; Cook, Reddy, and Hurrell 1995, p. 803.
79. Garn 1992, pp. 54–55.
80. Stuart-Macadam 1992, pp. 42–43.
81. Fairgrieve and Molto 2000, p. 329
82. Hanson and Winberg 1972, p. 845; Brothwell and Brothwell 1998, p. 188.

It is important to be aware of the limitations of using stable nitrogen isotope analysis to reconstruct weaning practices in past populations. As the cause of death cannot normally be determined, it is possible that the infants examined in this study died as a result of weaning, meaning that the pattern ascertained from the isotope evidence would not reflect the "normal" and successful weaning pattern of this population.[83] Pathological conditions, including nutritional stress, can also alter the stable nitrogen isotope signature in bone collagen.[84] Under-representation and/or inaccurate age estimation of infants in a skeletal sample may further hinder observation of the weaning pattern. In addition, it is impossible to discriminate between infants who were never breastfed from those who were breastfed for only a short period of time, as the precise lag time, or collagen turnover rate, in children remains unknown. Finally, biases in the ancient literary texts must be acknowledged, since most treatises from the Greco-Roman period were written by elite males and therefore may not be entirely representative of the general population.

In summary, this study contributes to our knowledge of infant feeding practices in classical antiquity and illustrates the value of incorporating multiple lines of evidence, including biochemical, skeletal, archaeological, and ancient literary sources, to gain further insight into child-rearing practices in the ancient Greek world.

ACKNOWLEDGMENTS

We would like to thank Kristina Panayotova (Institute of Archaeology, Sofia, Bulgaria) and Dimitar Nedev (Director of the Archaeological Museum, Sozopol, Bulgaria) for providing access to the skeletal samples utilized in this study. We would also like to acknowledge Henry Schwarcz and Martin Knyf (School of Geography and Earth Sciences, McMaster University), Jocelyn Williams (Department of Anthropology, Trent University), and M. Anne Katzenberg (Department of Archaeology, University of Calgary) for their comments, suggestions, and technical advice. Data on the infant feeding vessels from Apollonia were kindly provided by Margarit Damyanov. We would also like to thank Mike Richards and Anastasia Papathanasiou for inviting us to contribute to this volume and the three anonymous reviewers for their comments. Thanks should also be given to Stephen Gardiner (Department of Geography, Trent University) for the map of the Kalfata necropolis. This project was supported by an Ontario Graduate Scholarship (C.K.) and a Social Sciences and Humanities Research Council of Canada grant (#410–2005–0452) (A.K.).

83. Katzenberg, Herring, and Saunders 1996, pp. 192–193.
84. White and Armelagos 1997, pp. 192–193; Katzenberg and Lovell 1999, p. 322; Fuller et al. 2005, pp. 2503–2505.

REFERENCES

Ambrose, S. H. 1990. "Preparation and Characterization of Bone and Tooth Collagen for Isotopic Analysis," *JAS* 17, pp. 431–451.

Angel, J. L. 1966. "Porotic Hyperostosis, Anemias, Malarias, and Marshes in the Prehistoric Eastern Mediterranean," *Science* 153, pp. 760–763.

———. 1971. *The People of Lerna: Analysis of a Prehistoric Aegean Population* (*Lerna* II), Princeton.

Badalamenti, F., G. D'Anna, J. K. Pinnegar, and N. V. C. Polunin. 2002. "Size-Related Trophodynamic Changes in Three Target Fish Species Recovering from Intensive Trawling," *Marine Biology* 141, pp. 561–570.

Bender, M. M. 1968. "Mass Spectrometric Studies of Carbon 13 Variations in Corn and Other Grasses," *Radiocarbon* 10, pp. 468–472.

Blakey, M. L., T. E. Lelise, and J. P. Reidy. 1994. "Frequency and Chronological Distribution of Dental Enamel Hypoplasia in Enslaved African Americans: A Test of the Weaning Hypothesis," *American Journal of Physical Anthropology* 95, pp. 371–383.

Bocherens, H., and D. Drucker. 2003. "Trophic Level Isotopic Enrichment of Carbon and Nitrogen in Bone Collagen: Case Studies from Recent and Ancient Terrestrial Ecosystems," *International Journal of Osteoarchaeology* 13, pp. 46–53.

Bourbou, C., B. T. Fuller, S. J. Garvie-Lok, and M. P. Richards. 2013. "Nursing Mothers and Feeding Bottles: Reconstructing Breastfeeding and Weaning Patterns in Greek Byzantine Populations (6th–15th Centuries A.D.) Using Carbon and Nitrogen Stable Isotope Ratios," *JAS* 40, pp. 3903–3913.

Bourbou, C., and S. Garvie-Lok. 2009. "Breast-Feeding and Weaning Patterns in Byzantine Times: Evidence From Human Remains and Written Sources," in *Becoming Byzantine: Children and Childhood in Byzantium,* ed. A. Papaconstantinou and A M. Talbot, Washington, D.C., pp. 65–83.

Bourbou, C., and M. Richards. 2007. "The Middle Byzantine Menu: Paleodietary Information from Isotopic Analysis of Humans and Fauna from Kastella, Crete," *International Journal of Osteoarchaeology* 17, pp. 63–72.

Braun, T. 1991. "Ancient Mediterranean Food," in *The Mediterranean Diets in Health and Disease,* ed. G. A. Spiller, New York, pp. 10–55.

Brothwell, D., and P. Brothwell. 1998. *Food in Antiquity: A Survey of the Diet of Early Peoples,* 3rd ed., Baltimore.

Buikstra, J. E., and D. H. Ubelaker, eds. 1994. *Standards for Data Collection from Human Skeletal Remains. Proceedings of a Seminar at the Field Museum of Natural History, Organized by Jonathan Haas* (Arkansas Archeological Survey Research Series 44), Fayetteville.

Chisholm, B. S., D. E. Nelson, K. A. Hobson, H. P. Schwarcz, and M. Knyf. 1983. "Carbon Isotope Measurement Techniques for Bone Collagen: Notes for the Archaeologist," *JAS* 10, pp. 355–360.

Cook, J. D., M. B. Reddy, and R. F. Hurrell. 1995. "The Effect of Red and White Wines on Nonheme-Iron Absorption in Humans," *American Journal of Clinical Nutrition* 61, pp. 800–804.

Corruccini, R. S., J. S. Handler, and K. P. Jacobi. 1985. "Chronological Distribution of Enamel Hypoplasias and Weaning in a Caribbean Slave Population," *Human Biology* 57, pp. 699–711.

Dalby, A. 2003. *Food in the Ancient World, From A to Z,* London.

Danov, K. H. 1948. "Kam starata istoriya na Sozopol, Zapadnoto Chernomorie, I Trakiya (On the Ancient History of Sozopol, the Western Pontic Coast, and Thrace)," *Izvestiya na Bulgarskoto Istorichesko Druzhestvo* 22–24, pp. 172–183.

Deines, P. 1980. "The Isotopic Composition of Reduced Organic Carbon," in *Handbook of Environmental Isotope Geochemistry,* ed. P. Fritz and J. C. Fontes, Amsterdam, pp. 329–406.

DeNiro, M. J. 1985. "Post-Mortem Preservation and Alteration of *in vivo* Bone Collagen Isotope Ratios in Relation to Palaeodietary Reconstruction," *Nature* 317, pp. 806–809.

DeNiro, M. J., and S. Epstein. 1978. "Influence of Diet on the Distribution of Carbon Isotopes in Animals," *Geochimica et cosmochimica acta* 42, pp. 495–506.

———. 1981. "Influence of Diet on the Distribution of Nitrogen Isotopes in Animals," *Geochimica et cosmochimica acta* 45, pp. 341–351.

Dettwyler, K. A., and C. Fishman. 1992. "Infant Feeding Practices and Growth," *Annual Review of Anthropology* 21, pp. 171–204.

Dewey, K. G., R. J. Cohen, L. L. Rivera, and K. H. Brown. 1998. "Effects of Age of Introduction of Complementary Foods on Iron Status of Breast-Fed Infants in Honduras," *American Journal of Clinical Nutrition* 67, pp. 878–884.

Dupras, T. L., H. P. Schwarcz, and S. I. Fairgrieve. 2001. "Infant Feeding and Weaning Practices in Roman Egypt," *American Journal of Physical Anthropology* 115, pp. 204–212.

Dupras, T. L., and M. Tocheri. 2007. "Reconstructing Infant Weaning Histories at Roman Period Kellis, Egypt, Using Stable Isotope Analysis of Dentition," *American Journal of Physical Anthropology* 134, pp. 63–74.

Fairgrieve, S. I., and J. E. Molto. 2000. "Cribra Orbitalia in Two Temporally Disjunct Population Samples from the Dakhleh Oasis, Egypt," *American Journal of Physical Anthropology* 111, pp. 319–331.

Feeley, R. M., R. R. Eitenmiller, J. B. Jones Jr., and H. Barnhart. 1983. "Copper, Iron, and Zinc Contents of Human Milk at Early Stages of Lactation," *American Journal of Clinical Nutrition* 37, pp. 443–448.

Fidanza, F. 1979. "Diets and Dietary Recommendations in Ancient Greece and Rome and the School of Salerno," *Progress in Food and Nutrition Science* 3, pp. 79–99.

Fildes, V. A. 1981. "The Early History of the Infant Feeding Bottle," *Nursing Times* 77, pp. 128–129, 168–170.

———. 1986. *Breasts, Bottles, and Babies: A History of Infant Feeding,* Edinburgh.

———. 1988. *Wet Nursing: A History from Antiquity to the Present,* Oxford.

Flecker, H. 1932. "Roentgenographic Observations of the Times of Appearance of Epiphyses and Their Fusion with the Diaphyses," *Journal of Anatomy* 67, pp. 118–164.

Fogel, M. L., N. Tuross, and D. W. Owsley. 1989. "Nitrogen Isotope Tracers of Human Lactation in Modern and Archaeological Populations," *Annual Report of the Director, Geophysical Laboratory, Carnegie Institution for Science, Washington, D.C., 1988–1989,* pp. 111–117.

French, V. 1988. "Birth Control, Childbirth, and Early Childhood," in *Civilization of the Ancient Mediterranean: Greece and Rome,* ed. M. Grant and R. Kitzinger, New York, pp. 1355–1362.

Friedli, H., H. Lötscher, H. Oeschger, U. Siegenthaler, and B. Stauffer. 1986. "Ice Core Record of the $^{13}C/^{12}C$ Ratio of Atmospheric CO_2 in the Past Two Centuries," *Nature* 324, pp. 237–238.

Fuller, B. T., J. L. Fuller, D. A. Harris, and R. E. M. Hedges. 2006. "Detection of Breastfeeding and Weaning in Modern Human Infants with Carbon and Nitrogen Stable Isotope Ratios," *American Journal of Physical Anthropology* 129, pp. 279–293.

Fuller, B. T., J. L. Fuller, N. E. Sage, D. A. Harris, T. C. O'Connell, and R. E. M. Hedges. 2005. "Nitrogen Balance and $\delta^{15}N$: Why You're Not What You Eat during Nutritional Stress," *Rapid Communications in Mass Spectrometry* 19, pp. 2497–2506.

Fuller, B. T., T. I. Molleson, D. A. Harris, L. T. Gilmour, and R. E. M. Hedges. 2006. "Isotopic Evidence for Breastfeeding and Possible Adult Dietary Differences From Late/Sub-Roman Britain," *American Journal of Physical Anthropology* 129, pp. 45–54.

Fuller, B. T., M. P. Richards, and S. A. Mays. 2003. "Stable Carbon and Nitrogen Isotope Variations in Tooth Dentine Serial Sections from Wharram Percy," *JAS* 30, pp. 1673–1684.

Garn, S. M. 1992. "The Iron-Deficiency Anemias and Their Skeletal Manifestations," in *Diet, Demography and Disease: Changing Perspectives on Anemia,* ed. P. Stuart-Macadam and S. Kent, New York, pp. 33–61.

Garnsey, P. 1998. *Cities, Peasants, and Food in Classical Antiquity: Essays in Social and Economic History,* ed. with addenda by W. Scheidel, Cambridge.

———. 1999. *Food and Society in Classical Antiquity,* Cambridge.

Garvie-Lok, S. 2001. "Loaves and Fishes: A Stable Isotope Reconstruction of Diet in Medieval Greece" (diss. Univ. of Calgary).

Golden, M. 1990. *Children and Childhood in Classical Athens,* Baltimore.

Griffin, I. J., and S. A. Abrams. 2001. "Iron and Breastfeeding," *Pediatric Clinics of North America* 48, pp. 401–413.

Grivetti, L. E. 2001. "Mediterranean Food Patterns: The View from Antiquity, Ancient Greeks and Romans," in *The Mediterranean Diet: Constituents and Health Promotion,* ed. A.-L. Matalas, A. Zampelas, V. Stavrinos and I. Wolinsky, Boca Raton, pp. 3–30.

Hanson, L. A., and J. Winberg. 1972. "Breast Milk and Defence Against Infection in the Newborn," *Archives of Disease in Childhood* 47, pp. 845–848.

Herring, D. A., S. R. Saunders, and M. A. Katzenberg. 1998. "Investigating the Weaning Process in Past Populations," *American Journal of Physical Anthropology* 105, pp. 425–439.

Hershkovitz, I., B. M. Rothschild, B. Latimer, O. Dutour, G. Léonetti, C. M. Greenwald, C. Rothschild, and L. M. Jellema. 1997. "Recognition of Sickle Cell Anemia in Skeletal Remains of Children," *American Journal of Physical Anthropology* 104, pp. 213–226.

Jennings, S., O. Reñones, B. Morales-Nin, N. V. C. Polunin, J. Moranta, and J. Coll. 1997. "Spatial Variation in the ^{15}N and ^{13}C Stable Isotope Composition of Plants, Invertebrates and Fishes on Mediterranean Reefs: Implications for the Study of Trophic Pathways," *Marine Ecology Progress Series* 146, pp. 109–116.

Katz, P. B. 2007. "Educating Paula: A Proposed Curriculum for Raising a 4th-Century Christian Infant," in *Constructions of Childhood in Ancient Greece and Italy* (*Hesperia* Suppl. 41), ed. A. Cohen and J. B. Rutter, Princeton 2007, pp. 115–127.

Katzenberg, M. A. 1993. "Age Differences and Population Variation in Stable Isotope Values from Ontario, Canada," in *Prehistoric Human Bone: Archaeology at the Molecular Level,* ed. J. B. Lambert and G. Grupe, Berlin, pp. 39–62.

Katzenberg, M. A., D. A. Herring, and S. R. Saunders. 1996. "Weaning and Infant Mortality: Evaluating the Skeletal Evidence," *Yearbook of Physical Anthropology* 39, pp. 177–199.

Katzenberg, M. A., and N. C. Lovell. 1999. "Stable Isotope Variation in Pathological Bone," *International Journal of Osteoarchaeology* 9, pp. 316–324.

Katzenberg, M. A., and S. Pfeiffer. 1995. "Nitrogen Isotope Evidence for Weaning Age in a Nineteenth Century Canadian Skeletal Sample," in *Bodies of Evidence: Reconstructing History Through Skeletal Analysis,* ed. A. L. Grauer, New York, pp. 221–235.

Katzenberg, M. A., S. R. Saunders, and W. R. Fitzgerald. 1993. "Age Differences in Stable Carbon and Nitrogen Isotope Ratios in a Population of Prehistoric Maize Horticulturists," *American Journal of Physical Anthropology* 90, pp. 267–281.

Keenleyside, A., and K. Panayotova. 2005. "A Bioarchaeological Study of the Greek Colonial Population of Apollonia Pontica," *Archaeologia Bulgarica* 9, pp. 21–38.

———. 2006. "Cribra Orbitalia and Porotic Hyperostosis in a Greek Colonial Population (5th to 3rd Centuries B.C.) from the Black Sea," *International Journal of Osteoarchaeology* 16, pp. 373–384.

Keenleyside, A., H. Schwarcz, and K. Panayotova. 2006. "Stable Isotopic Evidence of Diet in a Greek Colonial Population from the Black Sea," *JAS* 33, pp. 1205–1215.

Keenleyside, A., H. Schwarcz, L. Stirling, and N. B. Lazreg. 2009. "Stable Isotopic Evidence for Diet in a Roman and Late Roman Population from Leptiminus, Tunisia," *JAS* 36, pp. 51–63.

King, H. 2001. *Greek and Roman Medicine,* London.

Krebs, N. F. 2000. "Dietary Zinc and Iron Sources, Physical Growth, and Cognitive Development of Breastfed Infants," *Journal of Nutrition* 130, pp. 358S–360S.

Kwok, C. S. 2007. "Baby Bones, Food, and Health: Stable Isotopic Evidence for Infant Feeding Practices in the Greek Colony of Apollonia (5th–2nd Centuries B.C.)" (M.A. thesis, Trent University).

Kwok, C. S., and A. Keenleyside. 2010. "Reaching Your Potential? Growth, Weaning Practices, Childhood Diet, and Health in the Ancient Greek Colony of Apollonia Pontica," in *Proceedings of the 18th European Meetings of the Paleopathology Association*, Vienna, p. 143.

Longin, R. 1971. "New Method of Collagen Extraction for Radiocarbon Dating," *Nature* 230, pp. 241–242.

Marino, B. D., and M. B. McElroy. 1991. "Isotopic Composition of Atmospheric CO_2 as Inferred from Carbon in C_4 Plant Cellulose," *Nature* 349, pp. 127–131.

Metsecky, J., C. Blair, and P. L. Ogra, eds. 1991. *Immunology of Milk and the Neonate* (Advances in Experimental Medicine and Biology 310), New York.

Millard, A. R. 2000. "An Evaluation of the Possible Use of Nitrogen Isotopes to Detect Milking in Cattle," in *Human Ecodynamics* (Symposia of the Association for Environmental Archaeology 19), ed. G. Bailey, R. Charles, and N. Winder, Oxford, pp. 51–59.

Minagawa, M., and E. Wada. 1984. "Stepwise Enrichment of ^{15}N along Food Chains: Further Evidence and the Relation between $\delta^{15}N$ and Animal Age," *Geochimica et Cosmochimica Acta* 48, pp. 1135–1140.

Moorrees, C. F. A., E. A. Fanning, and E. E. Hunt Jr. 1963a. "Age Variation of Formation Stages for Ten Permanent Teeth," *Journal of Dental Research* 42, pp. 1490–1502.

———. 1963b. "Formation and Resorption of Three Deciduous Teeth in Children," *American Journal of Physical Anthropology* 21, pp. 205–213.

Nakamura, K., D. A. Schoeller, F. J. Winkler, and H. L. Schmidt. 1982. "Geographical Variations in the Carbon Isotope Composition of the Diet and Hair in Contemporary Man," *Biomedical Mass Spectrometry* 9, pp. 390–394.

Nedev, D., and K. Panayotova. 2003. "Apollonia Pontica (End of the 7th–1st Centuries B.C.)," in *Ancient Greek Colonies in the Black Sea* (Publications of the Archaeological Institute of Northern Greece 4), ed. D. V. Grammenos and E. K. Petropoulous, Thessaloniki, pp. 95–155.

Ortner, D. J. 2003. *Identification of Pathological Conditions in Human Skeletal Remains,* 2nd ed., San Diego.

Papathanasiou, A., C. S. Larsen, and L. Norr. 2000. "Bioarchaeological Inferences from a Neolithic Ossuary from Alepotrypa Cave, Diros, Greece," *International Journal of Osteoarchaeology* 10, pp. 210–228.

Pennycook, C. 2008. "A Stable Isotope Reconstruction of Byzantine and Frankish Greek Diet in the Valley of Stymphalos" (M.A. thesis Univ. of Alberta).

Prowse, T. L. 2001. "Isotopic and Dental Evidence for Diet from the Necropolis of Isola Sacra (1st–3rd Centuries A.D.), Italy" (diss. McMaster University).

Prowse, T. L., S. R. Saunders, H. P. Schwarcz, P. Garnsey, R. Macchiarelli, and L. Bondioli. 2008. "Isotopic and Dental Evidence for Infant and Young Child Feeding Practices in an Imperial Roman Skeletal Sample," *American Journal of Physical Anthropology* 137, pp. 294–308.

Richards, M. P., S. Mays, and B. T. Fuller. 2002. "Stable Carbon and

Nitrogen Isotope Values of Bone and Teeth Reflect Weaning Age at the Medieval Wharram Percy Site, Yorkshire, UK," *American Journal of Physical Anthropology* 199, pp. 205–210.

Rosenthal, R. 1936. "A Short Pictorial Review of the Evolution of Infant-Feeding Vessels up to the Beginning of the 19th Century," *Bulletin of the Medical Library Association* 25, pp. 89–94.

Schoeninger, M. J., and M. J. DeNiro. 1984. "Nitrogen and Carbon Isotopic Composition of Bone Collagen from Marine and Terrestrial Animals," *Geochimica et Cosmochimica Acta* 48, pp. 625–639.

Schurr, M. R. 1997. "Stable Nitrogen Isotopes as Evidence for the Age of Weaning at the Angel Site: A Comparison of Isotopic and Demographic Measures of Weaning Age," *JAS* 24, pp. 919–927.

Slade, H. B., and S. A. Schwartz. 1987. "Mucosal Immunity: The Immunology of Breast Milk," *Journal of Allergy and Clinical Immunology* 80, pp. 348–358.

Smith, B. H. 1991. "Standards of Human Tooth Formation and Dental Age Assessment," in *Advances in Dental Anthropology*, ed. M. A. Kelley and C. S. Larsen, New York, pp. 143–168.

Smith, B. N. and S. Epstein. 1971. "Two Categories of $^{12}C/^{13}C$ Ratios for Higher Plants," *Plant Physiology* 47, pp. 380–384.

Stloukal, R. I., and H. Hanáková. 1978. "Die Länge der Längsknochen altslawischer Bevölkerungen unter besonderer Berücksichtigung von Wachtstumsfragen," *Homo* 29, pp. 53–69.

Stravopodi, E., S. K. Manolis, S. Kousoulakos, V. Aleporou, and M. P. Schultz. 2009. "Porotic Hyperostosis in Neolithic Greece: New Evidence and Further Implications," in *New Directions in the Skeletal Biology of Greece* (*Hesperia* Suppl. 43, OWLS 1), ed. L. A. Schepartz, S. C. Fox, and C. Bourbou, Princeton, pp. 257–270.

Stuart-Macadam, P. 1991. "Anemia in Roman Britain: Poundbury Camp," in *Health in Past Societies: Biocultural Interpretations of Human Skeletal Remains in Archaeological Contexts* (*BAR-IS* 567), ed. H. Bush and M. Zvelebil, Oxford, pp. 101–113.

———. 1992. "Porotic Hyperostosis: A New Perspective," *American Journal of Physical Anthropology* 87, pp. 39–47.

Sundick, R. I. 1978. "Human Skeletal Growth and Age Determination," *Homo* 29, pp. 228–249.

Triantaphyllou, S., M. P. Richards, C. Zerner, and S. Voutsaki. 2008. "Isotopic Dietary Reconstruction of Humans from Middle Bronze Age Lerna, Argolid, Greece," *JAS* 35, pp. 3028–3034.

Tuross, N., and M. L. Fogel. 1994. "Stable Isotope Analysis and the Subsistence Patterns at the Sully Site," in *Skeletal Biology in the Great Plains: Migration, Warfare, Health, and Subsistence,* ed. D. W. Owsley and R. L. Jantz, Washington, D.C., pp. 283–289.

Ubelaker, D. H. 1989. "The Estimation of Age at Death from Immature Human Bone," in *Age Markers in the Human Skeleton,* ed. M. Y. İşcan, Springfield, Ill., pp. 55–70.

van Klinken, G. J. 1999. "Bone Collagen Quality Indicators for Palaeodietary and Radiocarbon Measurements," *JAS* 26, pp. 687–695.

Vaughan, L. A., C. W. Weber, and S. R. Kemberling. 1979. "Longitudinal Changes in the Mineral Content of Human Milk," *The American Journal of Clinical Nutrition* 32, pp. 2301–2306.

Venedikov, I. 1963. *Apoloniya: Razkopkite v nekropola na Apoloniya prez 1947–1949,* Sofia.

Walker, P. L., R. R. Bathurst, R. Richman, T. Gjerdrum, and V. A. Andrushko. 2009. "The Causes of Porotic Hyperostosis and Cribra Orbitalia: A Reappraisal of the Iron-Deficiency-Anemia Hypothesis," *American Journal of Physical Anthropology* 139, pp. 109–125.

Wapler, U., E. Crubézy, and M. Schultz. 2004. "Is Cribra Orbitalia Synonymous with Anemia? Analysis and Interpretation of Cranial Pathology in Sudan," *American Journal of Physical Anthropology* 123, pp. 333–339.

Waterlow, J. C. 1989. "Diet of the Classical Period of Greece and Rome," *European Journal of Clinical Nutrition* 43 (Suppl. 2), pp. 3–12.

Waters-Rist, A. L., and M. A. Katzenberg. 2010. "The Effect of Growth on Stable Nitrogen Isotope Ratios in Subadult Bone Collagen," *International Journal of Osteoarchaeology* 20, pp. 172–191.

White, C. D., and G. J. Armelagos. 1997. "Osteopenia and Stable Isotope Ratios in Bone Collagen of Nubian Female Mummies," *American Journal of Physical Anthropology* 103, pp. 185–199.

Wright, L. E., and H. P. Schwarcz. 1998. "Stable Carbon and Oxygen Isotopes in Human Tooth Enamel: Identifying Breastfeeding and Weaning in Prehistory," *American Journal of Physical Anthropology* 106, pp. 1–18.

———. 1999. "Correspondence between Stable Carbon, Oxygen and Nitrogen Isotopes in Human Tooth Enamel and Dentine: Infant Diets at Kaminaljuyú," *JAS* 26, pp. 1159–1170.

CHAPTER 10

Bread, Oil, Wine, and Milk: Feeding Infants and Adults in Byzantine Greece

by Chryssi Bourbou and Sandra Garvie-Lok

In the last few decades, skeletal analyses in Greece have been broadened beyond their previous focus on prehistoric and classical populations. Studies of human skeletal material from previously under-examined time periods (including the Byzantine era, 6th–12th centuries A.D.) are now contributing to the reconstruction of health and disease patterns in the past.[1] Although these studies have added considerably to our knowledge of the biological profile of Byzantine populations, few of them have dealt with the reconstruction of dietary and weaning patterns for these eras.[2] Although small in number, these studies provide interesting information on the everyday diet of the Byzantines and their attitudes toward breastfeeding and weaning. In this paper, we will bring together recent stable isotope data on diet and weaning patterns for Byzantine Greece, suggest areas in which further work needs to be done, and point to specific research questions that remain to be addressed.

WRITTEN IN THE SOURCES: EVERYDAY DIET IN BYZANTIUM

In dealing with more recent populations, we enjoy a great advantage: the existence of documentary evidence against which the biological data can be projected and compared. Since the pioneering work of Koukoules[3] almost 60 years ago covering various aspects of everyday life in Byzantium, including diet, Byzantine dietary patterns have been a matter of ongoing interest for researchers.[4] The literary sources from which dietary information may be

1. Garvie-Lok 2001, p. iii; Bourbou 2004, 2010; Tritsaroli 2006; Bourbou and Tsilipakou 2009. For a thorough review of the bioarchaeology of early Byzantine (6th–7th centuries A.D.) populations from Greece, see Bourbou 2004, pp. 31–39.

2. Garvie-Lok 2001, p. iii; Bourbou and Richards 2007, pp. 69–70; Bourbou and Garvie-Lok 2009, pp. 78–82.

3. Koukoules 1952, pp. 136–140.

4. See, e.g., Kaplan 1992; Mango, Dagron, and Greatrex 1995; Kislinger 1997; Braudel 1998; Bober 1999; Flandrin, Montanari, and Sonnenfeld 1999. Special events have also been organized on the topic, such as the one-day meeting on *"Food and Cooking in Byzantium"* (Museum of Byzantine Culture, Thessaloniki 2001) and the 37th Spring Symposium of Byzantine Studies on *"Eat, Drink and Be Merry (Luke 12:19): Production, Consumption and Celebration of Food and Wine in Byzantium"* (University of Birmingham, 2003).

drawn include historic and social commentaries, travelers' tales, regulations for the taxation and sale of foods, records of supplies purchased for troops or ships' crews, physicians' comments on diet, and the dietary regimes of monasteries. Study of these texts has contributed to our knowledge of the most frequently consumed products, methods of preparation, cooking and preservation of foods, and various recipes ranging from simple meals to luxury banquets. In addition to literary evidence, artistic representations of meals can provide useful information on the subject—although this is largely limited to insights into the tableware and comestibles that were used, rather than detailed information on Byzantine gastronomy and diet. The pictures that tell us about Byzantine food are mainly religious images, or illustrations accompanying manuscripts on a variety of relevant topics, and one must interpret them with caution due to the artistic conventions, stereotypes, and symbolism inherent in these depictions.[5] Images of earthly fruits, agricultural activities, animal husbandry, game, fish, and wild fowl were also common in the mosaics and wall paintings of churches, palaces, and houses, especially in the 5th and 6th centuries A.D. Although complicated again by issues of symbolism, they illustrate the bounty of nature as well as the food sources available to the Byzantines.[6]

Before briefly outlining the main foodstuffs consumed by the Byzantines according to these sources, it is useful to comment on their dining and fasting habits. Normally, two daily meals were taken: *ariston* or *geuma* was the first meal of the day, sometime late in the morning or before noon; the second meal, *deipnon,* was consumed early in the evening, usually before sunset.[7] Fasting played a significant role in daily diet, since for a considerable portion of the year (170–180 days), the Byzantines had to stay away from specific foodstuffs—especially meat.[8] On this topic, it is useful to consider early Christian attitudes toward fasting, so as better to understand the dietary habits of both sexes. The early Christians adopted fasting as a religious practice from Jewish tradition. By the 4th century A.D., fasting had become a highly praised form of behavior that signified the struggle for holiness and perfection.[9] For common Christians, the figure of the self-mortifying ascetic was an ideal to admire—especially for monks and virgins. Jerome (A.D. 345–420), advising young women on the rules according to which virgins of the church ought to live (Letter 107.10), urges extreme frugality in food and the avoidance of wine: "Let the food be vegetables and wheaten bread and occasionally a little fish . . . let her meals always leave her hungry."[10] Not only are women encouraged never to satisfy their hunger, but they are also advised to separate themselves from general society, where many women "are intemperate as to the amount of food they take" (Letter 22.17), and where "women care for nothing but their belly and its adjacent members" (Letter 22.29). With such close association between

5. For a thorough discussion on the use of artistic sources for studying Byzantine food, see Maguire 1987, 2005; Anagnostakis and Papamastorakis 2005.

6. For a thorough review, see Maguire 1987, 2005.

7. Koder 2005, pp. 17–19.

8. Koder (2005, p. 19) notes that besides the usual fasting days of Wednesday and Friday, the Byzantines had to fast between the fifteenth of November and Christmas Day (40 days), from Ash Monday to Holy Saturday (50 days), and from the first to the fifteenth of August.

9. Grimm 1995.

10. Grimm 1995, p. 234.

fasting and sexual abstinence in early Christian tradition, it is possible that men and women of early Christian Greece ate different kinds and quantities of food. As fasting rules targeted certain animal products in the diet, these specific products should be considered carefully when reconstructing Byzantine diets.[11]

According to many researchers, the base of the Byzantine diet was grain,[12] with bread serving as the primary dietary staple.[13] Various types of bread, differing in quality, are attested in the sources (προφούρνια, ψωμὶν ἀφρατίτσιν, ἄσπρον σεμιδαλάτον). Social discrimination appears to have been marked, as bread of lower quality (μεσοκάθαρον, τριπτούτσικον, ἄρτος ὁ ξηρός) was left to the poor. Also popular were rusks (παξιμᾶς)—a dried bread product.[14] Grains in primary use appear to have been wheat and barley, with the grain of choice being wheat. Another Byzantine grain crop was millet, the importance of which seems to have varied by region and economic status.[15] In Roman times, millet was considered a food of the poor.[16] The Byzantine physician Simeon Seth's opinion of millet as an inferior food[17] and the fact that millet was taxed at a lower rate than other grains in Ottoman Greece[18] suggest that a low esteem was attached to millet throughout these later periods. Other grains traditionally grown or used include oats, rye, and rice. Oats and rye were of minor significance, but rice became quite important in later times.[19]

Other field crops of historical importance included various legumes such as broad beans, chickpeas, peas, lentils, vetch, and lupin. Vetches were commonly used as animal feed or sown to enrich soil, while lupins, eaten by both humans and animals, became more important as human food when other crops failed.[20] Legumes like chickpeas and lentils were consumed by all classes, and elaborate recipes incorporating them—flavored with imported spices—survive in the historical record.[21] However, the importance of legumes to the Byzantine diet is less clear than that of grain: although some sources have suggested a primary importance for legumes, the detailed data available suggest that actual consumption levels were sometimes quite low.[22]

Oil and wine formed the other basic elements of the Byzantine diet and, along with wheat, these foodstuffs represent what Braudel calls the "essential trinity" of Mediterranean crops.[23] Historical sources attest to the continued importance of olive oil through the Byzantine period. In a lower-class diet—one based on grain and legumes—olives and olive oil would

11. See also Kislinger 1999, pp. 199–201. Monks formed a special segment of the population, for their fasting included more days throughout the year (see also Dembinska 1985, pp. 446–447).

12. E.g., Teall 1959, p. 98; Braudel 1966; Dalby 1996, p. 190; 2003, pp. 58–61; Kazhdan 1997, p. 46; Motsias 1998.

13. For the prices of cereals and other important foodstuffs such as oil, cheese, fish, meat, and wine, which not only are of interest to the researcher from a statistical point of view but have much to say on Byzantine standards of living, agricultural production, the laws of the market, etc., see Cheynet 2005.

14. Koder 2005, pp. 19–20.

15. Teall 1959, pp. 92, 99; see also Chap. 11.

16. Garnsey 1988, p. 51; Gallant 1991, p. 116.

17. Teall 1959, pp. 92, 99.

18. Matthaiou 1997, p. 217.

19. Matthaiou 1997, p. 63.

20. Dalby 1996, pp. 24–25; Motsias 1998.

21. Jeanselme and Oeconomos 1923, pp. 4–5.

22. See also Koder 1994.

23. Braudel 1966, p. 215.

have been a main source of fat and a considerable source of calories.[24] Wine is treated as a basic staple by Byzantine authors who refer to it, along with bread, as a dietary essential. The wines of Chios, Samos, and Crete were particularly famous.[25] As in the case of olive oil, wine would have provided a significant proportion of calories in lower-class diets.

Historical documents indicate that sheep, goats, pigs, and cattle were the main domesticated animals in the Byzantine countryside, although sheep and goats predominated.[26] Meats most highly valued in gourmet recipes of the time were suckling pig, lamb, and kid, along with pigeons and chickens. Some animals were sold in the markets during specific seasons (e.g., lamb for Easter). Meat of good quality was expensive. In addition to domesticated animals and birds, hunted and collected meat likely played a role in the diet.[27] The nutritional importance of meat in Byzantine Greece is problematic. It is reasonable to assume that the upper classes must have consumed significant amounts of meat, but consumption by the average person remains in doubt. Some sources suggest fairly regular meat consumption, while others refer to it only on special occasions.[28] Koder argues that consumption of meat was generally restricted.[29] Meat consumption during fasting was prohibited, while during summer and generally in hot climates, meat was considered dangerous since it could decompose quickly. A number of historical sources indicate that chicken and eggs were important,[30] but Koder argues that neither was frequently consumed.[31]

Dairy products represent a practical, efficient solution to the problems of the Mediterranean environment, because they can be exploited even in mountainous areas where other forms of subsistence are difficult.[32] Documentary evidence for the Byzantine period indicates a dominance of sheep and goat—at least in the case of small landholders, who constituted most of the rural population and produced their own dairy products.[33]

Milk was usually converted into other products prior to being consumed. The dairy food most frequently mentioned in the sources is cheese. Less frequent references are made to other dairy products including *trachanas,* a dried mixture of milk, cheese, and roasted grain. This was a traveler's food important to soldiers and shepherds, who prepared it for consumption by simply stirring it into hot water.[34] Yogurt and butter are also mentioned.[35]

Dairy products held special importance in the countryside, although they were also seen in urban diets: laborers, for example, are reported to have existed on both bread and cheese, but shepherds lived primarily on dairy products. Some cheeses were luxury foods (e.g., Vlach cheese), while others appear to have been cheap enough to have been included in lower status diets.[36]

24. Braudel 1966; Dalby 1996, p. 89; Koder 2005.
25. Koder 2005, p. 26.
26. Kazhdan 1997, pp. 52–56. Koder (2005, p. 22) argues that no beef was consumed.
27. Motsias 1998.
28. Dalby 1996, p. 196; Kazhdan 1997, p. 55.
29. Koder 2005, p. 22.
30. Dembinska 1985, p. 446; Kazhdan 1997, p. 56.
31. Koder 2005, p. 22.
32. Braudel 1966, p. 34; Dalby 2003, pp. 72–74.
33. Kazhdan 1997, pp. 54–55.
34. Hill and Bryer 1995.
35. Motsias 1998.
36. Koder 2005, p. 20.

The relative importance of fish, shellfish, and other marine species in the Byzantine diet is difficult to assess from the historical sources but would have varied regionally and seasonally.[37] Marine resources were important, it seems, with fish especially popular. A 13th- or 14th-century satire *(Ψαρολόγος)* concerning a trial for betrayal even portrays the protagonists as fish.[38] Most of the marine foods known to be consumed during the Roman era are also mentioned in Byzantine times, and historical sources attest to more than 50 different species.[39] Preserved fish (salted or dried) was likely to have been important since methods for preserving fish eggs as caviar (sturgeon eggs) or botargo (salted, pressed mullet roe) are included in the historical sources. Additionally, there is evidence for the use of fish sauce *(garum)*.[40]

Finally, fruits (e.g., grapes, figs, melon), vegetables (e.g., celery, lettuce, endive, cress, kohlrabi, turnips, eggplant, cabbage), and wild vegetables gathered seasonally were also important components of the Byzantine diet.[41] Vegetables were consumed fresh, dried, or preserved in salt and vinegar (ταριχευμένα δι'ὀξάλμης) like modern pickles.[42] Nuts (e.g., almonds, chestnuts, pistachios, walnuts, pine nuts) and honey appear also to have formed a significant portion of the diet.[43] Cane sugar was known in the Byzantine world, but it was used more as a medicine than as a condiment. The *Prodromic Poems,* for example, contain a reference to a sick man needing sugar from the market as part of his cure.[44]

FEEDING INFANTS IN BYZANTIUM

Byzantine physicians, who have written entire chapters on the care of newborn babies and particularly perinatal nutrition, provide us valuable insight into Byzantine practices of nursing and weaning infants.[45] The most eminent early Byzantine physician was Oribasius from Pergamum (4th century A.D.), whose work influenced later Byzantine as well as Arabic medicine.[46] Oribasius and the preceding medical scholars on whose ideas he drew argued that newborns should not be breastfed immediately after birth, since the mother's milk was unwholesome during this time.[47] Instead, they should be given food thought to be easily digested, such as clear honey of high quality, drops of a lukewarm solution of honey and water *(hydromel),* or honey mixed with goat's milk. After that initial period, newborns should be breastfed to the age of 2 years—ideally by a wet nurse, if the family's resources permitted.[48] Solid foods were commenced around the age of 6 months, after the baby's body could tolerate it. These early foods included

37. Braund 1995; Dagron 1995; Maniatis 2000; Dalby 2003, pp. 66–69.
38. Kislinger 2005, p. 52.
39. Dalby 1996, p. 189; Kislinger 2005, p. 52.
40. Dalby 1996, p. 189; Motsias 1998. For the preparation of garum, see Koder 2005, pp. 24–25.
41. Dalby 2003, pp. 74–77.
42. Koder 1994, 2005; Littlewood, Maguire, and Wolschke-Bulmahn 2002; Anagnostakis and Papamastorakis 2005.
43. Motsias 1998.
44. Eideneier 1991, poem IV, 450–451.
45. For more detail on the documentary evidence for Byzantine weaning practices, see Bourbou and Garvie-Lok 2009, pp. 70–75.
46. Beaucamp 1982, p. 549; Congourdeau 1993; Lascaratos and Poulakou-Rebelakou 2003, p. 186.
47. Temkin 1991; Lascaratos and Poulakou-Rebelakou 2003, p. 186.
48. Beaucamp 1982, pp. 553–555; Temkin 1991; Lascaratos and Poulakou-Rebelakou 2003, p. 188.

crumbs of bread softened with *hydromel,* milk, sweet or honeyed wine, grain soup or porridge, and eggs.[49]

Useful clues to Byzantine weaning practices can be also found in the recorded *Vitae* and *Miracles* of saints. The *Vita* of Basil the Younger (316.37) refers to a 4-year-old child who was still breastfeeding.[50] The *Vita* of Symeon Stylites the Younger indicates that the saint was still breastfeeding at age 2, and relates that when Symeon began to eat solid food, his mother gave him honey on bread with water to drink.[51] The *Vitae* of Saints Alypios and Michael Synkellos mention children who had been weaned by the age of 3,[52] while in the *Miracles* of St. Thekla (*Mir.* 24) a child just weaned is said to be old enough to walk and run—which suggests it was at least 18 months old.[53] Although there is some variation, these Byzantine references generally indicate ages for the weaning practice that are fairly late by current Western standards.

STABLE ISOTOPES IN DIET AND BONE

Stable isotope analysis of archaeological skeletal remains is an established, powerful method for the reconstruction of dietary behavior in past populations. Its general theoretical basis is covered elsewhere in this volume.[54] This section will discuss some issues specific to this paper, including the likely $\delta^{13}C$ and $\delta^{15}N$ values of some of the items in Byzantine diets and some of the theoretical issues involved in interpreting the Byzantine stable isotope data discussed here.

Likely $\delta^{13}C$ Values for Some Byzantine Dietary Items

Most food plants used by the Byzantines, including wheat, barley, legumes, olives, and grapes, are C_3 plants. Of these, one—the olive—produces a staple with a distinctive $\delta^{13}C$ value. Olive oil shows very low $\delta^{13}C$ values (-30‰ to -27‰) characteristic of lipids and could have had significant impact on the bulk $\delta^{13}C$ value of the diet if consumed in sufficient amounts.[55] The only two C_4 crop plants mentioned in historical sources on Byzantine diets are millet and sugarcane.

The $\delta^{13}C$ values of meat and milk from Byzantine domesticated animals would reflect their diet, with values of muscle tissue somewhat higher than the diet[56] and values of milk falling close to that of the diet.[57] Recent analyses of domesticated fauna from Greek middens of Byzantine or slightly later date have yielded average collagen $\delta^{13}C$ values of -21.5‰

49. Temkin 1991, pp. 108–113.
50. Viliskij 1911, p. 316.37.
51. Van den Ven 1962, chaps. 5, 6, p. 7.
52. Delehaye 1923, p. 149; Cunningham 1991, p. 46.
53. Dagron and Dupré la Tour 1978, p. 350.
54. See Chap. 2.
55. Spangenberg, Macko, and Hunziker 1998, p. 4181; Royer et al. 1999, p. 361.
56. See, e.g., Vogel 1978, p. 299; Tieszen et al. 1983, p. 33; Tieszen and Boutton 1989, p. 177; Hare et al. 1991, p. 282.
57. Data for dairy cattle show whole milk $\delta^{13}C$ to fall within about 3‰ of the animal's fodder (Minson, Ludlow, and Troughton 1975; Tyrrell et al. 1984, p. 128; Boutton et al. 1988, p. 2639; Metges, Kempe, and Schmidt 1990, p. 191). The $\delta^{13}C$ value of casein, the primary protein of milk, is typically a little higher than that of whole milk (Nakamura et al. 1982, p. 392; Kornexl et al. 1997, p. 22).

to -19.6‰ for sheep and goat, -20.3‰ to -19.6‰ for pig, and -21.4‰ to -20.2‰ for cow (Table 10.1). This indicates the consumption of C_3-based fodder and suggests that δ^{13}C values of milk and meat would have fallen in the C_3 range.

The other major source of potential δ^{13}C patterning in the Byzantine diet would have been the distinctive values of marine resources. A recent analysis of a number of modern Aegean food fish species yielded average δ^{13}C values of -17.0‰ for small, low trophic level fish (sardine and anchovy) and -15.2‰ for larger species of higher trophic level (Table 10.2). Although recent industrial pollution may have altered Aegean food web δ^{13}C values somewhat, it can still be assumed that marine resources would have contributed an enriched δ^{13}C signal to Byzantine diets.

Likely δ^{15}N Values for Some Byzantine Dietary Items

Recent analyses of domesticated fauna from some Greek Byzantine-era middens have recorded collagen δ^{15}N values for sheep and goat ranging from 4.0‰ to 6.5‰, for cattle ranging from 3.5‰ to 6.0‰, and for pig ranging from 3.6‰ to 6.1‰ (Table 10.1). The δ^{15}N values of these animals' meat and milk protein would have been similar.[58] Subtracting a 4‰ trophic level enrichment from these collagen values suggests low δ^{15}N values in the 0‰ to 2‰ range for non-leguminous plants such as grains.[59] Lentils, beans, and other leguminous field crops would have had the lowest δ^{15}N values, around 0‰.

A recent analysis of a number of modern food fish species from the Aegean yielded average δ^{15}N values of 6.5‰ for small, low trophic level fish (sardine, anchovy) and 10.5‰ for larger, higher trophic level fish (Table 10.2). As with δ^{13}C values, the impact of recent industrial pollution (especially fertilizer-laden runoff from agricultural areas) may have altered the Aegean food web of δ^{15}N values. However, it remains a safe assumption that marine resources, especially those of high trophic level, would have contributed an enriched ^{15}N signal to Byzantine diets.

Inferring Past Diet from Archaeological Human Remains

The fundamental assumptions involved in interpreting collagen δ^{13}C values are reviewed elsewhere in the volume.[60] Approaches to carbonate δ^{13}C values are more variable. Mammalian bone and mineral carbonate δ^{13}C values are roughly 9‰ to 15‰ higher than bulk diet δ^{13}C, depending on the species in question.[61] Much of this variation may reflect interspecies

58. Steele and Daniel 1978, p. 8; DeNiro and Epstein 1981, p. 344; Hare et al. 1991, p. 282; Kornexl et al. 1997, p. 22. Although we use these values as general estimates of Byzantine Greek milk and meat δ^{15}N here, we recognize that local plant δ^{15}N values can vary substantially and that there may have been significant regional variation in milk and meat δ^{15}N.

59. As discussed below, the use of domesticate δ^{15}N values to estimate the δ^{15}N values of crop plants is somewhat problematic, and these values may not be accurate.

60. See Chap. 2.

61. DeNiro and Epstein 1978, p. 503; Ambrose and Norr 1993, p. 26; Tieszen and Fagre 1993, p. 155; Balasse, Bocherens, and Mariotti 1999, p. 596; Ayliffe et al. 2004, p. 21; Passey et al. 2005, p. 1467.

TABLE 10.1. STABLE ISOTOPE VALUES FOR DOMESTICATED FAUNA FROM BYZANTINE SITES*

Sample	*Common Name*	*Site*	*δ¹³C (‰ VPDB)*	*δ¹⁵N (‰ AIR)*	*Collagen Preservation*			
					C:N	*% C*	*% N*	*% Yield*
SHEEP/GOAT								
SAGF01	Sheep/Goat	Athens	-19.9	3.3	3.2	43.1	15.7	13.0
SAGF04	Sheep/Goat	Athens	-19.8	11.7	3.3	43.1	15.4	13.6
SAGF07	Sheep/Goat	Athens	-19.2	3.9	3.3	43.8	15.6	14.2
SAGF09	Sheep/Goat	Athens	-20.1	8.0	3.2	44.2	15.4	11.9
SAGF13	Sheep/Goat	Athens	-19.9	9.3	3.2	43.4	15.7	7.8
SAGF14	Sheep/Goat	Athens	-19.7	3.9	3.2	43.2	15.8	14.1
SAGF15	Sheep/Goat	Athens	-19.5	4.2	3.3	43.6	15.8	9.5
SAGF18	Sheep/Goat	Athens	-19.4	5.6	3.2	40.7	14.6	16.3
SAGF21	Sheep/Goat	Athens	-19.0	7.3	3.2	42.3	15.0	15.0
SAGF22	Sheep/Goat	Athens	-19.9	3.7	3.3	44.3	15.7	19.0
		Mean for Athens Sheep/Goat	*-19.6*	*6.1*	–	–	–	–
SCOF01	Sheep/Goat	Corinth	-19.8	5.3	3.4	41.8	14.2	3.1
SCOF02	Sheep/Goat	Corinth	-19.8	3.5	3.2	43.7	15.3	4.5
SCOF04	Sheep/Goat	Corinth	-19.9	4.0	3.3	44.2	15.5	11.2
SCOF05	Sheep/Goat	Corinth	-19.1	3.1	3.3	41.1	14.4	2.3
		Mean for Corinth Sheep/Goat	*-19.6*	*4.0*	–	–	–	–
CBYF1	Sheep	Kastella	-20.6	5.9	3.4	41.3	14.3	1.6
CBYF6	Goat	Kastella	-18.7	5.3	3.3	37.6	13.2	1.9
CBYF7	Goat	Kastella	-21.1	3.8	3.3	36.0	12.6	2.2
CBYF8	Goat	Kastella	-20.3	4.6	3.3	39.5	14.0	2.1
CBYF9	Goat	Kastella	-20.4	4.4	3.3	38.0	13.6	1.6
CBYF10	Goat	Kastella	-19.8	5.0	3.3	37.0	13.2	4.9
		Mean for Kastella Sheep/Goat	*-20.2*	*4.8*	–	–	–	–
SMYF02	Sheep	Mytilene	-21.5	6.2	3.3	46.3	15.9	4.2
SMYF03	Sheep	Mytilene	-21.2	6.3	3.4	46.1	15.3	4.4
SMYF04	Sheep/Goat	Mytilene	-21.5	5.2	3.4	46.4	16.1	3.6
SMYF12	Sheep/Goat	Mytilene	-21.7	8.4	3.3	46.6	16.4	7.7
		Mean for Mytilene Sheep/Goat	*-21.5*	*6.5*	–	–	–	–
PIG								
SAGF02	Pig	Athens	-20.9	6.4	3.2	42.9	15.8	14.7
SAGF05	Pig	Athens	-20.3	4.5	3.2	43.3	15.6	12.6
SAGF06	Pig	Athens	-18.9	9.1	3.2	42.9	15.6	8.8
SAGF16	Pig	Athens	-19.6	4.0	3.3	43.8	15.6	3.3
SAGF20	Pig	Athens	-20.8	6.4	3.3	43.2	15.8	12.2
SAGF23	Pig	Athens	-21.1	6.2	3.2	43.4	15.8	10.3
		Mean for Athens Pig	*-20.3*	*6.1*	–	–	–	–
SCOF07	Pig	Corinth	-19.6	3.6	3.3	42.8	15.3	9.8
		Mean for Corinth Pig	*-19.6*	*3.6*	–	–	–	–

TABLE 10.1—*Continued*

Sample	*Common Name*	*Site*	*δ¹³C (‰ VPDB)*	*δ¹⁵N (‰ AIR)*	*Collagen Preservation*			
					C:N	*% C*	*% N*	*% Yield*
CBYF4	Pig	Kastella	-20.6	7.1	3.4	39.4	13.6	0.1
CBYF3	Pig	Kastella	-21.0	4.2	3.4	8.6	3.0	0.5
CBYF5	Pig	Kastella	-20.5	5.8	3.3	40.8	14.5	3.1
		Mean for Kastella Pig	*-20.7*	*5.7*	–	–	–	–
SMYF05	Pig	Mytilene	-21.1	5.7	3.2	43.7	15.7	4.2
SMYF06	Pig	Mytilene	-20.4	6.0	3.3	43.8	15.6	5.7
		Mean for Mytilene Pig	*-20.8*	*5.9*	–	–	–	–
Cow								
SAGF08	Cow	Athens	-21.1	4.6	3.3	44.2	15.8	15.9
SAGF12	Cow	Athens	-19.9	6.3	3.2	44.0	15.8	9.4
SAGF17	Cow	Athens	-19.9	6.7	3.2	44.0	15.5	16.3
SAGF19	Cow	Athens	-19.9	6.4	3.3	43.2	15.9	8.4
		Mean for Athens Cow	*-20.2*	*6.0*	–	–	–	–
SCOF06	Cow	Corinth	-20.4	4.2	3.3	42.4	15.1	7.9
		Mean for Corinth Cow	*-20.4*	*4.2*	–	–	–	–
SMYF15	Cow	Mytilene	-21.4	3.5	3.4	47.0	16.0	5.5
		Mean for Mytilene Cow	*-21.4*	*3.5*	–	–	–	–

*Data are from Garvie-Lok 2001 and Bourbou and Richards 2007. Athens = Athenian Agora; Mytilene = acropolis of Mytilene, Lesbos. Analytical methods used are the same as those described for human archaeological samples. Although the Athens and Mytilene samples extend into the early modern era, no significant temporal trend in $\delta^{13}C$ or $\delta^{15}N$ is seen (Garvie-Lok 2001).

differences in digestive physiology.[62] Here we follow a number of recent archaeological studies[63] in using a value of 13‰, which is consistent with values measured in recent controlled feeding experiments on swine—an animal often considered a good model for human digestion.[64] When interpreting collagen and carbonate $\delta^{13}C$ values, we assume that in most situations, human collagen $\delta^{13}C$ values primarily reflect the $\delta^{13}C$ value of high-protein items in the diet, while carbonate $\delta^{13}C$ values reflect the overall $\delta^{13}C$ value of the diet.

As reviewed elsewhere in this volume,[65] the collagen $\delta^{15}N$ values of human bone should fall about 3‰–5‰ above dietary $\delta^{15}N$ and reflect the relevant trophic level, with $\delta^{15}N$ values substantially higher than local fauna indicating a heavy reliance on animal products. As stable isotope studies of European agricultural groups have accumulated, however, several have produced results that are unexpected in this model.[66] These groups are characterized by very high human $\delta^{15}N$ values at least one trophic level above

62. Hedges and van Klinken 2000, pp. 223–224; Hedges 2003, pp. 67–70; Passey et al. 2005, p. 1467.

63. See, e.g., Prowse et al. 2004, p. 267; Keenleyside, Schwarcz, and Panayotova 2006, p. 1208.

64. Passey et al. 2005, p. 1467.

65. See Chap. 2.

66. For review and theoretical comments on the following, see van Klinken, Richards, and Hedges 2000, pp. 55–57; Hedges and Reynard 2007, pp. 1241–1243.

TABLE 10.2. STABLE ISOTOPE VALUES FOR SELECT MODERN AEGEAN FAUNA*

Sample	*Species*	*Common Name*	*$\delta^{13}C$ (‰ VPDB)*	*$\delta^{15}N$ (‰ AIR)*
SAT-07	*Engraulis encrasicolus*	European anchovy	-18.3	5.2
SAT-08	*Engraulis encrasicolus*	European anchovy	-17.1	7.1
SAT-09	*Engraulis encrasicolus*	European anchovy	-18.1	6.6
SAT-10	*Engraulis encrasicolus*	European anchovy	-16.6	6.4
SAT-17	*Sardina pilchardus*	Pilchard	-16.7	7.1
SAT-18	*Sardina pilchardus*	Pilchard	-16.8	5.9
SAT-19	*Sardina pilchardus*	Pilchard	-16.4	7.7
SAT-20	*Sardina pilchardus*	Pilchard	-15.8	6.1
		Mean for Smaller Fish	*-17.0*	*6.5*
SAT-01	*Mullus surmuletus*	Red mullet	-14.8	8.5
SAT-02	*Mullus surmuletus*	Red mullet	-16.1	8.4
SAT-05	*Trachurus mediterraneus*	Horse mackerel	-15.5	8.1
SAT-11	*Micromestitius poutassou*	Blue whiting	-14.8	17.9
SAT-14	*Diplodus sargus*	White sea bream	-12.1	10.6
SAT-16	*Sparus aurata*	Gilt-head sea bream	-17.9	9.7
		Mean for Larger Fish	*-15.2*	*10.5*

*Data are from Garvie-Lok 2001. Measurements are on collagen prepared from fresh bone using the method outlined for archaeological samples from Nemea, Petras, and Servia, with the addition of a defatting treatment of 100% diethyl ether or 2:1 chloroform:methanol.

those of local domesticated fauna, a finding suggestive of a diet in which most or all dietary protein was derived from animal products.[67] Such a finding is unexpected when we consider other lines of evidence that indicate heavy dietary reliance on a grain staple by many European agricultural groups.

In some cases, these high $\delta^{15}N$ values may reflect the consumption of high-nitrogen aquatic resources, producing a mixed diet that had a high overall $\delta^{15}N$ value relative to the $\delta^{15}N$ values of local domesticates.[68] Hedges and Reynard, however, have argued that aquatic resource consumption alone is unlikely to have raised $\delta^{15}N$ values to this degree at more than a few sites.[69] High $\delta^{15}N$ values may instead be explained through another factor such as the consumption of omnivorous animals such as swine that may have unexpectedly high $\delta^{15}N$ values or the use of animal manure as fertilizer, which could produce grain $\delta^{15}N$ values that are unexpectedly high in comparison to the tissue $\delta^{15}N$ values of domesticated animals fed on local plants.[70] At present, this debate has no clear resolution. It can be

67. See, e.g., van Klinken, Richards, and Hedges 2000, p. 55; Privat, O'Connell, and Richards 2002, p. 785; Polet and Katzenberg 2003, p. 527; Prowse et al. 2004, p. 263; Müldner and Richards 2005, p. 42; Dürrwachter et al. 2006, p. 43; Jay and Richards 2006, p. 655; Keenleyside, Schwarcz, and Panayotova 2006, p. 1209; Richards, Fuller, and Molleson 2006, p. 125.

68. See, e.g., Privat, O'Connell, and Richards 2002, p. 785; Polet and Katzenberg 2003, pp. 527–528; Prowse et al. 2004, p. 270; Müldner and Richards 2005, p. 44; Keenleyside, Schwarcz, and Panayotova 2006, p. 1211; Richards, Fuller, and Molleson 2006, p. 126.

69. Hedges and Reynard 2007, p. 1244.

70. Van Klinken, Richards, and Hedges 2000, p. 57; Privat, O'Connell, and Richards 2002, p. 785; Bogaard et al. 2007, pp. 340–341; Hedges and Reynard 2007, p. 1247.

said, however, that unexpectedly high $\delta^{15}N$ values are common in European agricultural groups. Furthermore, contrary to initial expectations, these high $\delta^{15}N$ values may not necessarily indicate an extremely high proportion of animal protein in past diets.

STABLE ISOTOPE EVIDENCE FOR BYZANTINE DIETS

Relatively few Byzantine-era Aegean populations have to date been studied using stable isotope analysis. Although much research remains to be done, these studies do offer some valuable preliminary information (Table 10.3).[71]

Bone collagen and carbonate from three Middle Byzantine sites were analyzed as part of a wider study on diet in medieval and early modern Greece.[72] These samples were excavated at Nemea (12th–13th centuries A.D.), Petras (12th–13th centuries A.D.), and Servia (11th to ca. 15th centuries A.D.). To isolate collagen, samples from these sites were demineralized by soaking whole chunks of bone in repeated changes of 1% HCl solution.[73] This was followed by a 20-hour 0.125M NaOH treatment to remove humic and fulvic acids.[74] Isotopic values of the collagen were measured at the University of Calgary Department of Physics and Astronomy using a Finnegan Mat TracerMat mass spectrometer coupled to an NA 1500 elemental analyzer. For carbonate analysis, samples were ground and treated with 2% sodium hypochlorite and 0.1M acetic acid as described in detail elsewhere.[75] The prepared samples were analyzed on a VG Sira mass spectrometer at the University of Calgary Department of Physics and Astronomy. Sample stable isotope values and preservation indicators are presented in Table 10.3.

Bone collagen from the Middle Byzantine cemetery of Kastella (11th century A.D.) was studied as part of a general examination of health and diet at that site. Human bone collagen was extracted from the Kastella samples following a modified Longin procedure,[76] as outlined in detail elsewhere,[77] with the addition of an ultrafiltration step.[78] Isotopic values for the Kastella samples were measured using a Carlo Erba elemental analyzer coupled to a Thermo Finnigan Delta Plus XL mass spectrometer at the Isotope Laboratory, Department of Archaeological Sciences, University of Bradford.[79] Sample stable isotope values and preservation indicators are presented in Table 10.3.

Lastly, a small data set is available for Byzantine burials at Abdera. In this case, Agelarakis analyzed collagen and carbonate from four individuals as part of a wider study on the diet and health of two groups from the site—one dating to the 6th–9th centuries A.D. and the other to the 12th–13th centuries A.D.[80] Sample stable isotope values for these individuals are presented in Table 10.3.

Together, these studies offer information on the diets of Byzantine communities from both inland (Servia, Nemea) and coastal (Abdera, Kastella, Petras) areas in which some variation in available resources likely existed. Despite this variation, the stable isotope values of the human remains from the various sites are broadly similar (Table 10.3). Collagen shows mean $\delta^{13}C$ values ranging from -19.2‰ (Petras) to -18.5‰ (Abdera) and mean $\delta^{15}N$ ranging from 8.6‰ (Nemea) to 9.7‰ (Abdera). Bone

71. Data on Byzantine diet are also available for the early Byzantine site of Gortyn in Crete (Fornaciari, Ceccanti, and Menicagli 1988, pp. 403–416). However, as the method used in that study was trace element analysis, the data are not reviewed here.

72. Garvie-Lok 2001, pp. 386–398.

73. Sealy 1986, p. 51.

74. DeNiro and Epstein 1981, p. 343.

75. Garvie-Lok, Varney, and Katzenberg 2004, pp. 766, 774.

76. Longin 1971, p. 241.

77. Richards and Hedges 1999.

78. Brown et al. 1988 p. 176.

79. Bourbou and Richards 2007, p. 67.

80. Agelarakis 1989, 1992–1993.

TABLE 10.3. STABLE ISOTOPE VALUES FOR HUMANS FROM BYZANTINE SITES*

Site	*Sample*	*Sex*	*Age***	*Tissue*	$\delta^{13}C_{coll}$ *(‰ VPDB)*	$\delta^{15}N_{coll}$ *(‰ AIR)*	$\delta^{13}C_{carb}$ *(‰ VPDB)*	*Collagen Preservation*			
								C:N	*% C*	*% N*	*% Yield*
Kastella	CBYA1	F	ca. 25 y	rib	-18.7	9.4	–	3.3	45.2	15.8	6.2
Kastella	CBYA2	M	30–35 y	rib	-18.6	9.5	–	3.3	46.1	16.4	6.5
Kastella	CBYA3	M	Ad	rib	-18.5	10.6	–	3.3	40.1	14.1	3.0
Kastella	CBYA4	M	ca. 50 y	rib	-18.7	8.9	–	3.3	39.2	13.9	1.6
Kastella	CBYA5	M	Ad	rib	-18.2	10.9	–	3.3	30.2	10.8	1.4
Kastella	CBYA6	M	ca. 31 y	rib	-18.9	8.2	–	3.3	20.0	7.1	1.6
Kastella	CBYA7	M	ca. 44 y	rib	-19.1	9.0	–	3.3	40.0	14.2	2.2
Kastella	CBYA8	M	ca. 39 y	rib	-19.1	8.8	–	3.4	45.5	15.7	1.0
Kastella	CBYA9	F	35–40 y	rib	-19.1	9.1	–	3.3	47.1	16.5	3.0
Kastella	CBYA10	F	ca. 24 y	rib	-19.1	8.7	–	3.3	42.0	14.8	3.4
Kastella	CBYA11	F	50–60 y	rib	-18.9	8.6	–	3.3	44.7	15.8	6.5
Kastella	CBYA12	F?	Ad	rib	-19.2	7.8	–	3.4	40.1	13.9	2.2
Kastella	CBYA13	M	ca. 35 y	rib	-19.4	8.6	–	3.3	44.1	15.4	4.3
Kastella	CBYA14	M	35–40 y	rib	-18.4	10.1	–	3.4	36.8	12.8	0.5
Kastella	CBYA15	M?	Ad	rib	-19.0	8.5	–	3.3	37.0	13.2	5.9
Kastella	CBYS1	?	7 y ± 24 mo	rib	-19.0	8.9	–	3.3	40.3	14.4	5.6
Kastella	CBYS2	?	2 y ± 8 mo	rib	-18.0	12.1	–	3.2	40.8	14.7	6.3
Kastella	CBYS3	?	0.5 mo–1.5 y	rib	-18.3	11.6	–	3.3	33.0	11.7	3.0
Kastella	CBYS4	?	10 y ± 30 mo	rib	-18.8	7.9	–	3.2	37.2	13.4	4.0
Kastella	CBYS5	?	15 y ± 36 mo	rib	-19.2	10.7	–	3.3	31.6	11.1	0.7
Kastella	CBYS6	?	12 y ± 36 mo	rib	-19.3	8.8	–	3.3	37.5	13.3	2.2
Kastella	CBYS7	?	15 y ± 36 mo	rib	-18.5	9.1	–	3.3	32.3	11.6	3.3
Kastella	CBYS8	?	4.5–5.5 y	rib	-18.7	8.3	–	3.2	26.5	9.8	1.9
Kastella	CBYS9	?	15 y ± 36 mo	rib	-19.1	8.2	–	3.3	42.0	14.8	3.0
Kastella	CBYS10	?	7.5–8.5 y	rib	-18.7	8.5	–	3.3	38.2	13.7	2.6
Kastella	CBYS11	?	4 y ± 12 mo	rib	-18.4	7.5	–	3.4	46.5	15.8	1.4
				Mean for Kastella	*-18.8*	*9.2*	–	–	–	–	–
				SD for Kastella	*0.4*	*1.2*	–	–	–	–	–
Nemea	SNE-30	M	20–30 y	rib	-19.6	8.3	-11.4	3.2	43.1	15.6	7.2
Nemea	SNE-31	?	5–10 y	rib	-19.2	7.5	–	3.3	44.1	15.4	4.9
Nemea	SNE-32	?	16–32 mo	fibula	-19.0	8.6	-9.6	3.3	44.3	15.6	7.1
Nemea	SNE-33	M	40–44 y	rib	-18.9	8.9	-10.0	3.3	44.5	15.5	5.6
Nemea	SNE-34	F	>50 y	rib	-19.0	7.7	-12.0	3.2	44.3	16.1	16.3
Nemea	SNE-36	M	Ad	rib	-18.6	9.0	-11.9	3.2	44.5	15.9	9.5
Nemea	SNE-38	M	30–34 y	rib	-18.6	9.2	-10.8	3.3	45.2	16.1	11.7
Nemea	SNE-39	?	Ad	rib	-19.2	7.9	-10.8	3.3	44.9	15.7	5.6
Nemea	SNE-40	M	>35 y	rib	-19.3	8.9	-11.7	3.3	43.5	15.3	7.2
Nemea	SNE-41	M	25–39 y	rib	-19.2	8.4	–	3.3	43.6	15.3	4.2
Nemea	SNE-42	F	30–45 y	rib	-19.1	9.2	–	3.3	41.4	14.7	3.5
Nemea	SNE-43	M	18–35 y	rib	-19.2	8.9	-10.0	3.3	42.1	14.8	5.7
Nemea	SNE-45	M	Ad	rib	-18.7	9.2	-12.6	3.3	44.9	15.7	9.1
				Mean for Nemea	*-19.0*	*8.6*	*-11.1*	–	–	–	–
				SD for Nemea	*0.3*	*0.6*	*1.9*	–	–	–	–

TABLE 10.3—*Continued*

Site	*Sample*	*Sex*	*Age***	*Tissue*	$\delta^{13}C_{coll}$ (‰ VPDB)	$\delta^{15}N_{coll}$ (‰ AIR)	$\delta^{13}C_{carb}$ (‰ VPDB)	*Collagen Preservation*			
								C:N	*% C*	*% N*	*% Yield*
Petras	SPE-02	?	4–5 y	rib	-20.2	8.6	-14.0	3.2	45.2	16.8	19.4
Petras	SPE-03	M	35–50 y	rib	-18.7	10.8	–	3.4	44.3	15.0	5.6
Petras	SPE-04	M	35–50 y	rib	-18.7	10.7	–	3.4	45.5	15.6	4.1
Petras	SPE-05	F	18–35 y	rib	-18.9	9.6	-10.6	3.3	45.4	16.1	12.5
Petras	SPE-06	M	18–35 y	rib	-19.4	8.9	-11.9	3.3	45.5	15.9	9.2
Petras	SPE-07	?	6–7 y	rib	-19.7	8.9	-11.3	3.3	47.0	16.7	13.6
Petras	SPE-08	M	18–35 y	rib	-19.2	9.5	–	3.4	45.5	15.6	5.4
Petras	SPE-09	F	>35 y	rib	-19.5	9.2	-12.4	3.2	45.6	15.9	15.3
Petras	SPE-10	M	35–50 y	rib	-18.9	8.5	–	3.4	44.5	15.5	10.3
Petras	SPE-11	F	Ad	femur	-19.3	9.1	–	3.3	44.6	15.6	6.7
Petras	SPE-12	M	>35 y	rib	-19.5	9.4	-12.3	3.2	45.3	16.4	21.2
Petras	SPE-13	F	18–35 y	rib	-19.5	9.1	–	3.4	45.1	15.4	8.8
Petras	SPE-14	?	5.5–6.5 y	femur	-18.9	9.6	–	3.3	43.5	15.4	4.2
Petras	SPE-15	M	18–35 y	rib	-18.8	9.8	-12.5	3.3	44.4	15.6	16.7
Petras	SPE-16	?	30–42 mo	rib	-18.6	12.7	-12.0	3.3	47.1	16.7	18.6
Petras	SPE-17	F	Ad	rib	-19.4	8.9	-12.4	3.3	44.9	16.1	10.9
Mean for Petras					*-19.2*	*9.6*	*-12.2*	–	–	–	–
SD for Petras					*0.4*	*1.0*	*0.9*	–	–	–	–
Servia	SSE-01	M	Ad	femur	-18.1	8.7	–	3.3	46.7	16.6	7.7
Servia	SSE-02	?	8–11 y	femur	-19.0	9.2	–	3.4	43.9	14.9	4.5
Servia	SSE-03	?	3–7 y	tibia	-18.6	9.5	-11.6	3.2	43.6	15.7	13.5
Servia	SSE-04	?	15–20 y	tibia	-19.2	9.3	–	3.4	45.3	15.5	5.1
Servia	SSE-05	F	Ad	tibia	-18.8	8.6	–	3.4	47.4	16.4	11.3
Servia	SSE-06	F	18–35 y	tibia	-18.8	9.0	–	3.3	46.0	16.2	10.0
Servia	SSE-08	F	35–50 y	femur	-19.0	8.4	-12.3	3.2	46.0	16.8	21.2
Servia	SSE-18	M	Ad	femur	-18.0	9.6	-12.1	3.2	47.2	17.2	17.8
Servia	SSE-33	M	Ad	tibia	-18.7	9.7	–	3.3	43.5	15.2	2.2
Servia	SSE-34	F	Ad	tibia	-19.0	8.2	-12.1	3.3	43.6	16.0	20.0
Servia	SSE-35	M	Ad	femur	-18.6	8.5	-12.4	3.2	44.9	16.4	16.3
Servia	SSE-36	F	35–50 y	rib	-18.7	8.5	-12.3	3.2	45.6	16.2	21.2
Servia	SSE-37	?	ca. 12 mo	rib	-17.6	11.3	-11.0	3.3	44.9	15.9	14.0
Servia	SSE-38	?	Ad	tibia	-18.6	8.3	-11.8	3.2	47.0	16.9	10.2
Servia	SSE-39	F	18–35 y	rib	-19.1	8.6	-11.7	3.3	45.5	15.9	20.4
Servia	SSE-42	?	14–18 y	femur	-18.6	7.3	-10.6	3.2	46.2	16.9	18.6
Servia	SSE-43	F	>35 y	femur	-18.8	8.4	-11.2	3.2	45.8	16.6	8.9
Servia	SSE-44	M	Ad	rib	-19.0	9.0	-12.4	3.2	44.0	16.3	20.3
Mean for Servia					*-18.7*	*8.9*	*-11.8*	–	–	–	–
SD for Servia					*0.4*	*0.8*	*0.6*	–	–	–	–
Abdera	6	?	Ad	–	-18.4	10.4	-13.8	–	–	–	–
Abdera	7	?	Youth	–	-18.8	9.8	-12.9	–	–	–	–
Abdera	8	?	Ad	–	-19.1	8.3	-12.6	–	–	–	–
Abdera	9	?	Infant	–	-17.5	10.2	-11.8	–	–	–	–
Mean for Abdera					*-18.5*	*9.7*	*-12.8*	–	–	–	–
SD for Abdera					*0.7*	*0.9*	*0.8*	–	–	–	–

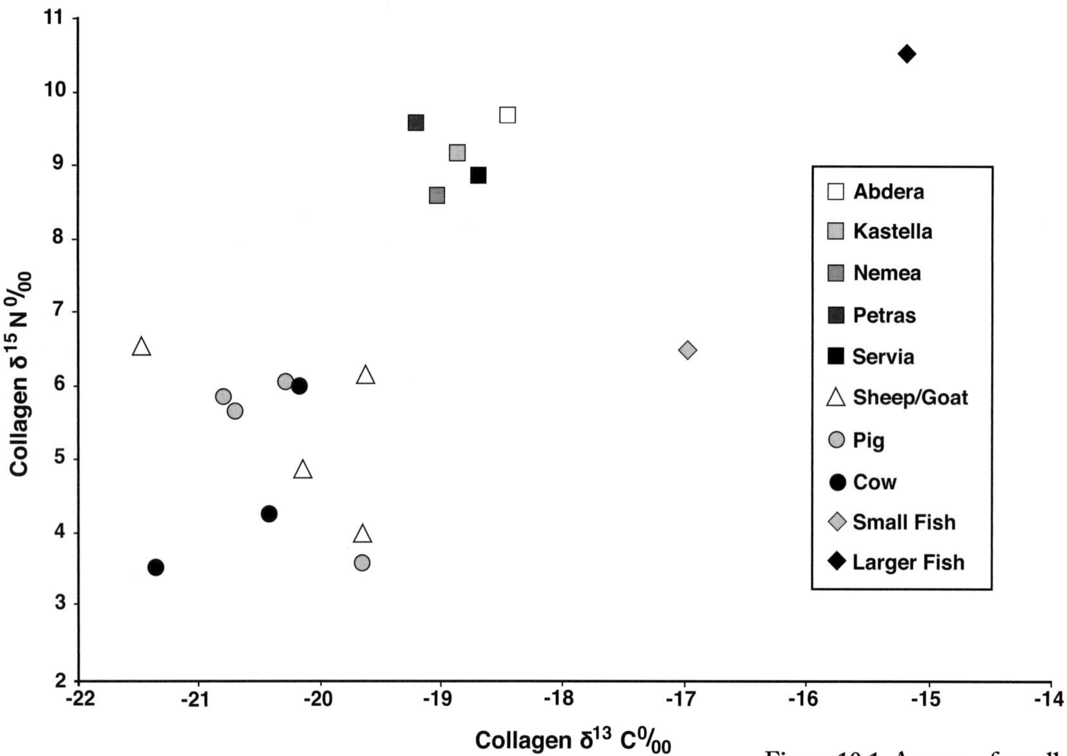

Figure 10.1. Averages for collagen $\delta^{13}C$ and $\delta^{15}N$ compared to fish and domesticate values from Byzantine sites

carbonate values vary slightly more, ranging from -12.8‰ (Abdera) to -11.1‰ (Nemea). These human collagen values, when plotted with values for some Byzantine-era domesticated fauna and modern Aegean fish, fall in a tight cluster with $\delta^{13}C$ values somewhat higher than the domesticates and $\delta^{15}N$ values markedly higher than the domesticates (Fig. 10.1).

If we consider the diverse foods available to Byzantine populations, and the wide range of likely $\delta^{13}C$ and $\delta^{15}N$ values estimated above for these resources, the $\delta^{13}C$ and $\delta^{15}N$ differences between these five human populations are quite small. This result seems to suggest that a general "Byzantine diet" existed, which may have varied somewhat by location but was broadly similar between communities. Such a scenario agrees with the documentary sources reviewed above, which stress a fundamental reliance on a trio of staples—grain, oil, and wine—that was supplemented, depending on class and other variables, by other foods.

The bone carbonate values offer insight into the nature of this trio of staples. Figure 10.2 compares these values to estimates of bulk $\delta^{13}C$ for some Byzantine foods that likely made a substantial contribution to total dietary calories because of their high carbohydrate or lipid content.[81] For each site, the data point on the right in Figure 10.2 represents the average human bone carbonate value. The data point to the left is 13‰ lower and represents a rough estimate of average bulk diet according to current models of human bone carbonate $\delta^{13}C$ values discussed above. Estimated bulk diet $\delta^{13}C$ for all sites falls in the range expected for dependence on a C_3 grain staple. Substantial dependence on a C_4 grain staple such as millet is not indicated at any site. It should be noted, however, that more small-scale

81. Values for C_3 and C_4 grains are based on modern literature averages for C_3 and C_4 plants. The value for olive oil is based on Greek olive oil values (Spangenberg, Macko, and Hunziker 1998, p. 4181; Royer et al. 1999, p. 361). All plant values are adjusted by 1.5‰ to compensate for changes in the $\delta^{13}C$ of atmospheric CO_2 since the Industrial Revolution (Friedli et al. 1986, p. 238). Data in the literature indicate that the bulk $\delta^{13}C$ value of an animal's milk will be close to that of its feed (Minson et al. 1975; Tyrrell et al. 1984, p. 128; Boutton et al. 1988, p. 2639; Metges et al. 1990, p. 191). Based on this, the bulk $\delta^{13}C$ value for dairy foods is estimated here to be 5‰ lower than the average collagen $\delta^{13}C$ value obtained for goats and sheep from the sites listed in Table 10.1.

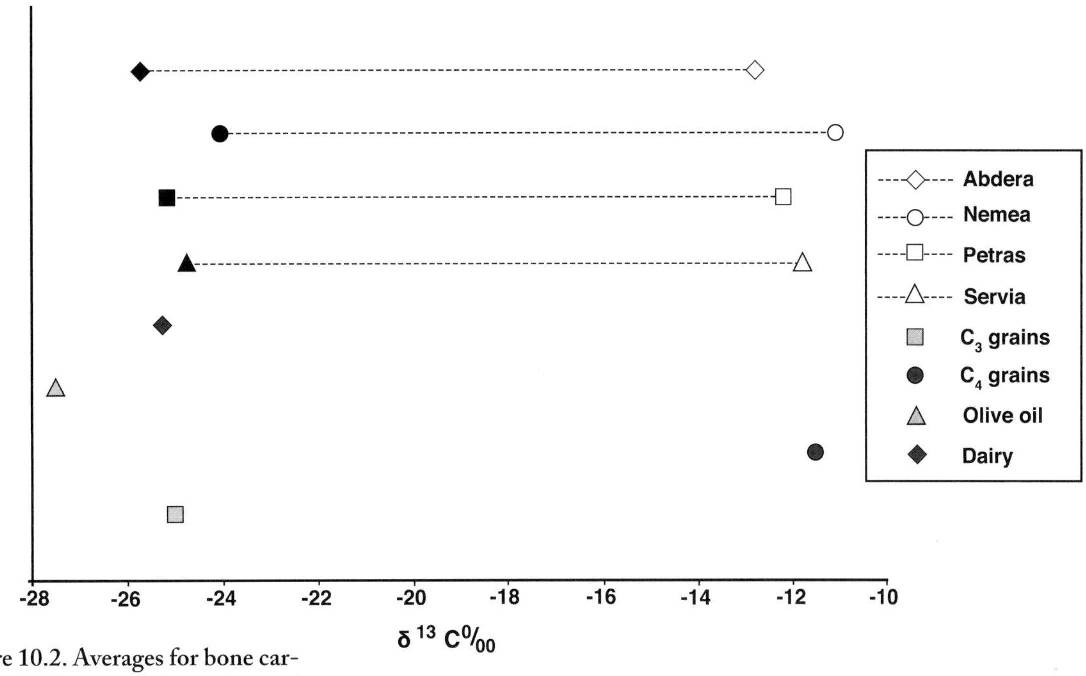

Figure 10.2. Averages for bone carbonate δ^{13}C compared to estimated bulk δ^{13}C values of some foods from Byzantine sites

use of millet would be compatible with the human values—especially since it could be offset by consumption of olive oil, which would be expected to draw carbonate δ^{13}C values down somewhat.

Variation in the consumption of resources such as millet and olive oil might explain the minor site-to-site differences seen in average carbonate δ^{13}C. Another potential source of variation may lie in the consumption of oily fish. The estimated bulk dairy δ^{13}C value used here is similar to the C_3 plant value and reflects the generally C_3-based diet of goats and sheep sampled to date for Byzantine sites. Within this scenario, addition of dairy foods would not be expected to shift bulk diet δ^{13}C away from the C_3 grain value.

The nature of the protein component of diets at the subject sites is less clear. The δ^{15}N values obtained are similar to those seen for many other European agricultural populations with respect to their substantial elevation above domesticated fauna.[82] Collagen δ^{13}C values fall somewhat above those of the domesticates and substantially below those of the marine resources. These results may be indicative of a general diet in which most dietary protein came from domesticated fauna, and very little from staple grains or marine resources. As already discussed above, however, other explanations for similar patterns have been advanced: for example, the inclusion of marine resources in the diet. In this scenario, marine resource consumption would contribute to elevated δ^{15}N levels, while δ^{13}C values suggesting a largely land-based diet would be explained by the body's use of non-protein (fat and lipid) carbon in the synthesis of non-essential amino acids.[83]

A closer look at the data on an individual level (Figs. 10.3, 10.4) offers some support for this idea.[84] Figure 10.3 shows data for the coastal sites of Abdera, Kastella, and Petras. Here a number of individuals can be seen to stand apart from the majority, with a tendency toward the higher δ^{13}C and δ^{15}N values expected for the consumption of marine resources. This

82. Van Klinken, Richards, and Hedges 2000, p. 55; Privat, O'Connell, and Richards 2002, p. 785; Polet and Katzenberg 2003, p. 527; Prowse et al. 2004, p. 263; Müldner and Richards 2005, p. 42; Dürrwachter et al. 2006, p. 43; Jay and Richards 2006, p. 655; Keenleyside, Schwarcz, and Panayotova 2006, p. 1209; Richards, Fuller, and Molleson 2006, p. 125.

83. Prowse et al. 2004, p. 270; Keenleyside, Schwarcz, and Panayotova 2006, p. 1211.

84. Garvie-Lok 2001, pp. 398–403; Bourbou and Richards 2007, pp. 67–68. To eliminate the confounding effects of breastfeeding, individuals aged less than three years are not shown in these figures.

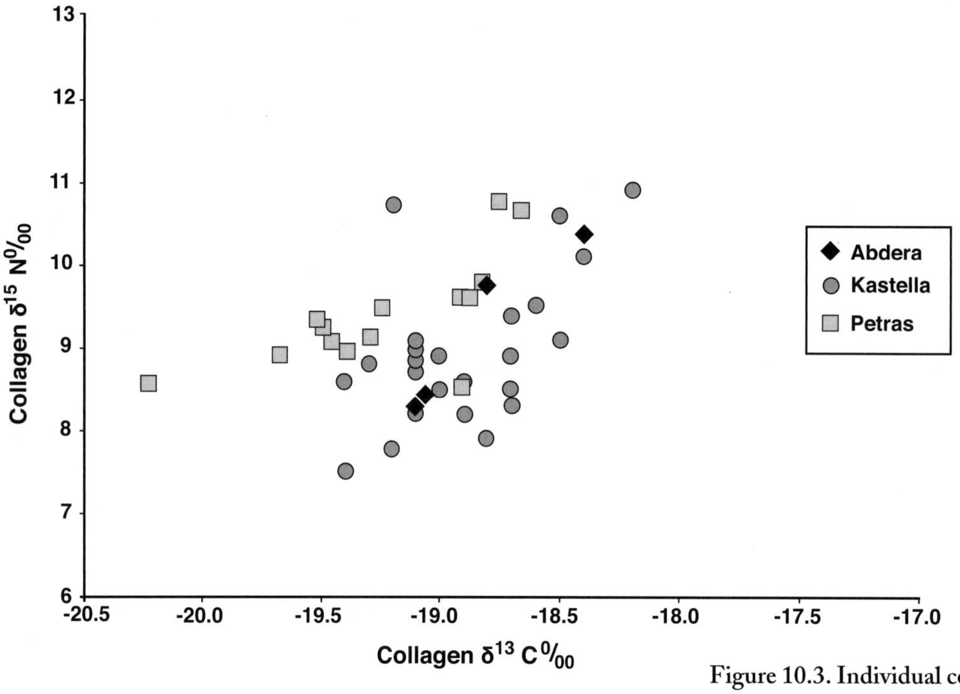

Figure 10.3. Individual collagen $\delta^{13}C$ and $\delta^{15}N$ values for coastal Byzantine sites

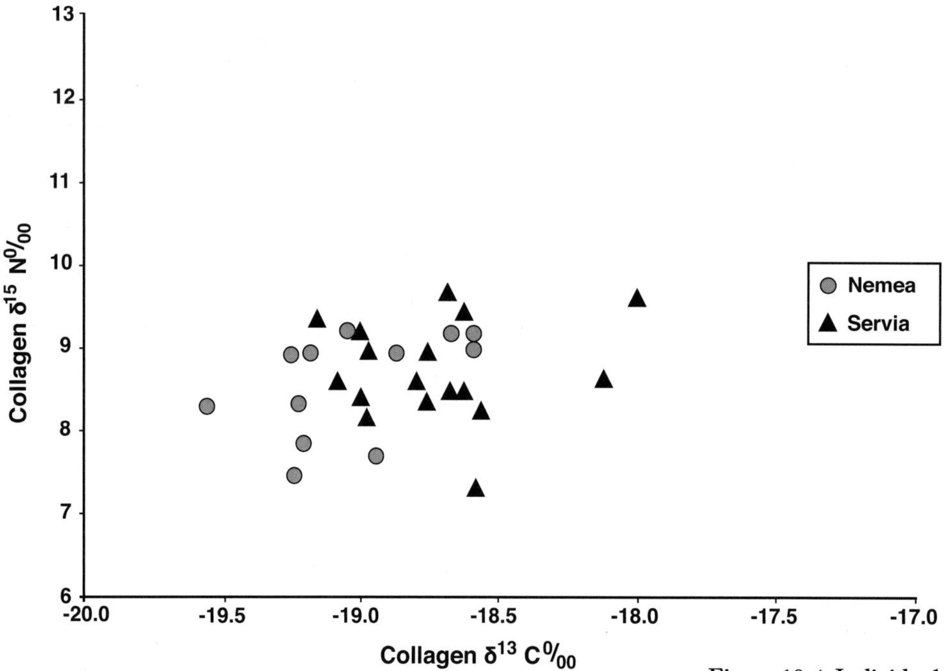

Figure 10.4. Individual collagen $\delta^{13}C$ and $\delta^{15}N$ values for inland Byzantine sites

pattern, argued elsewhere[85] to reflect some use of marine resources at these sites (a reasonable conclusion given their location), does not appear at the inland sites of Nemea and Servia (Fig. 10.4). Although two individuals at Servia do have higher $\delta^{13}C$ values, they do not exhibit higher $\delta^{15}N$ values similar to those found at the coastal sites. Average collagen $\delta^{15}N$ values at these inland sites of Nemea and Servia (8.6‰ and 8.9‰, respectively) are slightly lower than those at the coastal sites (9.2‰ to 9.6‰), which also points to greater use of marine resources on the coast than inland.

It is worth noting here that a similar pattern has been found for another set of Greek sites far removed in time from the Byzantine period. Papathanasiou has published stable isotope data for three coastal and three inland Neolithic Greek sites.[86] Two of the coastal sites, Franchthi and Kephala, exhibit mean $\delta^{15}N$ values elevated somewhat above mean $\delta^{15}N$ values seen elsewhere at the inland sites. Furthermore, these two sites contain a number of individuals with values tending toward higher levels of collagen $\delta^{13}C$ and $\delta^{15}N$, which are consistent with use of marine resources. Papathanasiou concludes that although none of the Neolithic groups she studied had a strong dependence on marine resources, the diets of the Franchthi and Kephala populations did include significant amounts of such foods.[87] The major differences in time, customs, and technology separating the Neolithic and Byzantine periods disallow any direct parallels between Neolithic and Byzantine diets, but the similar pattern exhibited by the stable isotope data is interesting. It suggests that coastal populations in both periods supplemented an essentially land-based diet with a small amount of marine resources.[88]

Although intra-site patterning of collagen values suggests some marine exploitation occurred at coastal sites, it seems unlikely that use of marine resources alone can explain the generally high collagen $\delta^{15}N$ values seen in Byzantine remains. The $\delta^{15}N$ values for inland sites are lower than those on the coast, but they are nevertheless still quite high: at 8.6‰ to 8.9‰, they are elevated over Byzantine faunal values reported here by up to 5‰. Despite an apparent proximity to the coast, transport of goods between the coast and the inland, upland sites of Nemea and Servia would have been fairly arduous in Byzantine times. Some dried, salted marine resources may have been imported, but it seems unlikely that enough of these foodstuffs would have been consumed to eclipse the $\delta^{15}N$ values of local grain and domesticates—especially in light of Hedges' and Reynard's recent argument that addition of marine resources to a land-based diet should not raise $\delta^{15}N$ values to a great extent in agricultural populations.[89]

Recent work by Bogaard and colleagues suggests that high $\delta^{15}N$ values in some European agricultural communities may reflect elevated grain $\delta^{15}N$ due to the manuring of fields.[90] This model, however, may not apply to the Byzantine agricultural system. Although domesticated animals did graze on cleared fields and would thus have deposited some manure there, they were often kept for much of the year on lands remote from cultivated fields, which made their dung unavailable for fertilizer during this period.[91]

Other explanations, nevertheless, do exist, which fit well with documentary evidence for the Byzantine diet. It has been suggested, for example, that high $\delta^{15}N$ values of European agricultural populations may in some

85. Agelarakis 1992–1993; Garvie-Lok 2001, pp. 438–439; Bourbou and Richards 2007, p. 69.

86. Papathanasiou 2003, pp. 318–319; see also Chap. 3.

87. Papathanasiou 2003, p. 321.

88. See also Chap. 11.

89. Hedges and Reynard 2007, p. 1244.

90. Bogaard et al. 2007, pp. 340–341.

91. Kazhdan 1997, pp. 52–54.

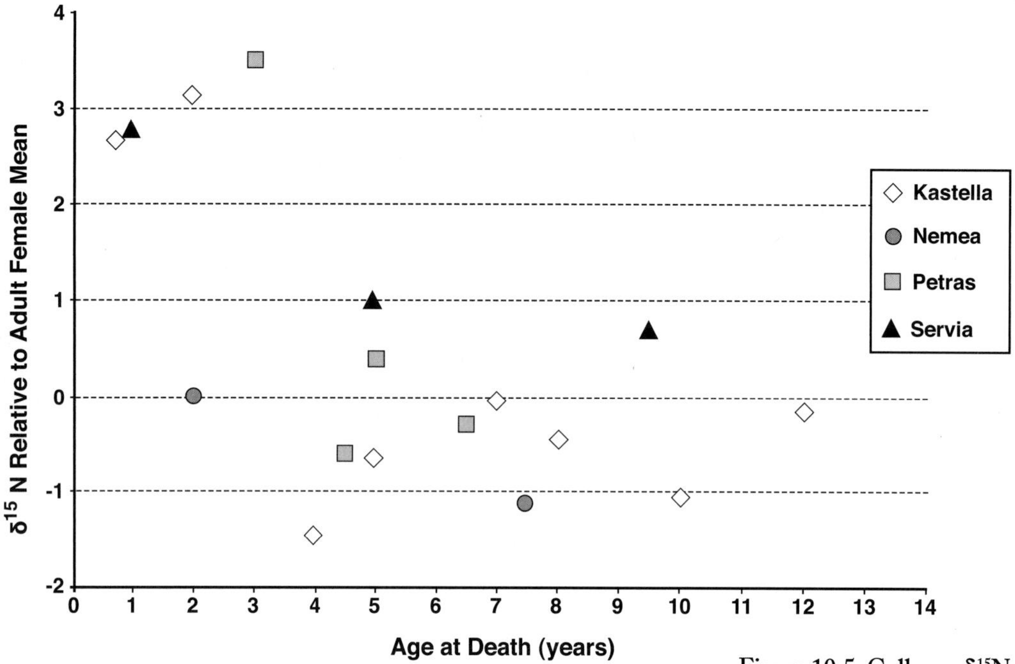

Figure 10.5. Collagen δ^{15}N values of juveniles relative to adult female Byzantine site means

cases reflect a very heavy reliance on dairy foods.[92] This scenario agrees with the importance of dairy foods in the Byzantine diet as attested in the documentary sources. Given the inexpensive nature of some dairy foods and their general availability, use of dairy foods is an attractive explanation for the generally high δ^{15}N values seen in Byzantine groups studied to date. It is also possible that high δ^{15}N values in Byzantine populations reflect consumption of dietary items of unexpectedly high δ^{15}N: for example, chickens, chicken eggs, and suckling animals.[93]

Stable isotope data also provide some tentative information on weaning practices in Byzantine Greece.[94] Collagen δ^{15}N values can be used to reconstruct weaning patterns in archaeological populations due to the fact that nursing infants have higher δ^{15}N than their mothers.[95] A nursing infant's tissue δ^{15}N values are the same as those of its mother at birth, then rise rapidly as tissue synthesized on a diet of milk replaces tissue formed *in utero,* and eventually fall again as milk begins to be supplemented with solid foods. Figure 10.5 shows juvenile δ^{15}N values for Kastella, Nemea, Petras, and Servia. These are expressed in terms of per mil (‰) departure from each site's mean adult female δ^{15}N value. Although few data points are available, a clear δ^{15}N elevation can be observed in most individuals with estimated ages at death of 3 years and younger. This result is consistent with the protracted period of breastfeeding described in documentary sources reviewed above. It is also consistent with late weaning ages suggested by stable isotope data for Roman populations from Isola Sacra, near Rome (1st–3rd centuries A.D.),[96] the Dakhleh oasis of Egypt (ca. A.D. 250–450),[97] and Queenford Farm in Britain (late 4th–mid 6th centuries A.D.).[98] Despite wide geographic separation of these populations and presumed local cultural differences, all three studies suggest a similar weaning pattern, with some individuals nursing to the age of 3 years. Fuller and

92. Hedges and Reynard 2007, p. 1248.

93. Garvie-Lok 2001, pp. 455–456; as foraging chickens consume many insects, their flesh and eggs show δ^{15}N values above those of herbivores; see Schwarcz, Dupras, and Fairgrieve 1999, p. 633.

94. For a more complete discussion of these results, see Bourbou and Garvie-Lok 2009, pp. 78–83.

95. Fogel, Tuross, and Owsley 1989, p. 114; Fuller et al. 2006a, p. 291. For detailed reviews, see Katzenberg, Herring, and Saunders 1996, pp. 187–193; Fuller et al. 2006a, p. 279; 2006b, pp. 46–47.

96. Prowse et al. 2004, p. 262.

97. Dupras, Schwarcz, and Fairgrieve 2001, p. 210.

98. Fuller et al. 2006b, pp. 47–50.

colleagues[99] have pointed out a contrast between this apparent Roman weaning pattern and that seen at the medieval British site of Wharram Percy (10th–16th centuries A.D.),[100] where stable isotope values suggest that children had begun the weaning process at earlier ages—between 18 months and 2 years. They suggest this contrast reflects a stronger influence by the physician Soranus and other medical scholars among these earlier Roman populations, who appear to have followed the physicians' advice for a gradual weaning process, while later medieval populations chose to wean their children at a younger age. It is interesting to note, then, that the Greek Byzantine data are consistent with a later weaning onset similar to that found among Roman populations. If these data are supported by further, more extensive studies on Byzantine populations, a case may be made that differences in weaning practices existed between Byzantine and medieval western European populations—with Byzantine groups having maintained traditions carried over from the Roman period.[101]

CONCLUSION

Historical sources are clear in portraying a Byzantine diet based on grain (primarily wheat and barley), oil, and wine, supplemented with legumes, dairy products, meat, and marine resources. How much of these supplements would have been consumed, however, is not always so clear. Some scholars have argued that legumes were a key source of protein—at least for the poor. A number of dairy products were also affordable enough to be an important component in the diet of all socioeconomic classes. Meat may have been a common food for the wealthy classes, but it has been argued that the average person ate little of it. Marine resources were valued and are often mentioned in the texts, but it also remains unclear how much of these were consumed, especially since their use would have varied by region and season.

Despite the small number of isotopic studies conducted on Byzantine Greek material to date, stable isotope analysis provides insight into some of the aforementioned dietary issues. The $\delta^{13}C$ and $\delta^{15}N$ values of the five Byzantine populations discussed here are quite tightly clustered, which indicates that all of these groups may have eaten a broadly similar diet. Carbonate $\delta^{13}C$ values, which indicate that most calories came from C_3 plants or C_3-feeding animals, are consistent with a diet based on the "Mediterranean trinity" of wheat, oil, and wine.

Stable isotope analysis indicates for at least these five Byzantine populations that millet was not a major resource—although low-level exploitation of millet cannot be eliminated and may in part account for minor $\delta^{13}C$ differences documented between populations. Collagen $\delta^{13}C$ values are in the range usually associated with C_3 terrestrial resources, while $\delta^{15}N$ values are high relative to those of domesticated animals from Byzantine sites. This pattern does not support a general dependence on low $\delta^{15}N$ legumes, since substantial legume consumption is inconsistent with the high collagen $\delta^{15}N$ values seen for all sites. Instead, it suggests that significant amounts of animal protein were included in the Byzantine diet. At coastal sites, a

99. Fuller et al. 2006b, p. 52.

100. Richards, Mays, and Fuller 2002, p. 209.

101. Also compare Chaps. 7, 9.

portion of this protein appears to have come from marine resources. Yet the fact that high $\delta^{15}N$ values are also seen at inland sites suggests that another factor is also at work. A reasonable explanation may lie in high consumption of dairy products, which are known from textual sources to have been generally available across classes and regions.

Isotopic studies conducted to date also offer insight into the diet of the very young—suggesting Byzantine infants were weaned relatively late—and point to possible cultural differences concerning weaning practices between Byzantines and contemporary western Europeans.

The isotopic studies discussed above raise as many questions as they answer and encourage further exploration into Byzantine diet. Studies of archaeological remains from a wider variety of sites could provide insight into regional variation in the use of millet and confirm the low importance of legumes in the Byzantine diet suggested by previous studies. Where possible, the study of human remains from religious foundations may be particularly interesting for this latter point, since the fasting rules applied to religious orders were extremely strict and would have encouraged consumption of legumes.

Some of the uncertainty surrounding collagen $\delta^{13}C$ and $\delta^{15}N$ values in Byzantine groups studied to date will be addressed by current research that is attempting to identify general aspects of $\delta^{15}N$ patterning in recent European populations. Further work on Byzantine-era Greek groups, however, should also help to clarify matters. As isotopic data for coastal and inland Byzantine populations accumulate, it will be interesting to observe whether the pattern argued here to reflect minor use of marine resources is consistent with coastal populations in general and whether the apparent lack of a marine signature in inland populations is also typical. Finally, additional studies on juvenile remains should be undertaken to shed further light on weaning patterns in Byzantine Greece and to allow testing of the difference in weaning ages proposed here between Byzantine and Western populations.

ACKNOWLEDGMENTS

Funding for the Nemea, Petras, and Servia analyses was provided by the Social Sciences and Humanities Research Council of Canada, the Wiener Laboratory, and the University of Calgary. Sandra Garvie-Lok thanks Ethne Barnes, Stephen Miller, Nikki Tsilipakou, and Metaxia Tsipopoulou for granting access to the skeletal material from those sites. Funding for the Kastella analysis was provided by the Centenary Bursary of the British School at Athens. Chryssi Bourbou would like especially to thank Anagnostis Agelarakis (Adelphi University) for sharing with the authors stable isotope data for the Abdera collection.

REFERENCES

Agelarakis, A. 1989. "The Paleopathological Evidence at Polystylon, Abdera," *Byzantinische Forschungen* 14, pp. 7–26.

———. 1992–1993. "Excavations at Polystylon (Abdera) Greece: Aspects of Mortuary Practices and Skeletal Biology," *ArchDelt* 47–48, A′, pp. 293–308.

Ambrose, S. H., and L. Norr. 1993. "Experimental Evidence for the Relationship of the Carbon Isotope Ratios of Whole Diet and Dietary Protein to Those of Bone Collagen and Carbonate," in *Prehistoric Human Bone: Archaeology at the Molecular Level,* ed. J. B. Lambert and G. Grupe, Berlin, pp. 1–37.

Anagnostakis, I., and T. Papamastorakis. 2005. "'. . . and Radishes for Appetizers': On Banquets, Radishes, and Wine," in Papanikola-Bakirtzi 2005, pp. 147–174.

Ayliffe, L. K., T. E. Cerling, T. Robinson, A. G. West, M. Sponheimer, B. H. Passey, J. Hammer, B. Roeder, M. D. Dearing, and J. R. Ehleringer. 2004. "Turnover of Carbon Isotopes in Tail Hair and Breath CO_2 of Horses Fed an Isotopically Varied Diet," *Oecologia* 139, pp. 11–22.

Balasse, M., H. Bocherens, and A. Mariotti. 1999. "Intra-Bone Variability of Collagen and Apatite Isotopic Composition Used as Evidence of a Change of Diet," *JAS* 26, pp. 593–598.

Beaucamp, J. 1982. "L'allaitement: Mère ou nourrice?," *JÖBG* 32, pp. 546–559.

Bober, P. P. 1999. *Art, Culture, and Cuisine: Ancient and Medieval Gastronomy,* Chicago.

Bogaard, A., T. H. E. Heaton, P. Poulton, and I. Merbach. 2007. "The Impact of Manuring on Nitrogen Isotope Ratios in Cereals: Archaeological Implications for Reconstruction of Diet and Crop Management Practices," *JAS* 34, pp. 335–343.

Bourbou, C. 2004. *The People of Early Byzantine Eleutherna and Messene (6th–7th Centuries A.D.): A Bioarchaeological Approach,* Athens.

———. 2010. *Health and Disease in Byzantine Crete (7th–12th Centuries A.D.),* Farnham.

Bourbou, C., and S. Garvie-Lok. 2009. "Breastfeeding and Weaning Patterns in Byzantine Times: Evidence from Human Remains and Written Sources," in *Becoming Byzantine: Children and Childhood in Byzantium,* eds. A. Papaconstantinou and A. M. Talbot, Washington, D.C., pp. 65–84.

Bourbou, C., and M. P. Richards. 2007. "The Middle Byzantine Menu: Palaeodietary Information from Isotopic Analysis of Humans and Fauna from Kastella, Crete," *International Journal of Osteoarchaeology* 17, pp. 63–72.

Bourbou, C., and A. Tsilipakou. 2009. "Investigating the Human Past of Greece during the 6th–7th Centuries A.D.," in *New Directions in the Skeletal Biology of Greece* (*Hesperia* Suppl. 43, OWLS 1), ed. L. A. Schepartz, S. C. Fox, and C. Bourbou, Princeton, pp. 121–136.

Boutton, T. W., H. F. Tyrrell, B. W. Patterson, G. A. Varga, and P. D. Klein. 1988. "Carbon Kinetics of Milk Formation in Holstein Cows in Late Lactation," *Journal of Animal Science* 66, pp. 2636–2645.

Braudel, F. 1966. *La Méditerranée et le monde méditerranéen à l'époque de Philippe II,* 2nd ed., 2 vols., Paris.

———. 1998. *Ιστορία της διατροφής: Προσεγγίσεις της σύγχρονης ιστοριογραφίας,* trans. K. Tsichli-Aroni, Athens.

Braund, D. 1995. "Fish from the Black Sea: Classical Byzantium and the Greekness of Trade," in Wilkins, Harvey, and Dobson 1995, pp. 162–170.

Brown, T. A., D. E. Nelson, J. S. Vogel, and J. R. Southon. 1988. "Improved Collagen Extraction by Modified Longin Method," *Radiocarbon* 30, pp. 171–177.

Cheynet, J.-C. 2005. "La valeur marchande des produits alimentaires dans l'Empire byzantin," in Papanikola-Bakirtzi 2005, pp. 31–46.

Congourdeau, M.-H. 1993. "Metrodora et son oeuvre," in *Maladie et société à Byzance* (Collectanea 3), ed. E. Patlagean, Spoleto, pp. 57–96.

Cunningham, M. B., ed. 1991. *The Life of Michael the Synkellos* (Belfast Byzantine Texts and Translations 1), Belfast.

Dagron, G. 1995. "Poissons, pêcheurs et poissonniers de Constantinople," in *Constantinople and Its Hinterland. Papers from the Twenty-Seventh Spring Symposium of Byzantine Studies, Oxford, April 1993* (Publications of the Society for the Promotion of Byzantine Studies 3), ed. C. Mango, G. Dagron, G. Greatrex, Aldershot, pp. 57–73.

Dagron, G., and M. Dupré la Tour, eds. 1978. *Vie et miracles de sainte Thècle: Texte grec, traduction et commentaire* (Subsidia hagiographica 62), Brussels.

Dalby, A. 1996. *Siren Feasts: A History of Food and Gastronomy in Greece,* London.

———. 2003. *Flavours of Byzantium,* Totnes.

Delehaye, H. 1923. *Les saints stylites* (Subsidia hagiographica 14), Brussels.

Dembinska, M. 1985. "Diet: A Comparison of Food Consumption between Some Eastern and Western Monasteries in the 4th–12th Centuries," *Byzantion* 55, pp. 431–462.

DeNiro, M. J., and S. Epstein. 1978. "Influence of Diet on the Distribution of Carbon Isotopes in Animals," *Geochimica et cosmochimica acta* 42, pp. 495–506.

———. 1981. "Influence of Diet on the Distribution of Nitrogen Isotopes in Animals," *Geochimica et cosmochimica acta* 45, pp. 341–351.

Dupras, T. L., H. P. Schwarcz, and S. I. Fairgrieve. 2001. "Infant Feeding and Weaning Practices in Roman Egypt," *American Journal of Physical Anthropology* 115, pp. 204–212.

Dürrwächter, C., O. E. Craig, M. J. Collins, J. Burger, and K. W. Alt. 2006. "Beyond the Grave: Variability in Neolithic Diets in Southern Germany?" *JAS* 33, pp. 39–48.

Eideneier, H., trans. and ed. 1991. *Ptochoprodromos: Einführung, kritische Ausgabe, deutsche Übersetzung, Glossar* (Neograeca medii aevi 5), Cologne.

Flandrin, J.-L., M. Montanari, and A. Sonnenfeld. 1999. *Food: A Culinary History from Antiquity to the Present,* New York.

Fogel, M. L., N. Tuross, and D. Owsley. 1989. "Nitrogen Isotope Tracers of Human Lactation in Modern and Archaeological Populations," *Annual Report of the Director, Geophysical Laboratory, Carnegie Institution for Science, Washington, D.C., 1988–1989,* pp. 111–117.

Fornaciari, G., B. Ceccanti, and E. Menicagli. 1988. "Ricerca degli elementi guida della nutrizione e di alcuni metalli pesanti mediante spettroscopia ad assorbimento atomico," in *Gortina* I, ed. A. di Vita, Roma, pp. 403–416.

Friedli, H., H. Lötscher, H. Oeschger, U. Siegenthaler, and B. Stauffer. 1986. "Ice Core Record of the $^{13}C/^{12}C$ Ratio of Atmospheric CO_2 in the Past Two Centuries," *Nature* 324, pp. 237–238.

Fuller, B. T., J. L. Fuller, D. A. Harris, and R. E. M. Hedges. 2006a. "Detection of Breastfeeding and Weaning in Modern Human Infants with Carbon and Nitrogen Stable Isotope Ratios," *American Journal of Physical Anthropology* 129, pp. 279–293.

Fuller, B. T., T. I. Molleson, D. A. Harris, L. T. Gilmour, and R. E. M. Hedges. 2006b. "Isotopic Evidence for Breastfeeding and Possible Adult Dietary Differences from Late/Sub-Roman Britain," *American Journal of Physical Anthropology* 129, pp. 45–54.

Gallant, T. W. 1991. *Risk and Survival in Ancient Greece: Reconstructing the Rural Domestic Economy,* Stanford.

Garnsey, P. 1988. *Famine and Food Supply in the Graeco-Roman World: Responses to Risk and Crisis,* Cambridge.

Garvie-Lok, S. J. 2001. "Loaves and Fishes: A Stable Isotope Reconstruction of Diet in Medieval Greece" (diss. Univ. of Calgary).

Garvie-Lok, S. J., T. L. Varney, and M. A. Katzenberg. 2004. "Preparation of Bone Carbonate for Stable Isotope Analysis: The Effects of Treatment Time and Acid Concentration," *JAS* 31, pp. 763–776.

Grimm, V. 1995. "Fasting Women in Judaism and Christianity in Late Antiquity," in Wilkins, Harvey, and Dobson 2005, pp. 225–240.

Hare, P. E., M. L. Fogel, T. W. Stafford Jr., A. D. Mitchell, and T. C. Hoering. 1991. "The Isotopic Composition of Carbon and Nitrogen in Individual Amino Acids Isolated from Modern and Fossil Proteins," *JAS* 18, pp. 277–292.

Hedges, R. E. M. 2003. "On Bone Collagen: Apatite-Carbonate Isotopic Relationships," *International Journal of Osteoarchaeology* 13, pp. 66–79.

Hedges, R. E. M., and L. M. Reynard. 2007. "Nitrogen Isotopes and the Trophic Level of Humans in Archaeology," *JAS* 34, pp. 1240–1251.

Hedges, R. E. M., and G. J. van Klinken. 2000. "'Consider a Spherical Cow . . .'—On Modeling and Diet," in *Biogeochemical Approaches to Paleodietary Analysis* (Advances in Archaeological and Museum Science 5), ed. S. H. Ambrose and M. A. Katzenberg, New York, pp. 211–241.

Hill, S., and A. Bryer. 1995. "Byzantine Porridge: Tracta, Trachanás, and Tarhana," in Wilkins, Harvey, and Dobson 2005, pp. 44–54.

Jay, M., and M. P. Richards. 2006. "Diet in the Iron Age Cemetery Population at Wetwang Slack, East Yorkshire, UK: Carbon and Nitrogen Stable Isotope Evidence," *JAS* 33, pp. 653–662.

Jeanselme, E., and L. Oeconomos. 1923. *Aliments et recettes culinaires des Byzantins*, Anvers.

Kaplan, M. 1992. *Les hommes et la terre à Byzance du VIe au XIe siècle: Propriété et exploitation du sol* (Publications de la Sorbonne, Série Byzantina Sorbonensia 10), Paris.

Katzenberg, M. A., D. A. Herring, and S. R. Saunders. 1996. "Weaning and Infant Mortality: Evaluating the Skeletal Evidence," *Yearbook of Physical Anthropology* 39, pp. 177–199.

Kazhdan, A. 1997. "The Peasantry," in *The Byzantines,* ed. G. Cavallo, Chicago, pp. 43–73.

Keenleyside, A., H. Schwarcz, and K. Panayotova. 2006. "Stable Isotopic Evidence of Diet in a Greek Colonial Population from the Black Sea," *JAS* 33, pp. 1205–1215.

Kislinger, E. 1997. "I cristiani d'Oriente: Regole e realtà alimentari nel mondo bizantino," in *Storia dell'alimentazione,* ed. J.-L. Flandrin and M. Montanari, Rome, pp. 250–265.

———. 1999. "Christians of the East: Rules and Realities of the Byzantine Diet," in *Food: A Culinary History from Antiquity to the Present,* ed. J.-L. Flandrin, M. Montanari, and A. Sonnenfeld, New York, pp. 194–206.

———. 2005. "Τρώγοντας και πίνοντας εκτός σπιτιού," in Papanikola-Bakirtzi 2005, pp. 47–60.

Koder, J. 1994. "Ο κηπουρός και η καθημερινή κουζίνα στο Βυζάντιον," in *Όψεις της Βυζαντινής κοινωνίας* 2, Athens.

———. 2005. "Η καθημερινή διατροφή στο Βυζάντιον με βάση τις πηγές," in Papanikola-Bakirtzi 2005, pp. 17–30.

Kornexl, B. E., T. Werner, A. Rossmann, and H. L. Schmidt. 1997. "Measurements of Stable Isotope Abundances in Milk and Milk Ingredients: A Possible Tool for Origin Assessment and Quality Control," *Zeitschrift für Lebensmittel-Untersuchung und Forschung, A,* 205, pp. 19–24.

Koukoules, P. 1952. *Βυζαντινών βίος και πολιτισμός,* vol. 5, Athens.

Lascaratos, J., and E. Poulakou-Rebelakou. 2003. "Oribasius (Fourth Century) and Early Byzantine Perinatal Nutrition," *Journal of Pediatric Gastroenterology and Nutrition* 36, pp. 186–189.

Littlewood, A., H. Maguire, and J. Wolschke-Bulmahn, eds., 2002. *Byzantine Garden Culture,* Washington, D.C.

Longin, R. 1971. "New Method of Collagen Extraction for Radiocarbon Dating," *Nature* 230, pp. 241–242.

Maguire, H. 1987. *Earth and Ocean. The Terrestrial World in Early Byzantine Art* (Monographs on the Fine Arts 43), University Park, Pa.

———. 2005. "A Fruit Story and an Aviary: Images of Food in Houses, Palace, and Church," in Papanikola-Bakirtzi 2005, pp. 133–146.

Mango, C., G. Dagron, and G. Greatrex, eds. 1995. *Constantinople and Its Hinterland. Papers from the Twenty-Seventh Spring Symposium of Byzantine Studies, Oxford, April 1993* (Publications of the Society for the Promotion of Byzantine Studies 3), Altershot.

Maniatis, G. C. 2000. "The Organizational Setup and Functioning of the Fish Market in Tenth-Century Constantinople," *DOP* 54, pp. 13–42.

Matthaiou, A. 1997. *Aspects de l'alimentation en Grèce sous la domination ottomane: Des réglementations au discours normatif* (Studien zur Geschichte Südosteuropas 14), Frankfurt.

Metges, C., K. Kempe, and H.-L. Schmidt. 1990. "Dependence of the Carbon-Isotope Contents of Breath Carbon Dioxide, Milk, Serum, and Rumen Fermentation Products on the $\delta^{13}C$ Value of Food in Dairy Cows," *British Journal of Nutrition* 63, pp. 187–196.

Minson, D. J., M. M. Ludlow, and J. H. Troughton. 1975. "Differences in Natural Carbon Isotope Ratios of Milk and Hair from Cattle Grazing Tropical and Temperate Pastures," *Nature* 256, p. 602.

Motsias, C. 1998. *Τι έτρωγαν οι Βυζαντινοί,* Athens.

Müldner, G., and M. P. Richards. 2005. "Fast or Feast: Reconstructing Diet in Later Medieval England by Stable Isotope Analysis," *JAS* 32, pp. 39–48.

Nakamura, K., D. A. Schoeller, F. J. Winkler, and H. L. Schmidt. 1982. "Geographical Variations in the Carbon Isotopic Composition of the Diet and Hair in Contemporary Man," *Biomedical Mass Spectrometry* 9, pp. 390–394.

Papanikola-Bakirtzi, D., ed. 2005. *Food and Cooking in Byzantium. Proceedings of the Symposium "On Food in Byzantium," Thessaloniki Museum of Museum Culture, 4 November 2001,* Athens.

Papathanasiou, A. 2003. "Stable Isotope Analysis in Neolithic Greece and Possible Implications on Human Health," *International Journal of Osteoarchaeology* 13, pp. 314–324.

Passey, B. H., T. F. Robinson, L. K. Ayliffe, T. E. Cerling, M. Sponheimer, M. D. Dearing, B. L. Roeder, and J. R. Ehleringer. 2005. "Carbon Isotope Fractionation between Diet, Breath CO_2, and Bioapatite in Different Mammals," *JAS* 32, pp. 1459–1470.

Polet, C., and M. A. Katzenberg. 2003. "Reconstruction of the Diet in a Mediaeval Monastic Community from the Coast of Belgium," *JAS* 30, pp. 525–533.

Privat, K. L., T. C. O'Connell, and M. P. Richards. 2002. "Stable Isotope Analysis of Human and Faunal Remains from the Anglo-Saxon Cemetery at Berinsfeld, Oxfordshire: Dietary and Social Implications," *JAS* 29, pp. 779–790.

Prowse, T., H. P. Schwarcz, S. Saunders, R. Macchiarelli, and L. Bondioli. 2004. "Isotopic Paleodiet Studies of Skeletons from the Imperial Roman-Age Cemetery of Isola Sacra, Rome, Italy," *JAS* 31, pp. 259–272.

Richards, M. P., B. T. Fuller, and T. I. Molleson. 2006. "Stable Isotope Palaeodietary Study of Humans and Fauna from the Multi-Period (Iron Age, Viking, and Late Medieval) Site of Newark Bay, Orkney," *JAS* 33, pp. 122–131.

Richards, M. P., and R. E. M. Hedges. 1999. "How Chemical Analysis of Human Bones Can Tell Us the Diets of People Who Lived in the Past (and Other Contributions)," in *Minoans and Mycenaeans: Flavours of Their Time: National Archaeological Museum, 12 July–27 November 1999,* ed. Y. Tzedakis and H. Martlew, Athens, p. 214.

Richards, M. P., S. Mays, and B. T. Fuller. 2002. "Stable Carbon and Nitrogen Isotope Values of Bone and Teeth Reflect Weaning Age at the Medieval Wharram Percy Site,

Yorkshire, UK," *American Journal of Physical Anthropology* 119, pp. 205–210.

Royer, A., C. Gerard, N. Naulet, M. Lees, and G. J. Martin. 1999. "Stable Isotope Characterization of Olive Oils. I—Composition and Carbon-13 Profiles of Fatty Acids," *Journal of the American Oil Chemists' Society* 76, pp. 357–363.

Schwarcz, H. P., T. L. Dupras, and S. I. Fairgrieve. 1999. "^{15}N Enrichment in the Sahara: In Search of a Global Relationship," *JAS* 26, pp. 629–636.

Sealy, J. C. 1986. *Stable Carbon Isotopes and Prehistoric Diets in the Southwestern Cape Province, South Africa* (*BAR-IS* 293), Oxford.

Spangenberg, J. E., S. A. Macko, and J. Hunziker. 1998. "Characterization of Olive Oil by Carbon Isotope Analysis of Individual Fatty Acids: Implications for Authentication," *Journal of Agricultural and Food Chemistry* 46, pp. 4179–4184.

Steele, K. W., and R. M. Daniel. 1978. "Fractionation of Nitrogen Isotopes by Animals: A Further Complication to the Use of Variations in the Natural Abundance of ^{15}N for Tracer Studies," *The Journal of Agricultural Science* 90, pp. 7–9.

Teall, J. L. 1959. "The Grain Supply of the Byzantine Empire, 330–1025," *DOP* 13, pp. 87–139.

Temkin, O. 1991. *Soranus's Gynecology*, Baltimore.

Tieszen, L. L., and T. W. Boutton. 1989. "Stable Carbon Isotopes in Terrestrial Ecosystem Research," in *Stable Isotopes in Ecological Research* (Ecological Studies 68), ed. P. W. Rundel, J. R. Ehleringer, and K. A. Nagy, New York, pp. 167–195.

Tieszen, L. L., T. W. Boutton, K. G. Tesdahl, and N. A. Slade. 1983. "Fractionation and Turnover of Stable Carbon Isotopes in Animal Tissues: Implications for δ^{13}C Analysis of Diet," *Oecologia* 57, pp. 32–37.

Tieszen, L. L., and T. Fagre. 1993. "Effect of Diet Quality and Composition on the Isotopic Composition of Respiratory CO_2, Bone Collagen, Bioapatite, and Soft Tissues," in *Prehistoric Human Bon: Archaeology at the Molecular Level,* ed. J. B. Lambert and G. Grupe, Berlin, pp. 121–155.

Tritsaroli, P. 2006. "Pratiques funéraires en Grèce centrale à la période byzantine: Analyse à partir des données archéologiques et biologiques" (diss. Univ. de Paris).

Tyrrell, H. F., G. Pelletier, R. Chevalier, C. Hillaire-Marcel, and M. Gagnon. 1984. "Use of Carbon 13 as a Tracer in Metabolism Studies," *Canadian Journal of Animal Science* 64 (Suppl.), pp. 127–129.

van den Ven, P. 1962. *La vie ancienne de s. Syméon Stylite le jeune (521–592)* (Subsidia hagiographica 32), Brussels.

van Klinken, G. J., M. P. Richards, and R. E. M. Hedges. 2000. "An Overview of Causes for Stable Isotopic Variation in Past European Human Populations: Environmental, Ecophysiological, and Cultural Effects," in *Biogeochemical Approaches to Paleodietary Analysis* (Advances in Archaeological and Museum Science 5), ed. S. H. Ambrose and M. A. Katzenberg, New York, pp. 39–64.

Viliskij, S. G. 1911. *Žitie sv: Vasilija Novogo v russkoj literature,* Odessa.

Vogel, J. C. 1978. "Isotopic Assessment of the Dietary Habits of Ungulates," *South African Journal of Science* 74, pp. 298–301.

Wilkins, J., D. Harvey, and M. Dobson, eds. 1995. *Food in Antiquity,* Exeter.

CHAPTER 11

Summary: Patterns in the Carbon and Nitrogen Isotope Data through Time

by Anastasia Papathanasiou and Michael P. Richards

Greece has a rich archaeological record and has long been the focus of bioarchaeological research. It has led the way for the application of cutting-edge scientific methods, starting with studies by researchers such as J. Lawrence Angel, who looked at trace elemental analysis of bone as a dietary indicator in the 1960s and 1970s. More recently, Greek archaeological material has been the focus of residue analysis studies, which have applied novel methods to reconstruct the contents and use of pottery from many different time periods.[1] Now Greece has become a center for the application of stable isotope analysis of human remains to reconstruct diet. Starting with the first published study in the mid-1990s,[2] subsequent publications in the last decade, and now with the addition of the significant amount of new data presented in this volume, Greece is one of the most studied regions in the world for the application of palaeodietary isotopic analysis.

This then allows researchers to begin looking for long-term trends in the data that have been produced and to build up a comprehensive picture of human dietary changes and adaptations in prehistory and into the historic periods. In this final summary chapter, an initial examination of this body of data is presented, which will surely be added to after this volume is published, to look for long-term trends in the diets of prehistoric and historic populations in Greece.

LONG-TERM TRENDS IN THE ISOTOPIC DATA

The focus of this volume is mainly on bone collagen carbon and nitrogen isotope analysis. As discussed in Chapter 2, this method reflects the main protein sources in human diets over a period of years. As also discussed in Chapter 2, carbon isotope analysis can be particularly useful in Greece for investigation of the importance of marine versus terrestrial protein in human diets and the importance of millet. Nitrogen isotope analysis can also contribute to this, indicating the consumption of marine foods (which often have a much higher isotope ratio than terrestrial foods) as well as the importance of plant vs. animal protein in human diets.

Selected published human carbon and nitrogen isotope data from Greece have been collated and plotted, with the carbon and nitrogen isotope values plotted separately (Table 11.1, Figs. 11.1, 11.2).

1. Evershed et al. 2000.
2. Papathanasiou, Larsen, and Norr 1995.

TABLE 11.1. MEAN STABLE ISOTOPE VALUES FOR HUMANS FROM SELECTED SITES IN GREECE

Site	*Location*	*$\delta^{13}C \pm 1\sigma$ (‰ VPDB)**	*$\delta^{15}N \pm 1\sigma$ (‰ AIR)*	*N*	*Source*
Mesolithic Sites					
Franchthi Cave	Mainland	-21.2	6.4	1	Richards and Papathanasiou (unpublished data)
Neolithic Sites					
Alepotrypa	Mainland	-20.0 ± 0.4	7.2 ± 1.0	26	Papathanasiou, Larsen, and Norr (2000), Papathanasiou (2001, 2003)
Franchthi	Mainland	-18.7 ± 0.8	9.2 ± 1.8	11	Papathanasiou (2003)
Kephala	Mainland	-19.1 ± 1.2	9.2 ± 1.0	5	Papathanasiou (2003)
Theopetra	Mainland	-19.8 ± 0.9	7.4 ± 1.0	13	Papathanasiou (2003)
Tharrounia	Mainland	-20.0 ± 0.2	8.0 ± 0.7	20	Papathanasiou (2003)
Kouveleiki	Mainland	-19.8	8.1	2	Papathanasiou (2003)
Proskynas	Mainland	-19.8 ± 0.8	7.3 ± 1.6	8	Papathanasiou (2003)
Makriyalos	Mainland	-19.9 ± 0.3	6.8 ± 0.9	21	Papathanasiou (2003)
Gerani	Crete	-19.9 ± 0.8	9.2 ± 0.3	3	Richards and Hedges (2008)
Bronze Age Sites					
Mycenae chamber	Mainland	-19.3 ± 0.2	8.0 ± 1.0	11	Richards and Hedges (2008)
Mycenae A	Mainland	-18.6 ± 0.5	10.5 ± 1.1	9	Richards and Hedges (2008)
Mycenae B	Mainland	-19.4 ± 0.6	8.6 ± 1.9	9	Richards and Hedges (2008)
Pylos tholos	Mainland	-19.8 ± 0.4	9.6 ± 0.6	4	Papathanasiou et al. (2012)
Pylos chamber	Mainland	-19.8 ± 0.3	8.8 ± 0.6	26	Papathanasiou et al. (2012)
Sykia	Mainland	-19.4 ± 0.3	8.2 ± 0.7	6	Richards and Vika (2008)
Spaliareika	Mainland	-20.1 ± 0.3	8.4 ± 0.9	8	Richards and Vika (2008)
Kalamaki	Mainland	-20.0 ± 0.6	7.8 ± 0.8	15	Richards and Vika (2008)
Spathes	Mainland	-19.0 ± 0.5	7.8 ± 0.6	11	Triantaphyllou (Chap. 4)
Rymnio	Mainland	-17.6 ± 0.8	8.6 ± 1.2	4	Triantaphyllou (2001)
Korinos	Mainland	-19.2 ± 0.6	8.7 ± 1.4	4	Triantaphyllou (2001)
Makriyalos	Mainland	-19.6 ± 0.4	7.4 ± 0.7	6	Triantaphyllou (2001)
Perahora	Mainland	-19.3 ± 0.3	9.1 ± 0.6	19	Petroutsa (2007)
Koufovouno	Mainland	-19.8 ± 0.3	7.9 ± 0.4	5	Petroutsa (2007)
Tragana	Mainland	-19.3 ± 1.3	9.3 ± 1.0	4	Iezzi (Chap. 6)
Atalanti	Mainland	-20.5 ± 0.9	7.7 ± 0.8	4	Iezzi (Chap. 6)
Kolaka	Mainland	-18.3 ± 1.6	7.8 ± 1.7	4	Iezzi (Chap. 6)
Modi	Mainland	-19.3 ± 0.3	7.3 ± 1.3	4	Iezzi (Chap. 6)
Armenoi	Crete	-19.8 ± 0.3	7.8 ± 0.5	39	Richards and Hedges (2008)
Ayios Charalambos	Crete	-19.5 ± 0.1	8.5 ± 0.5	8	Richards (unpublished data)
Platyvola	Crete	-19.6 ± 0.4	8.1 ± 2.0	9	Richards (unpublished data)
Iron Age Sites					
Athens	Mainland	-19.2 ± 0.2	10.6 ± 0.8	12	Lagia (Chap. 8)
Makriyalos	Mainland	-18.9 ± 0.5	7.1 ± 0.6	14	Lagia (Chap. 8)
Treis Elies	Mainland	-16.9 ± 1.6	8.3 ± 0.7	6	Triantaphyllou (Chap. 4)

TABLE 11.1—*Continued*

Site	*Location*	*$\delta^{13}C \pm 1\sigma$ (‰ VPDB)**	*$\delta^{15}N \pm 1\sigma$ (‰ AIR)*	*N*	*Source*
Karitsa	Mainland	-16.7	9.6	2	Triantaphyllou (Chap. 4)
Kladeri	Mainland	-17.0 ± 0.6	9.3 ± 0.7	5	Triantaphyllou (Chap. 4)
BYZANTINE SITES					
Abdera	Mainland	-18.5 ± 0.7	9.7 ± 0.9	4	Bourbou and Garvie-Lok (Chap. 10)
Nemea	Mainland	-19.0 ± 0.3	8.6 ± 0.6	13	Bourbou and Garvie-Lok (Chap. 10)
Servia	Mainland	-18.7 ± 0.4	8.9 ± 0.8	18	Bourbou and Garvie-Lok (Chap. 10)
Kastella	Crete	-18.8 ± 0.4	9.2 ± 1.2	26	Bourbou and Garvie-Lok (Chap. 10)
Petras	Crete	-19.2 ± 0.4	9.6 ± 1.0	16	Bourbou and Garvie-Lok (Chap. 10)

*Collagen values only; does not include carbonate values.

As discussed in Chapter 2, human bone collagen carbon isotope values of close to -12‰ indicate that most dietary protein came from marine sources, while humans that have values closer to -20‰ have mainly terrestrial (C_3 based) dietary protein. As can be observed over time, most of the human samples included in this volume exhibit carbon isotope values (Fig. 11.1) that indicate their dietary protein was largely from terrestrial sources, as they have $\delta^{13}C$ values that are below -19‰ (a somewhat arbitrary cut-off point, assuming a terrestrial "end-point" bone collagen carbon isotope value of -19‰ based on large-scale studies of over 10,000 animals in prehistoric temperate Europe, and also in this volume). This is somewhat surprising, as many of these sites are located near the coast, but it is apparent that marine foods have never been an important staple in Greek diets, in either prehistoric or historic times. However, there are some very notable exceptions that have carbon isotope values indicating marine food consumption, and this is confirmed by their elevated nitrogen isotope values as is evidenced in Figure 11.2. As also discussed in Chapter 2, bone collagen nitrogen isotope values are usually approximately 3‰ higher than dietary protein. Since there are more steps in marine ecosystems, marine organisms generally have much higher nitrogen isotope values than organisms in terrestrial ecosystems.

Additionally, there are some individuals with $\delta^{13}C$ values higher than -19‰, but in these cases the associated $\delta^{15}N$ values are not elevated, which would indicate the consumption of marine foods. For these individuals, the higher carbon isotope values are most likely due to the consumption of a significant amount of dietary protein from C_4 plants, most likely millet (as also discussed in Chap. 2). If the trend is examined over time, the first clear signal indicating C_4 plant consumption appears in animals of the Neolithic period (in which case it would imply either millet or seaweed), which continues into the Bronze Age and is maintained at least into the Byzantine era.

An overview of the $\delta^{15}N$ values (Fig. 11.2) reveals variations in time that are likely due to differing amounts of animal and plant protein in human diets. Although there is likely very little fluctuation in environmental and animal isotope values during the time period plotted on the graph

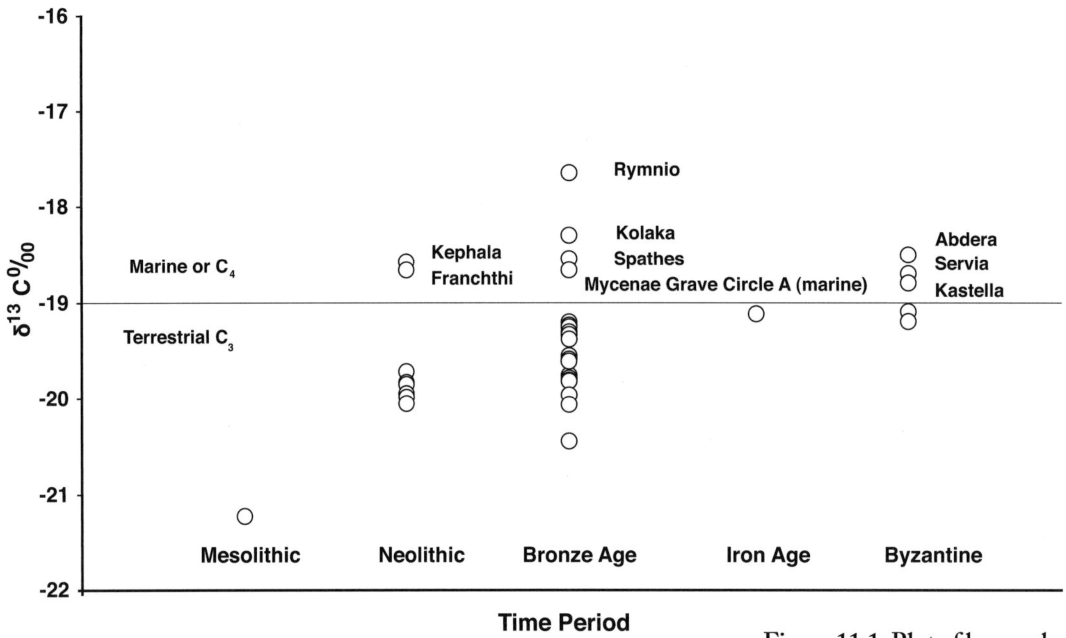

Figure 11.1. Plot of human bone collagen $\delta^{13}C$ values through time from various locations in Greece. Data are from the sources given in Table 11.1.

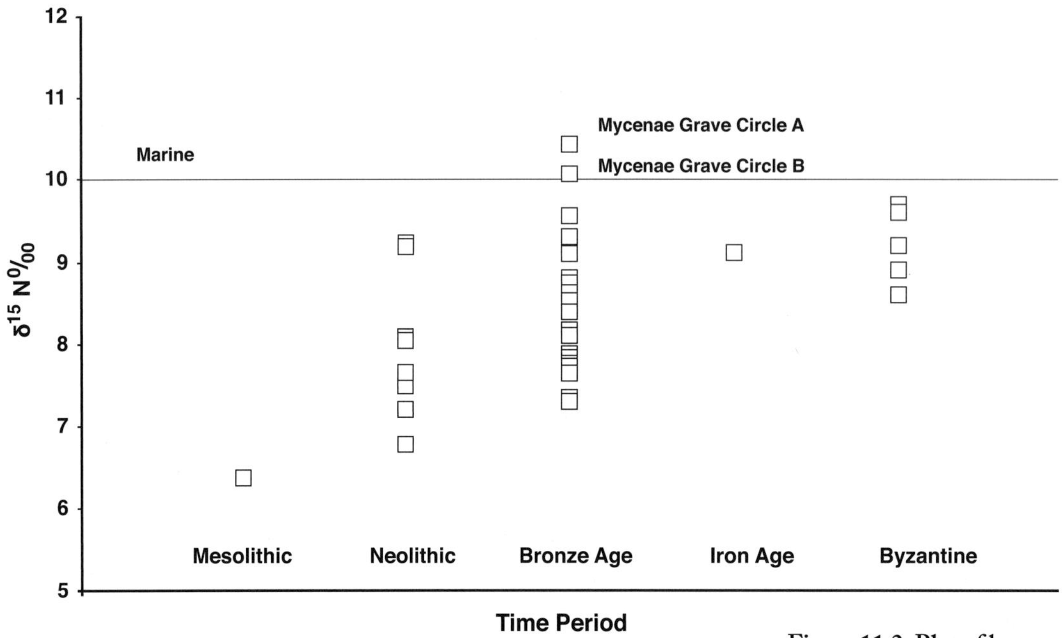

Figure 11.2. Plot of human bone collagen $\delta^{15}N$ values through time from various locations in Greece. Data are from the sources given in Table 11.1.

(compared to large fluctuations observed in the Pleistocene),[3] the data still need to be interpreted with caution, as the individual food webs for each site need to be established first before anything can be definitively concluded about the amounts of plant and animal protein in human diets. What can be noted, however, is that apart from a few exceptions, all of the human $\delta^{15}N$ values are lower than 10‰, so there are no indications for the consumption of foods with high $\delta^{15}N$ values such as from fish. Instead, the $\delta^{15}N$ values for most humans are consistent with a diet where the protein came from both animals and plants.

ARCHAEOLOGICAL INFERENCES

One of the key challenges in using isotopic analysis to interpret past diets is to link them with larger archaeological trends and changes, such as the transition to agriculture, the role of marine resources in prehistoric and historic diets, the documentation of gender and status differences in the access to resources, the tracing of fluctuations of animal protein consumption through time, the determination of weaning age, and the role of C_4 plants in human and animal diet through time.

Looking at the large dataset of isotope values now available for Greece, some long-term patterns and trends can be elucidated. Starting with the shift to agriculture during the Mesolithic–Neolithic transition, there are not many samples from Greek Mesolithic sites available for comparative purposes to confirm differences that may characterize this transition. The very few available samples as well as archaeological data imply significant reliance on terrestrial resources during the Greek Mesolithic, while recent isotopic studies from Mesolithic Italy have produced similar evidence.[4] During the Neolithic, even in its early periods, human diet is already very homogeneous, dominated by terrestrial resources, especially domesticated plant and animal species. Despite the existing variation in the available resources, dietary variation among Neolithic people is limited and is affected more by geographical location than by sex, age, or potential status differences. Although a trend might have existed in the Mesolithic, the appearance of Neolithic material culture is characterized by a complete embracement of agriculture,[5] which induced a strong emphasis on domesticated resources with a series of demographic and political consequences from adopting farming as a livelihood. This resulted in a dramatic population increase beyond the carrying capacity of local environments. Whether this happened as a result of the engagement of the society in the changing cultural networks, because this was dictated by ecological/climatic pressure, as an economic decision related to the productivity output versus the labor input, or for other "practical" purposes needs to be investigated.[6] However, this shift was so intense that even successful foraging communities worldwide were persuaded to eventually switch to agriculture. Furthermore, it was sustained, despite an association with a general decrease in human health, expressed as increased rates of carious lesions, extreme dental wear, decreased stature, and increased prevalence of anemic conditions.[7] The effects of this Neolithic Revolution

3. Richards and Hedges 2008.
4. Mannino et al. 2012.
5. Papathanasiou 2000, 2011; Papathanasiou and Richards 2011.
6. Richards and Schulting 2006; Bowles 2011.
7. DeNiro and Epstein 1981; Cohen and Armelagos 1984; Larsen 1997.

have set the main dietary directions for all subsequent prehistoric and historic cultural periods, as becomes evident in every chapter of this volume.

The supposedly important role of fishing in the Mediterranean diet and in Aegean prehistoric and historic archaeology is challenged by stable isotope analyses. All the contributors to this volume stated that they were surprised when their results implied insignificant seafood consumption. This holds true from the Neolithic to the Byzantine period, with the exception of the few Mycenaean high-status individuals.[8] Marine resources were used occasionally throughout all these time periods, but they were a peripheral aspect of subsistence. Namely, less than 10% of the overall diet comes from the consumption of marine protein. Fishing as well as harvesting of wild game was not an important economic activity. Instead they both were possibly retained for their symbolic value as representations of power and masculinity. Fish and game are most likely exotic, rare, and symbolic food items and not the everyday staples that a farming community would rely on for its subsistence and economy. The notion persists through time, and it is supported by historical evidence as well. The zooarchaeological evidence also points to the majority of the diet being composed of domesticated species at both coastal and inland sites. Additionally, textual evidence attests to the rare use of wild game and fish relative to domesticated plant and animal species.

Another extensive discussion in Aegean archaeology is about the practice of specialized pastoralism and extensive agriculture versus mixed farming. In this volume, both Papathanasiou (Chap. 3) and Triantaphyllou (Chap. 4) discuss a model of mixed farming, as the isotope data do not point toward diets where protein was mainly from animal consumption. Additionally, if we take into account secondary products that exhibit the same values as meat and land use logistics that indicate that pastoralism could not have been an efficient solution for the Mediterranean environment, it seems more parsimonious to support a mixed farming model dominated by domesticated plant species with the later inclusion of dairy products and occasionally meat. Without intensive dairying, livestock had limited potential as a staple food source, while mixed farming had many practical advantages: the combination of crops and livestock offered a more secure subsistence base and was less labor intensive.[9] The available evidence is compatible with the dominance of mixed farming during the Neolithic and the patchy adoption of extensive farming starting in the later part of the Bronze Age with the absence of specialized pastoralism prior to the historical era.[10] Even for Classical Athens, Sallares[11] mentions that "only an Athenian who was so wealthy . . . would have used the produce of land to support animals, beyond the bare minimum required for ploughing, if that same land could produce cereals, because animal husbandry is an extremely costly way of producing food, although its products are very prestigious, for the same reason."

Animal protein consumption fluctuates through time and is related to cultural changes by certain researchers. Taking a broad look at the nitrogen isotope values, it appears that animal protein consumption is high in the Early Neolithic, decreases during the Late Neolithic, increases again during the Bronze Age,[12] and drops again at the beginning of the Iron Age to increase once more during classical times, especially for elite

8. See Chap. 2.
9. Halstead 2008, p. 244.
10. Halstead 2008.
11. Sallares 1991, p. 311.
12. See Chaps. 3–5.

individuals.[13] It decreases during the Late Classical period[14] and stays low throughout the Byzantine period. This pattern of animal protein consumption correlates very well with cultural and socioeconomic shifts. The highest points of animal protein consumption coincide with periods of economic and cultural growth and prosperity, such as the Mycenaean civilization and Classical Athens. The lowest points are associated with the Late and Final Neolithic or the transitional Geometric and Hellenistic periods, which are characterized by population movements, poverty, war, and limited access to resources. Much better documentation and data are needed to support this trend, however. It is also important to remember that animal protein can appear in two forms: the more expensive meat and fish, and the more affordable dairy products, and later eggs, which cannot be distinguished isotopically. Poorer dental health and increased anemic conditions are related to low animal protein consumption.[15] Therefore, it is possible to correlate diet, animal protein consumption, human health, and the standard of living in general, and associate them with cultural and political events and periods.

Evidence also exists for sex and status differences in isotope values, especially for prehistoric Greece. No statistically significant differences are apparent throughout the entire Neolithic and the EBA/MBA, but during the LBA, for the first time, these differences emerge and become detectable and measurable through stable isotope analysis. The humans from Mycenae Grave Circle A[16] have carbon isotope values indicating marine food consumption, and this is indicated as well by their elevated nitrogen isotope values as can be seen in Figure 11.2. In this volume, Iezzi (Chap. 6) and Papathanasiou (Chap. 3) also observe statistically significant differences between males and females with males consuming more animal protein, and status differences with individuals buried in the elite tombs showing significantly higher animal and marine protein consumption. Similar differences are observed by Lagia (Chap. 8) for the cemetery of the Kerameikos in classical Athens. Differences in dietary practices between males and females in the Classical period are also documented in literary sources, such as different preparation procedures and meal times and differential access to certain animal parts.[17]

Most of the authors in this volume notice isotope evidence for occasional C_4 protein consumption by animals or humans. Although 800 plants that grow in Greece are recognized as C_4 plants, millet is the most probable one, as it is the only one that can be consumed in considerable amounts. The occurrence of millet, however, is rather sporadic and always constitutes a minor part of the overall diet even in historic times, with texts supporting the isotopic data. *Panicum miliaceum* and *Setaria italica* are mentioned by classical authors as being "the least productive of all cereals, and were only cultivated because they had an ecological niche all to themselves as there were no other cereals in Europe adapted to European summer day-lengths."[18]

Another issue that is addressed by stable isotope analysis is the weaning age and weaning diet of children, as it is closely related with child rearing practices in the ancient Greek world, early death, stunted development, and selective malnutrition. Kwok and Keenleyside[19] address this question, indicate that weaning began between 6 months and 1 year of age and was

13. See Chap. 8.
14. See Chap. 8.
15. See Chaps. 3, 4, 6, 9.
16. Richards and Hedges 2008.
17. Dalby 1996, pp. 57–92.
18. Sallares 1991 p. 363.
19. See Chap. 9.

completed between the ages of 2 and 4, and conclude that Roman practices were heavily influenced by Greek customs, observations consistent with literature. These practices remained standard throughout the Greco-Roman period until the Byzantine era.

Most researchers utilize terms such as "Byzantine diet" and "Neolithic diet" similar to the archaeologically known term "Mycenaean *koine*," suggesting that, although some regional variability may occur, the samples are clustered very tightly. Diet is a part of the society's engagement in cultural networks, and culture determines the food choices more than availability, nutritional value, or even the taste of foodstuffs. Isotopic analysis therefore points to a generalized Mediterranean diet starting as early as the Neolithic and continues with little change until recently. Variations do, however, exist through time and within each society as governed by socioeconomic parameters.

Another issue that arises from the above is that when more samples are available—as in the Bronze Age or the Classical Age—and when studies are oriented around specific research questions, the results are more revealing. Because data are very closely clustered in all time periods, internal variability is easily masked. When more site-specific targeted sampling occurs, subtle but important differences within a society are revealed, as well as differences between time periods.

As mentioned above, Greece is now one of the most isotopically studied areas of the world. In this volume, approximately 1,000 samples are presented or referred to from all time periods, a sample set that includes humans and animals, males and females, adults and subadults. The specific, individual, long-term diets add up to accurately represent the entire population that they came from. It is rare in archaeology to be able to quantify and scientifically prove clear differences using any kind of methodology. From the above studies, it is obvious that isotopic data can provide key evidence to our understanding of the past and its transitions and internal variability, and that it can answer questions about the relationship between diet and the flourishing or the decline of the standard of living and culture.

CONCLUSIONS

Stable isotope analysis is a powerful tool for reconstructing past diets in Greece, and there are increasing numbers of studies that are applying this method, many of which are included in this volume. The method is well suited to this region, as it can readily identify the consumption of marine foods and millet, both of which have been important foods in Greece's past. As the dataset grows, it is possible to look at long-term dietary changes to see how people have changed the way they obtained food over time in Greece. In the future, more data will allow us to better see these trends, and the addition of new isotope measurements, such as sulfur isotopes[20] and the measurement of the isotope ratios of single amino acids from collagen, will allow us to build up a comprehensive picture of the diets of humans in prehistoric and historic Greece.

20. Richards, Fuller, and Hedges 2001.

REFERENCES

Bowles, S. 2011. "Cultivation of Cereals by the First Farmers Was Not More Productive than Foraging," *Proceedings of the National Academy of Sciences of the United States of America* 108, pp. 4760–4765.

Cohen, M. N., and G. J. Armelagos, eds. 1984. *Paleopathology at the Origins of Agriculture,* Orlando.

Dalby, A. 1996. *Siren Feasts: A History of Food and Gastronomy in Greece,* London.

DeNiro, M. J., and S. Epstein. 1981. "Influence of Diet on the Distribution of Nitrogen Isotopes in Animals," *Geochimica et cosmochimica acta* 45, pp. 341–351.

Evershed, R. P., S. J. Vaughan, S. N. Dudd, and J. S. Soles. 2000. "Organic Residue, Petrographic and Typological Analysis of Late Minoan Lamps and Conical Cups from Excavations at Mochlos in East Crete, Greece," in *Palaeodiet in the Aegean. Papers from a Colloquium Held at the 1993 Meeting of the Archaeological Institute of America in Washington D.C.* (Wiener Laboratory Monograph 1), ed. S. J. Vaughan and W. D. E. Coulson, Oxford, pp. 37–54.

Halstead, P. 2000. "Land Use in Postglacial Greece: Cultural Causes and Environmental Effects," in *Landscape and Land Use in Postglacial Greece* (Sheffield Studies in Aegean Archaeology 3), ed. P. Halstead and C. Frederick, Sheffield, pp. 110–128.

———. 2008. "Between a Rock and a Hard Place: Coping with Marginal Colonisation in the Later Neolithic and Early Bronze Age of Crete and the Aegean," in *Escaping the Labyrinth: The Cretan Neolithic in Context* (Sheffield Studies in Aegean Archaeology 8), ed. V. Isaakidou and P. Tomkins, Oxford, pp. 229–257.

Larsen, C. S. 1997. *Bioarchaeology: Interpreting Behavior from the Human Skeleton* (Cambridge Studies in Biological Anthropology 21), Cambridge.

Mannino, M. A., G. Catalano, S. Talamo, G. Mannino, R. Di Salvo, V. Schimmenti, C. Lalueza-Fox, A. Messina, D. Petruso, D. Caramelli, M. P. Richards, and L. Sineo. 2012. "Origin and Diet of the Prehistoric Hunter-Gatherers on the Mediterranean Island of Favignana (Ègadi Islands, Sicily)," *Public Library of Science (PLOS) ONE* 7, p. e49802.

Papathanasiou, A. 2000. "The Reconstruction of the Theopetra Cave Population Diet from Stable Isotope Analysis of Human Bone," in *Theopetra Cave: Twelve Years of Excavation and Research 1987–1998. Proceedings of the International Conference, Trikala 6–7 November 1998,* ed. N. Kyparissi-Apostolika, Athens, pp. 119–127.

———. 2001. *Bioarchaeological Analysis of Neolithic Alepotrypa Cave, Greece* (*BAR-IS* 961), Oxford.

———. 2003. "Stable Isotope Analysis in Neolithic Greece and Possible Implications on Human Health," *International Journal of Osteoarchaeology* 13, pp. 314–324.

———. 2011. "Health, Diet, and Social Implications in Neolithic Greece from the Study of Human Osteological Material" in *Human Bioarchaeology of the Transition to Agriculture,* ed. R. Pinhasi and J. T. Stock, Chichester, pp. 87–106.

Papathanasiou, A., C. S. Larsen, and L. Norr. 1995. "A Bioarchaeological Analysis of a Neolithic Ossuary from Alepotrypa Cave, Diros, Greece," *American Journal of Physical Anthropology* 38, Suppl. 20, p. 167 (abstract).

———. 2000. "Bioarchaeological Inferences from a Neolithic Ossuary from Alepotrypa Cave, Diros, Greece," *International Journal of Osteoarchaeology* 10, pp. 210–228.

Papathanasiou, A., and M. P. Richards. 2011. "Ανθρωπολογικά κατάλοιπα από τις πρώιμες θέσεις Μαυροπηγής, Ξηρολίμνης και Ποντοκώμης της Αρχαιότερης Νεολιθικής στη Δυτική Μακεδονία," in *Το Αρχαιολογικό Έργο στην Άνω Μακεδονία,* Aiani, pp. 257–274.

Papathanasiou, A., L. A. Schepartz, M. P. Richards, and E. Malapani. 2012. "Bioarchaeological Evidence for Social Differentiation in the Health and Diet of Mycenaean Pylos," in *Proceedings of the 2nd ARCH-RNT Symposium, University of the Peloponnese, Kalamata 2011,* pp. 143–151.

Petroutsa, E. I. 2007. "Μελέτη της διατροφής σε πληθυσμούς της Ελλάδας της Εποχής του Χαλκού" (diss. Univ. of Athens).

Richards, M. P., B. F. Fuller, and R. E. M. Hedges. 2001. "Sulphur Isotopic Variation in Ancient Bone Collagen from Europe: Implications for Human Palaeodiet, Residence Mobility, and Modern Pollutant Studies," *Earth and Planetary Science Letters* 191, pp. 185–190.

Richards, M. P., and R. E. M. Hedges. 2008. "Stable Isotope Evidence of Past Human Diet at the Sites of the Neolithic Cave of Gerani; the Late Minoan III Cemetery of Armenoi; Grave Circles A and B at the Palace Site of Mycenae; and Late Helladic Chamber Tombs," in *Archaeology Meets Science: Biomolecular Investigations in Bronze Age Greece. The Primary Scientific Evidence, 1997–2003,* ed. Y. Tzedakis, H. Martlew, and M. K. Jones, Oxford, pp. 220–230.

Richards, M. P., and R. J. Schulting. 2006. "Touch Not the Fish: The Mesolithic–Neolithic Change of Diet and Its Significance," *Antiquity* 80, pp. 444–456.

Richards, M. P., and E. Vika. 2008. "Stable Isotope Results from New Sites in the Peloponnese: Sykia, Kalamaki, and Spaliareka," in *Archaeology Meets Science: Biomolecular Investigations in Bronze Age Greece. The Primary Scientific Evidence, 1997–2003,* ed. Y. Tzedakis, H. Martlew, and M. K. Jones, Oxford, pp. 231–235.

Sallares, R. 1991. *The Ecology of the Ancient Greek World,* Ithaca.

Triantaphyllou, S. 2001. *A Bioarchaeological Approach to Prehistoric Cemetery Populations from Central and Western Greek Macedonia* (*BAR-IS* 976), Oxford.

INDEX